HANDBOOK CONTENT MAP

Control Room Operator's Handbook

CONTROL ROOM OPERATOR'S HANDBOOK

AT-THE-READY CONTROL ROOM & OPERATIONS CENTER GUIDANCE

DOUGLAS H. ROTHENBERG
BEVERLY B. KUCH

PB
PENNWELL BOOKS

Disclaimer

The recommendations, advice, descriptions, and the methods in this book are presented solely for educational purposes. The author and publisher assume no liability whatsoever for any loss or damage that results from the use of any of the material in this book. Use of the material in this book is solely at the risk of the user.

PennWell Books, LLC
10050 E 52nd Street
Tulsa, Oklahoma 74146 USA

866.777.1814
sales@pennwellbooks.com
www.pennwellbooks.com

Publisher: Matthew Dresher

Images: gettyimages
iStock/Getty Images Plus - kodda
iStock/Getty Images Plus - gorodenkoff
E+ - halbergman
E+ - youngvet

Library of Congress Cataloging-in-Publication Data Available upon Request.

LCCN 2022008593

ISBN 9781955578042 (Hardcover)
ISBN 9781955578059 (eBook)

Printed in the United States of America

1 2 3 4 5 26 25 24 23 22

Dedications

To the men and women from all walks of life who leave their homes, families, and friends and step into a control room or operation center for their assigned shifts. They work with responsible professionalism and a keen sense of duty. We also recognize with respect and gratitude the entire surrounding supportive staff, who provide proper resources, teach, and advise them as they do the important work of making and distributing much of the things we value as a grateful society.

To my dearest wife, constant companion, champion, and bestest friend, Kasia Gustaw Rothenberg, who encourages my countless hours in the *scriptorium*, writing, editing, and just thinking about everything that goes into such a project of this unimaginable draw and humility. And for ever humoring me as I almost daily "torment" her with small indications, weak signals, and the greatest new idea of the minute.

Douglas H. Rothenberg

To my family, particularly my husband, Larry, and daughter, Heather, for their patience and support when my writing kept me locked away in my office for hours on end, and for their understanding of my inability to stay truly retired, at least not today.

Beverly Biel Kuch

Contents

Part 2
Operations

Module 16

Part 4

The control room

Module 17

Module 18

Module 19

Module 20

Module 25

Module 26

Part 6
Awareness

Module 27

Module 28

Part 7
Managing Situations

Module 29

Appendix 3

List of Figures & Tables

Preface

This is a unique handbook to use in your control room. At the ready are helpful reminders of things you already understand, things about which you may have heard but would like to know more, and new ideas of ways for working the many jobs you do all of the time on your shift. Everything here is intended to assist you to get the job done. It will touch on almost all of your tasks, duties, and responsibilities. As the creators of the Handbook, we have done the best we know how to bring together the collective experiences of engineers, operators, and controllers like you from around the world. We have worked hard to be clear, well organized, and respectful. We now offer the Handbook to you and your management. It is our hope that you find it a valuable resource.

The handbook is large, containing many individual topic modules. Some contain old information, some new. Some very new and useful. There is a simple yet important reason for the size—your job as operator or controller is not simple. You have a lot on your plate! Your every shift hour involves observing, deciding, scheduling, and all manner of managing you do for a really expensive plant or facility. For a resource to be helpful, it must cover and respect your broad range of responsibility. If it is actually going to be of value for you, it needs to cover what you need.

The underlying goal of this handbook is singular: safe and reliable operations. Inside you will find explanations about the intended purpose of the wide range of your everyday tools and activities. There are introductions to new ones. There are discussions to help you understand and appreciate how human nature affects what goes on in control rooms and operation centers. You will learn about new ways to get help when help is needed—and to avoid continuing to operate when operational success is threatened by doing so. We hope that all of this enables you to be more skillful, more effective, and more comfortable sitting in the operator or controller chair. When all is said, understood, and done, the desired result will be for you to gain and maintain a higher level of operational, personal, and enterprise safety. Financial success and pride of operation will follow. Above all, you should recognize that, as a highly trained and ready operator, you are a *Professional*!

DHR & BBK
Ohio USA

Acknowledgments

The authors gratefully acknowledge the generous contributions of the following friends and colleagues:

Chris Conlon and members of the engineering and operations teams, including Walter (Skip) Doucette, Byron Knoth, and Kevin Thomas, for proof-of-concept reviews and valuable critique, suggestions, and keen interest.

Joel Stein, for his faith in our work and active sponsorship for bringing it to reality.

Joseph Doemling, for a complete and insightful review of the material, valuable evaluation, and recommendations for enhancement and improvement.

Walt Boyes, for his steadfast support, enthusiastic endorsement, and plentiful advocacy for the work.

Matthew Dresher and the staff at PennWell Books. In particular to Matthew, for his constant enthusiasm for our book; and the many ways he provided encouragement, thoughtful comment, willingness to listen, and jointly shape the design and messages of the handbook. His constant advice to make it the best it can be was all we needed to go the extra miles.

Ann Cutler, for many thoughtful discussions of weak signals and what they might really mean for us as individuals searching for explanation and truth.

Steve Apple, Neil Baltes, Ardis Bartle, Frank Bradshaw, Steve Elwart, Van Harp, Joseph Kaulfersch, Jack Pankoff, Dan Scarberry, and Russel Treat, for reviewing portions of the manuscript and for the benefits of their advice and suggestions.

Momentum Press, for their generous permission to incorporate relevant materials from *Alarm Management for Process Control* (2009, 2018).

The International Society for Automation (ISA), for permission to incorporate relevant materials from *Situation Management for Process Control* (2019).

The many whose fingerprints can be found covering the prescient thinking and concepts of the seminal earlier works of the Alarm Management Task Force (Honeywell, ca 1991) and the Abnormal Situation Management Consortium (NIST, ca 2002). The energy, urgency, and promise of those activities have shaped much of the current work toward better and safer industrial manufacturing enterprise.

And, on a personal note (*from Rothenberg*), thank you to Beverly for stepping bravely into the roles of author and collaborator with such patience, competence, intelligence, style, and creativity. Much of the heart and accessibility of the work can be attributed to your efforts.

And a last personal note to Doug (*from co-author Kuch*): Thank you for giving me the opportunity to join you in this (ad)venture. One could not ask for a more knowledgeable collaborator. Your meticulous nature counterbalanced my sometimes less thorough style.

PART 1
THE BASICS

Part 1 opens with the goals of the handbook and suggestions for how you might consider using it.

It continues with an explanation of the handbook's modular format, designed to help you find what you wish to use. Its several unique features are described.

Next the types of enterprises you operate are explored for: leadership, philosophies, and types of control room personnel.

It closes with characteristics and important components of safety.

PART 1 — THE BASICS

Module 1: Getting Started

Process Operators oversee and manage the full production process of a manufacturing plant or other industrial facility. They monitor equipment to ensure the quality, efficiency, and safety of the plant.... Process Operators are skilled professionals who often work as part of a team and provide leadership.

—JobHero.com

Welcome to your handbook. Up front, we admit that it is large. Lots of pages. Lots of individual topic modules. Some old material. Some new material. Some very new and quite useful material. The reason for the size: your job as operator or controller is not simple. You have a lot on your plate! Your every hour involves observing, deciding, scheduling, and managing a really expensive plant or facility. If any resource is going to be of help for you, it must cover and respect that broad range of responsibility. This one is for you to explore, test, and use.

When you venture inside this handbook, you will find a cornucopia of things you might need to know to help make your job as operator and controller work better for your enterprise and be more comfortable for you. Even if your enterprise has none of these tools and practices, and you are sure that will not change anytime soon, what you can find here will help you to do better with what you have available to you.

If you are new to being an operator, this can guide you into the position and help you get your arms around the job to become more comfortable. If you are a seasoned operator, you can discover many ideas that would be helpful and provide confirmation of what you already know and do.

1.1 Introduction

To our knowledge, this handbook is a first of its kind. The *Operators' Handbook* is compact but extensive. It is a ready reference, with tools, advice, experience, and examples. It explains how to gain and sustain a responsible level of control room operational effectiveness and safety. It reinforces the concept of you as a professional. The content is built upon material adapted from the groundbreaking engineering reference book *Situation Management for Process Control*, published by the International Society of Automation (ISA) in 2019. [1.1]

Your handbook is designed to explain, amplify, and supplement your existing knowledge and practices. The goal is to reinforce your background as you work in the control room. Your site-based training, experience, policies, and procedures, as well as the guidelines given to you and the expectations placed upon you by your plant or operation (called enterprise in this handbook) take precedence and are primary. This handbook is extra. It is intended to give you options, not to tell you what to do. Please use the material as an aid and hands-ready assistant to (a) recognize gaps, small and potentially large, (b) assist filling in the gaps to keep things going right, and (c) provide new ways of meeting your job responsibility. What you find useful, review and discuss with your management. If you are new to operations, this material provides a supportive background and comprehensive introduction. Take the time to understand what is going on and what is needed to do your job well. If you are a seasoned operator, see how supportive this can also be for you. This material is not the only way to achieve effective situation management for good operation. If a particular idea, tool, or concept doesn't seem to work, skip it and use what already works. If you see gaps, ask your supervisor for advice.

> It is the sum total of the decisions and actions that the operator makes, within the capabilities of the enterprise, that determines whether or not the enterprise in his care operates safely and productively.

Situation management is the core message of this handbook. Recognize that the sum total of the decisions and actions that you make as operator, within the capabilities of your specific enterprise, determine whether or not the enterprise in your care operates safely and productively. It is your end-game mission.

Format

Parts

The material is presented in succinct modules that are grouped into seven parts. Each part is identified by a colored tab that is visible from the edge.

PART 1: THE BASICS—the basics of the handbook are explained with suggestions for reading. Cautions are given and special features are illustrated.

PART 2: OPERATIONS—discusses what goes on in a control room and who is there.

PART 3: KEY CONSIDERATIONS—tools and ideas are covered, keeping the focus on what needs to happen in a well-run control room to avoid distraction and wrong turns.

PART 4: THE CONTROL ROOM—the control room is defined and described along with successful control room requirements and management. A module is included on the Human Machine Interface (HMI).

PART 5: WEAK SIGNALS—weak signals, a powerful and effective new way to discover early problems, is introduced and fully explained.

PART 6: AWARENESS—the tools and methods are provided to see what is going on in the enterprise and decide its importance.

PART 7: MANAGING SITUATIONS—how to keep things going properly and responding when and where they are not.

ADDITIONAL MATERIAL: supporting material, appendices, glossary, and index

Dividers and Covers

There is a complete map of the contents in the very front of the book. Each module is identified, with the part in which it is included. This will easily aid you in finding the location of each module of interest. Each part is separated by a divider, with an overview of its contents (front side) and a complete listing of the included modules, with major sub-topics identified (reverse side). Each part is identified by an appropriately colored tab. All of this will help you easily and quickly find a topic.

Contents, Index, and Lists

The table of contents has been expanded to include major sub-parts of each module. This will assist you in seeing the organization and finding related topics for which you might be looking. There is an extensive index. Reference notes are listed at the end, since normally you will not need them for most work. A table of figures is included to help you locate any you need.

CueCards

CueCards are a special part of this handbook. You can pull out the corresponding cue card and use the information contained on it to keep focused and on course for a particular activity. Later on, perhaps your first day back on a rotation, pick a few out and use them as a handy review. This way, you can both ease into the shift and brush up on what those cards cover. Cue Card material (be it conversations, procedures, lists, etc.) is pulled from the text, where it is explained in context with the situation it covers. The handbook location for the material is noted on the appropriate cue card using the convention [**Mx.y**], where Mx identifies the modue, and .y identifies the appropriate section within the module. This information is also repeated in Appendix 3 for permanent reference. Specific Cue Card references are noted within the handbook text using the convention [**Q#**] where # identifies the specific card.

Audience

The form and format is specifically tailored for control room operators and controllers. We use the term *operator* for both. Others with operational responsibilities will also find the material quite useful. Operations teams will appreciate the breadth of coverage and the accessibility of the presentation. All are encouraged to seek and adopt strong tools and processes for doing a reliable job and to together work desired changes back into approved operations protocols for their site.

1.2 Situation Management

Situation management is an essential work process in an industrial control room or operation center. It effectively positions operators as a responsible component of successful enterprise performance. The handbook provides unified understanding and solutions by integrating the needs and the work process. Before, the many discussions on and approaches to important control room and operations center requirements were fragmented. While each approach might be quite insightful and independently competent, it was hard to combine the various ideas into a workable enterprise culture. They are brought together here. This coverage provides the capability for you to understand the operations situations and the methodology for managing them. This helps you understand the need, assess your particular situation, and effect useful change.

Situation management is not crisis management. Nor is it disaster management. For these, your operational role ceases when it is deemed likely that a safe state of the enterprise is not attainable, even though you initiate action(s) to move everything toward that safe state and ensure that the actions to do so have been taken. From this point forward, the problem is shifted to a crisis management responsibility or a disaster management responsibility. Both are critically important. Both are covered in an extensive array of other resources. Both are beyond the scope and intention of this book, notwithstanding the possibility that you may play a role in or perform duties for the furtherance of enterprise safety.

1.3 Suggestions for Reading

Most of the duties you perform and the ways you do it are touched upon here. You already have much of what you need to do that job well. This is in addition. Almost every activity in which you engage is addressed in one or more modules. And there are many modules. This is by design. Each focuses on a specific topic, skill, or activity. The handbook should make it easier, more effective, and more consistent. The advice and material is all there, all in one convenient place. This way, you can learn all about it quickly and without being distracted by "other" topics. When you are interested in looking for more information on a topic, we cross-reference it in the text using the convention [**Mx**], where **"x"** stands for the module number. Many threads are discussed in several places in the handbook. Each discussion has specific relevance for where it is located.

We are counting on your reading the entire module. That way, you can fully understand what it is about and ensure that you have seen the cautions and recommendations. We are also counting on you to not only find interesting and useful things, but at the same time to appreciate the limitations and guidance for its use. If you are a novice, keeping the handbook at the ready can be an efficient and knowledgeable way to support understanding your job training. New to the job or not, you can directly turn to a module when you are doing a specific task or facing a challenging problem. Or you can use a module to help you explain something to a colleague. For example, you can directly turn to the module on getting help [**M34**] to brush up on the suggestions to make sure that both of you are able to productively assist each other. On the other hand, when you are curious, you could open a module on doubt [**M15**]**,** multitasking [**M7**], or managing automation [**M20**]**,** just to remind you of the many different facets of doing your job well. This material can also be an excellent way to help you explain to others what goes on in your control room.

1.4 Cautions and Notices

Design and Safety Notice

All material described and illustrated in this handbook (control schemes, design suggestions, displays, diagrams, lists, and the like) have been designed to amplify and explain concepts and practices intended to help explain material to you. It is provided for training and understanding purposes only and not intended as specific models for design or implementation. Qualified, authorized members of the enterprise staff, who must act with full knowledge of their specific enterprise's configuration, process conditions, equipment, and applicable statutory practices and requirements, must retain the choice of what to design and implement and how it will be done.

Conflicts with Established Protocols or Statutory Requirements

No material contained herein is intended to circumvent the sole responsibility of your management to decide your specific operational policies and your role as operator.

No claim in this handbook is meant to imply or otherwise suggest that any item, example, technology, tool, or illustration is or would be a sufficient solution for responsible operation. No included material is meant to suggest compliance with any policy, standard, or legal or other statutory requirement.

Exclusion of Special Systems and Responsibilities

This handbook covers important aspects regarding the safety of personnel, the physical security of plants and enterprises, and related aspects that may in some way apply or relate to operators, control rooms, and operation centers. All of these topics require special knowledge and additional practices that are well beyond the scope and intention of this reference. No part of this coverage is intended to guide or inform as to proper design or use.

Terminology and Gender

Operators and Controllers

This guide applies equally to control room operators (*operators*) and operations controllers (*controllers*) in operations centers. For simplicity, the term operator is used to apply for all.

Gender

Your guide is intended to be gender neutral. The words *he, his,* and *him* are meant simply to be placeholders for individuals of all genders and identity.

1.5 Knowing What Is Right

How will you decide if the material contained in this handbook is worth reading and understanding? And how might you decide that it could work for your enterprise? The first question has an easy answer: contained in this handbook are principles and best practices that work. You will find answers to questions that you might not have ever asked, or even considered asking. Having seen them here, you may realize that the answers have a great deal of use for you. Without some of the approaches suggested, your operations might be more exposed to difficulty and unwanted impact. If you are open to learning and rethinking, here are insight, discussion, tools, and technology to put you on the path to success for a very long and productive career.

The next three tests will allow you to grasp the concept and keep an open mind to see if what you find here would make a difference for you. They are thought points intended to guide approaches and plans, not design requirements.

Test 1: Simplicity vs. Correctness

How can anyone identify what is the truth as opposed to what is not? How can we know whether a particular solution is the way to go? These questions have vexed thinkers and decision makers for as long as there has been thought and important things to be decided. These are easy questions to ask. As you might imagine, definitive tests do not exist. There is guidance. A test known as Occam's Razor [1.2] states that if there are two competing theories or explanations for something important, we believe the simpler one (of course, one's intellect is not ignored in the decision). [1.3]

Please accept that genuine solutions can be more complex. That is, good solutions can require many "moving parts." But the end result should fit together simply, naturally, and efficiently. In this handbook, we strive to keep it as simple as we can, while getting the points across.

Test 2: The Concept of Scale

The concept of scale means that if one's solution-focused actions are on the right track to solve a problem, then more of the same actions should solve the problem even better. If using this "more of the same" concept quickly gets out of hand and becomes an equal or worse burden than the original problem, it strongly suggests that you might be using an ineffective approach. It probably means that your approach was more about resolving only symptoms. You missed the root cause. Look again.

Test 3: Not by Subtraction Alone

Proper resolution of real problems almost always involves a combination of reducing unwanted aspects and adding in missing aspects. Therefore, as we approach a problem with too much of something—e.g., too much data on a display—work to reduce the "too much." Then, using the real need as guidance, find what is required for a proper solution and add it back in. For example, too much data on a display will involve a reduction of data. Adding back to identify the true problem will then compensate for the reduction. Addition would be the need for more effective maintenance, or better ways of recognizing abnormal plant operations and conditions, or more effective options for managing plant upsets and disaster avoidance. To summarize: after completing work to remove the "too much" parts, always look to find out what needs adding back and add it. Every solution requires "adding back."

1.6 Looking Ahead

You are just starting into the *Operators' Handbook*. Maybe someone has suggested it to you. Maybe you came across it incidentally. But you are here. As authors, we have invested into getting this ready. Our technical backgrounds are in chemical engineering, electrical engineering, process control, alarm management, and operator training. We have worked for major oil companies and run our own businesses. We have consulted for academia and a slew of other industrial enterprises. We have published technical books and professional articles and have spent time in leading-edge research and development. It has been our focus for these many years. This handbook is our way of "paying it forward" to you and the next generations of the industrial community.

You may have already heard or know about many of the topics in this handbook. Other topics or material that is here you will not find anywhere else, at least not soon. This new material provides important tools that are either missing now or present but not well understood. The power of this handbook is its solid understanding of why and how these topics and materials fit together and work. As you read, you might be of two minds. One part may think that a couple of engineers cannot really know what you need in order to do your job. After all, you have been operating for years—and doing pretty well at that, thank you very much! And to an extent, that is correct. On the other hand, another part of you may be thinking, how come it took so long to get a useful operator guide out? Wow, if I would have had this early on, I cannot tell you how many close calls I could have missed, or how much clearer doing my job might have been.

Please remember that the design and intent of this handbook covers broad portions of your responsibilities: how shifts start and end, how to keep aware of what is going on when it is oh-so-easy to assume more than you can prove, how to know that you might need assistance even when it may seem that doing so might seem weak, as well as how to obtain that assistance from others. You will discover tools that clarify and supplement your existing skills. You will discover tools to do old work in a better way. You will find new tools and concepts that can change the way you find and manage certain situations. All of this should lead you to feel more confident, better equipped, and more comfortable. Those are all good for you.

It is time to get started.

Module 2: Enterprise

We almost never fail the leadership test during a crisis that we didn't already fail before. Responsibility is developed before it is needed—it is practiced when it is.

—Lasky [2.1]

The Enterprise is your workground. When you know more about what it is and what drives its design and management, you can make your way to a more comfortable understanding of how your operating fits into everything else.

2.1 Goals, Roles, and Culture

You sit in the control room of your enterprise as the operator. An enterprise is the organized entity for making, storing, or transporting goods and services for sale. It may be owned and controlled by any type of organization, including businesses, non-profits, and government agencies. A well-designed enterprise includes proper physical equipment (including maintenance), operational procedures and protocols, a willing and able professional staff, and a rich culture that melds it all together into a reliable and safe entity. It is often extensive and complex, with lots of moving parts—or it can be small. Responsible organizations accept, that despite your best efforts as a response to problems, things can go wrong. They strive to identify and resolve problems and issues before they escalate.

This module provides a view into the enterprise within which you work. This discussion can be important for you. You will be able to see how your enterprise "fits you." Yes, this is a turnabout view. It is very likely that most of your thinking about this is the reverse: how you fit into the enterprise. When you know more about how responsible enterprises support their operators, you can better understand how to work together. Working well together is a comfortable and worthy goal.

Types of Enterprises

An enterprise can be a collection of physical equipment and energy and resources usually used to manufacture or produce something. It is often called a plant or operation. Things like petroleum refineries, chemical plants, electrical power generation and transmission lines, food processing, and the like come to mind. In general, they cannot operate without direct operator supervision—your supervision. This may be because of their complexity. There may be requirements of public safety. It may be a critical public resource—for example, highway transportation networks, pipelines, electronic data networks, air traffic, security monitoring systems, and so on. Some may be able to operate without supervision, but supervision brings an important level of performance not attainable without it.

The scale of plants and operations is of little concern unless they are very small or very, very large. The very small ones are for special purposes. Those purposes generally require very special operational management. The very, very large ones can be so impactful to a community or a locale that special or critical operational practices and regulations would govern many aspects of their activity.

Here is what to look for:

Leadership

Leadership requires capable leaders who are knowledgeable, honest, and communicative. "We almost never fail the leadership test during a crisis that we didn't already fail before. Responsibility is developed before it is needed—it is practiced when it is." [2.1] Responsible leadership requires the desired goals be reachable and sustainable. Effective leaders share the workload and the prizes. Their methods must be fair and effective. Different hands do different work, but the work is done. Managers also must do their hard work and share equitably the success.

Walking the Walk

Leaders are expected to lead in integrity, responsibility, honesty, and success. Leaders "walk the walk." This is an expression that captures it all. Leaders will get down into the trenches and see what you see, feel what you feel, and understand what they ask you to contribute. They are also expected to make the hard decisions thoughtfully and carefully—just as they expect their personnel to pay careful attention to their duties and responsibilities.

These brief illustrations get to the heart of this message: [2.2]

> Whenever Wal-Mart (WMT) founder Sam Walton traveled on business, he rented the same compact economy cars and stayed in the same inexpensive hotels as his employees.
>
> Ray Kroc picked up the wastepaper in the parking lot whenever he visited a McDonald's (MCD) to show cleanliness was a continual job for everyone—even the CEO.
>
> During the early years at Charles Schwab (SCHW), whenever the customer service phone lines got really busy, founder 'Chuck' Schwab dropped everything and answered calls along with everyone else at the company who had a stock broker's license.

Talking the Talk

Communication is a great tool [**M6**]. Effective leaders use it consistently and often. Talk is easy, because leaders can use the enterprise infrastructure (newsletters, memoranda, town hall meetings) to talk "top down" to their organization. But "talk" is not "communication." Effective talk is the truth couched in a responsible way. Effective communication begins with listening. Responsible organizations develop powerful mechanisms for all stakeholders to communicate—to talk and to listen.

Step-up Leadership

Enterprise leadership is not only about senior and junior leaders. It is different and important. You as a member of the operations team may suddenly find yourself in a situation that requires leadership. While in the control room while on shift, your duties include the *in charge rule* [**M34.2**] that requires you to always make sure someone is in charge of operations when, at the moment, you are not. Step up leadership applies when you are not leading but part of a team or even a loose group and you recognize that no one else is leading and immediate leadership is necessary for responsible action. At that moment, informally, and respectfully you assume a temporary leadership role to either actually lead the group or recruit a leader who will. Later in this handbook you will find a discussion on how to handle situations when no one is acting. **[M16.5]** Diffusion of Responsibility **[M34.2].**

Understanding the Enterprise

A perfect understanding of an enterprise is not essential for proper operation. However, every operational piece that is used needs to be very well understood and specifically accounted for in operational procedures and training. This goes way beyond familiarity. This is about having a level of expertise that permits everyone to recognize proper operation, understand the nuisances of operation going astray, and competently manage situations that require intervention. This understanding must be embedded into the fabric of the enterprise to ensure continuous improvement and continuity between stewards.

Control Personnel

Every control room or operations center also has a "field." The field is the physical location(s) that contains the actual working equipment, other components, materials, and energy. The field may begin at the doorstep of the control room or be at a far distance. Operators who reside in the specific control room for their operational area manage the operational area partially or totally. They are the board operators (also referred to as control room operators). When things go wrong in the field, and understanding and/or remedy is beyond the timely physical reach of the control room operator, the work in the field is done by others. The term "field operator" describes that field-resident person (or one specifically designated to visit the field) who performs much of the routine work there. Of course, there may be others in the field including maintenance, inspection, delivery, visitors, supervisors, and the like.

Let us review the key personnel who have active roles in the control room and related plant areas.

Control Room or Board Operator

This is the primary person responsible for ensuring the process parameters are maintained within limits. Normally, this individual manages from a purpose-built operator console within a control room. This individual is expected to be "operator qualified" by possessing the necessary required knowledge and skill to recognize abnormal operating conditions and react appropriately. For some industries, this qualification is established by legal requirement.

The control room or board operator:

- is in charge during shift.
- focuses on coordination with the schedule and other operations or work.
- retains overall responsibility for all operations within his span of management.
  - No informal passing of tasks or duties.

Field Operator

Field operators are the extension of the control room operator's eyes, ears, hands, and feet. They are responsible for ensuring the outside-the-control-room process equipment is monitored and any hands-on operations and adjustments made. Normally these individuals act under the direction and authority of a qualified board operator. They are present at the operating equipment as a "long arm" of the operator [**M4**].

Field operators:

- focus on the immediate time frame.
- may perform general tasks (such as being eyes and ears) but are usually devoted to a specific plant component or localized situation.

- possess responsibility for competently observing, reporting, and recommending.
  - Hands-on actions must proceed only with the agreement of and under the strategic control of the control room operator.

Shift Supervisor

The shift supervisor is the individual responsible for providing direction, guidance, and feedback to operations personnel on shift who are performing all manner of operator duties. This includes a measure of training, fitness and competency evaluation, and remediation; and all manner of appropriate administrative actions necessary for a proper function of a shift.

2.2 Long Term vs. Short Term

We know that long term is better than short term. Remember the parable of the race between the tortoise and the hare. Both begin at the same starting point, leaving at the same time. The first one to cross the finish line is the winner. The hare leaves in a flash, gets far along the course and then pauses, exhausted. Believing that he has traveled too far and too fast to be overtaken, he takes a nap. Meanwhile the tortoise adopts a pace that, while slow and plodding, is steady. He passes the sleeping hare and wins the race. The sustainable long-term plan delivers.

To demonstrate long-term sustainability in a more relevant way, we consider two hypothetical, identical companies: Company A and Company B. Each reports to the same top executive, but with different accountability goals. Each company starts out in the same region, with the same resources. Each makes the same product using the same suppliers, cost, and profit structure. Their single difference is that Company A's performance is judged on a strict yearly basis: every year at year's end. Company B's performance will only be judged at the end of the first five-year period and every five years thereafter. Figure 2.2.1 shows the progress, year after year.

In Figure 2.2.1 you can see that Company A (orange line) has excellent performance, increasing in value more than 50% by the end of the first year. Company B (blue line) has dropped about 50% in value over exactly the same time. Company A gets a gold star. Company B gets a pass for now since the top executive upstairs is not looking yet. What might be happening? Company A's results at the end of year one are understandable. This boss is going all out, got cooperation from his entire team to go for the maximum, pushed the production machinery for their highest rates, priced product to sell, and got results to match. Company B operated very differently. It planned and invested in new or improved production equipment and methods, better sales training, and enhanced marketing systems. Quite predictably, its sales team had less time to sell and their equipment received lots of attention and more than a few tweaks—but at the expense of production numbers and, of course, sales.

In year two, Company A has increased value, but less than in their first year. Company B has started reversing its decreasing trend. Dig a bit deeper. Company A achieved less the second year than the first likely because its sales team needed to catch their breath; the machinery needed maintenance due to working overtime. Pushing at maximum works only for a while. On the other hand, Company B started seeing better results using their new approaches and skills, along with improved supplier agreements, and they were able to keep their machinery working more reliably. You can see how this is shaping up!

At the end of five years, the difference in policy is telling. Not only has the long-term company (Company B) dramatically exceeded the results of the short-term one (Company A), but also the short-term one appears to be headed for collapse. We could see this coming since it was pretty clear that it had likely spent considerable, if not all, of its reserves in very earlier years to meet those short-term goals.

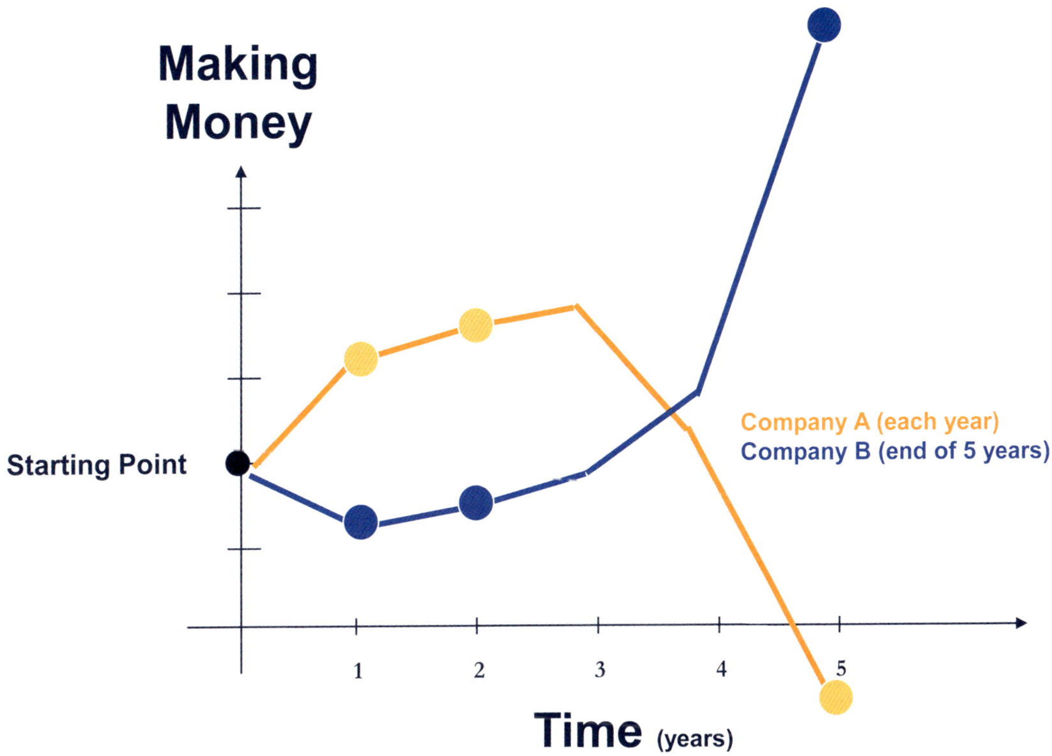

Figure 2.2.1. Two competing companies, over five years, showing a clear and dramatic loss by pursuing short-term goals over long-term goals

Valuable Message

All long-term strategies will lose in the game of competition if judgments are made in the short term. Your enterprise can work towards sustainability and stay there only if everyone and everything is working for the long term. Otherwise, what looks okay at the moment will succumb to all manner of operational drift and all manner of immediacies of little value.

> All long-term strategies will lose in the game of competition if judgments are made in the short term. Always! Every time!

Your job as operator is to decide whether you are working, or someone else is asking you to work, for the long term or the short term. If it looks like short term, you need to understand what to compare against what for a fair comparison. Then you can decide the responsible way forward: question and get advice so you are not making decisions that seem to work now but will cost much more later on.

2.3 Operational Drift

Operational drift always results from small unresolved problems. Small changes are a part of everyday life. We often do not know they are there; or if we do, they quickly fall under the radar. Yet, small changes in the control room can be a big problem if they later lead to important things we do not want to miss. The sticky problem area is in short-term progress versus long-term goals:

- Conflict between making production (and profit) vs. operating completely safely
- Number and quality of the messages about operational activities and safety
- Forgiving small irregularities in favor of waiting until they became strong enough to call for attention

These problems can come from anywhere. No matter where they come from, if not fixed, they will drift. The gradual changes will always build up. When they do, they almost never are pretty. Watch for them. Attend to them. They will be covered in Part 5 (*Weak Signals*).

2.4 Mindful Organizations

A mindful organization seems like something made up to sound politically correct. This is not the case here! A mindful organization is one that has achieved a sustained level of successful operation. You may be more familiar with it being called a high-reliability organization. Much of the success is attributable to the inherent strength of its entire design. This is so important that no organization can aspire to achieve real long-term success without it. In this handbook, a large number of useful tools and ideas will be presented. At the end of the day, they are going to be largely a waste of time unless your organization itself is healthy and fit—highly reliable.

2.5 Enterprise Capabilities

Enterprises are conceived by entrepreneurs, designed by engineers and architects, and constructed by builders and fabricators. Specifically, others on the enterprise team are responsible for the following:

- Design and build of the plant or enterprise
- Safety, health, and environmental awareness and attainment
- Regulatory compliance
- Financial conditions to be met by the enterprise
- Markets, marketing, delivery objectives, products, product quality, and production scheduling (including end-of-operation decision)
- Staffing levels, competency, remuneration, and retention
- Enterprise maintenance, replacement, and enhancements

Management intends to do a proper job, and for the most part they do. But we must not assume that they got it right from the start, or that they fixed it later and now it is good to go. Despite best efforts, or as a response to irresponsible actions, things go wrong. Responsible organizations accept this and strive to identify and resolve problems and issues before they escalate. This means in the boardroom, the engineering and design, the operational policies and procedures, and all of the rest.

Nothing in the above list is your responsibility. You do have an essential role. You are the primary agent to keep everything going properly during your duty shift. Guidance and assistance for what you do and how you do it form the bulk of this handbook.

Module 3: Safety

Safety is the most important responsibility and criterion for guiding your job. So let us find out what it really means for you in the control room.

3.1 Safety

Your entire handbook is written with the goal of safe operation. Everything works to help you understand what to do, to do it well, and to make sure it works. When you operate safely, costs stay in line and your job stays valuable. "Safety is the state of being 'safe,' the condition of being protected against physical, social, financial, emotional, occupational, and other types of consequences of failure, damage, error, accidents, harm, or any other event that is undesirable. Safety can also be defined as the control of recognized hazards to achieve an acceptable level of risk." [3.1] Safety is not the job of your safety department. It is not the responsibility of your safety manager. It is the job of your management. It is the job of senior management. [3.2] And it is a job with the responsibility to deliver; the requirements to demand what is needed to be safe; the determination of all needed actions to meet the requirements; and the commitment to gain lasting compliance to ensure that what is needed is evaluated, supplemented where necessary, and maintained to the full. Having said that, it is everyone's job to act and operate safely. It must be part of the DNA of every individual and every responsible organization.

> The bottom line is a complete culture, from the highest manager to the furthest employee, to understand safety, accept safety as the only appropriate operational parameter, and to be provided with all necessary means to achieve the full measure of safe and responsible operation.

3.2 Characteristics of Safety

Enterprise safety must be gained first and foremost by its design and build. Atop this, safety is completely dependent upon how people and equipment function together to meet the operational requirements necessary for safe operations. It is gained by having sufficient technology and operational excellence. It requires an engineering community constantly tasked with participating in evaluating the fitness of the enterprise to operate as designed and whether the design is safe. It must be constantly supplemented by consistent practices that ensure the enterprise is properly run. It must be fully supported by all leadership to ensure the highest needed standards are met. It also requires continuous evaluation of the competency of the equipment operating in all production campaigns. The constant presence of sufficient experienced, trained, guided, and confident operations leadership and technology personnel must be assured.

3.3 Components of Safety

Safety means different things to different people. Unfortunately, actually being safe does not respect your meaning. Safety cannot be defined into being, and danger cannot be kept at bay by what anyone thinks about safety. Safety can only be gained by actually working hard to be safe. Being safe has four distinct aspects. It depends upon the unrelenting adherence to all of them, all of the time. No single one is more important than any other. None are optional.

Personal Protection

Individuals will wear and use the proper personal protection equipment (PPE) whenever needed. Each enterprise will understand the risks and exposure and decide what to use, how to use it, and when to use it.

Safe Activities

Everyone must go about their work in a safe and responsible manner. Everyone must use safe work practices, such as wearing the proper PPE, notifying the proper personnel when working in specific areas, using proper tools and equipment, climbing ladders properly, entering confined spaces properly, following proper lockout/tagout policies, using the permit-to-work system faithfully, and all of the rest. No shortcuts are safe. No shortcuts should be permitted.

Process Safety Equipment

The process and equipment being operated must have been designed for safety. It must contain all reasonable engineering and operational safeguards needed to keep things safe even if someone makes a mistake, something breaks, or something gets out of control. Operating interlocks are one way to achieve this. Over-temperature, over-pressure, high- or low-level shutdowns, dikes to contain spills, and vessels designed to withstand fire are also ways—and there are many others. All are needed.

Process Safe Operations

This is where it all comes together. It is where your part as operator is important. Even though you are wearing and using the right equipment and doing your work correctly, and the plant has the requisite safe operations safeguards, you must conduct all of your activities in a safe and effective manner. Following operating procedures is a part. Not inventing ways to handle situations where no appropriate procedure exists or the existing procedures do not appear to be working is a part. Effective training and practice is a part.

3.4 Operator vs. Independent Layer of Protection

This aspect of being safe is about whether you as an operator should be considered a formal layer of protection (see below) to ensure safe operations. The concern is straightforward—people being people, they cannot be counted on to be perfect enough, to be right enough, and to be effective enough all of the time. We best not place lives, fortunes, and our environment to the actions of an individual alone. Yes, of course, good operators protect lives, fortunes, and the environment every day. That is your job. However, the modern industrial society knows to backstop any operator error with effective protective defenses. Responsible enterprises implement best practices in situation management without shortcuts.

> No enterprise should depend primarily upon operator action to provide the necessary full degree of safe operational outcomes. Other formal protections must be in place to limit or otherwise mitigate the effects of abnormal operation.

This discussion is to better understand your bottom-line responsibility. We bring it up here so that you and your operations management can be on the same page with respect to accountability. This is about how far operator action alone (your action) is expected to formally protect operational success. Certainly, your enterprise depends upon your proper actions and responses. You are expected to do your best. In turn, operators must be able to rely upon specific formal independent layers of protection to backstop situations that are not or cannot be successfully managed by their action alone. Therefore, no responsible enterprise should depend primarily upon operator action to provide the necessary full degree of safe operational outcomes. All enterprises that can develop abnormal operations that pose a degree of danger to people, the community, and the availability of essential goods and services must have independent, proven, and ever-present safeguards and protections.

Layer of Protection

Let us look at protection layers first.

> "Protection layers [(PL)] typically involve special process designs, process equipment, administrative procedures, the basic process control system (BPCS) and/or planned responses to imminent adverse process conditions; and these responses may be either automated or initiated by human actions." [3.3]

Next look at independent layers.

> "Independent layers of protection for which each layer helps to reduce the overall level of risk of the process. The inner layers help to prevent a hazardous event (explosion, fire, etc.) from ever occurring, and the outer layers are designed to help lessen the consequences of an event once it has occurred." [3.4]
>
> "[Independent Protection Layer (IPL)]: A system or subsystem specifically designed to reduce the likelihood or severity of the impact of an identified hazardous event by a large factor, i.e., at least by 100-fold reduction in likelihood. An IPL must be independent of the protections layers associated with the identified hazardous event, as well as dependable, and auditable."[3.5]

Bottom Line

Responsible operator performance depends on your enterprise having provided a safe design for the equipment, proven safe operating procedures, qualified operators, effective alarm systems, reliable maintenance practices, and the full measure of safe protection equipment. Certainly, you, the operator, are important. Just as vital, you must be able to depend upon the rest of the enterprise doing its part well.

3.5 Guidelines, Standards, and Practices

Guidelines, standards, and best practices provide invaluable benefit by informing all of us of what has been understood, accepted, tried out by those before, and works.

They are useful because generally it is not possible to evaluate the full extent of a design until it is built and tried. In this way, understanding the experience of others enables us to move our efforts from "trial and error" into the arena of "very likely to work and be of real benefit." Hence, guidelines, standards, and practices form the cornerstone of good engineering design and operation. They may be prescriptive, outcome-based, or a combination of both.

Prescriptive means that they describe a specific procedure or steps to follow; and they must be followed no matter what. Outcome-based means that they describe an outcome or requirement but without providing specific ways or procedures for producing the required results. In cases where there may be a reasonable number of different acceptable ways to achieve an outcome, the outcome-based approach works extremely well. Even for this, a specific way is usually preselected by the enterprise. In other cases, where it is either difficult to select the correct approach in the heat of the moment, or where undue prejudicial influences exist for a selection, the prescriptive approach is best.

When it comes down to the design of a specific level of company or corporate policy, almost everything that is set out internally by that company is done prescriptively.

Guideline

A guideline is a principle used to make a judgment, determine a policy, or choose a course of action. Guidelines may be large in scope (e.g., for shift handover or electronic logging) or limited (e.g., if 8-hour shifts are used, be sure to change shift times to a later time slot rather than an earlier one). They offer a wealth of things to consider. There is a common wisdom that, when using guidelines, it is best to adopt a consistent approach rather than pick and choose different parts from different sources. Taken as a whole, guidelines fill in the gap between having no guidance or recommendations vs. other, more formal options.

Standard

A standard is an established criterion or set of rules that has been defined by an entity that governs approved, acceptable practices. The main purposes of standards are to promote safety, reliability, productivity, and efficiency. Such practices may involve design, operation, construction, or other aspects. For standards that have been approved by governmental entities, within their applicable jurisdiction, all behavior subject to the standard is expected to comply. Enterprises that do not follow this practice cannot later assert that they exercised due care in the event of problems or issues. Standards that are not adopted by statute or law are considered important because they describe the extent that a duty of due diligence is expected.

Example standards-setting bodies are ANSI (American National Standards Institute) in the US, CSA (Canadian Standards Association) in Canada, ISO (International Standards Organization), IEC (International Electrotechnical Commission), and others.

It is important to observe that a standard that is required within an applicable legal jurisdiction can also be extremely useful guidance for all other jurisdictions where it is not a legal requirement. For those situations, consider it as a recommended practice. This benefit should not be underestimated.

Recommended Practice

A recommended practice is a way of doing things that comes directly from the available pool of resources (standards, guidelines, long experience, and such) and is considered the appropriate way to do things. In general, a recommended practice is more long-standing than a best practice and usually leads it in time. This means that standards and guidelines usually draw on the earlier experiences of best practices and recommended practices.

Best Practice

Best practice describes methodologies, processes, or techniques that consistently produce results superior to other alternative designs or approaches. A best practice is a way of doing things (for which the practice applies) that may come from the available pool of resources (standards, guidelines, recommended practices, long experience) and is considered, by those with sufficient experience and expertise to make such a judgment, the most appropriate practice. Unless it conflicts with applicable jurisdictional law, it would be the most appropriate practice to use. While they generally do not carry the weight or responsibility that a mandatory requirement may, oftentimes they evolve, in the absence of such requirements, as the standard of care that responsible organizations strive to follow.

Special Considerations

RAGAGEP

The acronym RAGAGEP stands for Recognized And Generally Accepted Good Engineering Practice. Recognized means that a sufficient number of those in the direct field of application have looked at it in a suitably substantial way. Generally Accepted means that there is consensus among those who did the recognizing that they agreed with what they saw. And Good Engineering Practice means that what they saw and agreed to is the right way to do the work. The strength of this category is that RAGAGEP carries the dual benefit of effectiveness and appropriate exercise of due diligence. In the United States, there is a generally found category in the Occupational Safety and Health Administration (OSHA) [3.6] regulations that has the legal force of including RAGAGEP requirements in certain disciplines (alarm management, for example) as carrying the weight of law.

BATNEEC

Best Available Technology Not Entailing Excessive Cost (BATNEEC) is most often used to decide the limitations when considering alternative remedies or technologies to use to control environmental exposure. The determination of "excessive" is not related to available funds, but rather to the extent of safety or environmental operational risk compared against the general enterprise cost and profit numbers. This means that a manufacturing operation using $20,000 a day in raw materials and utilizing capital equipment costing tens of millions of dollars cannot suggest that a $100,000 expense for needed technology is excessive (even if the enterprise is operating at a loss) when the likely risk of serious issues can result in life lost or extensive environmental damage or property losses. This principle is also known both as As Low As Reasonably Achievable (ALARA) and So Far As Is Reasonably Practicable (SFAIRP) in the United Kingdom as well as in New Zealand.

Duty of Care

Duty of care is defined as follows:

"A requirement that a person [or enterprise] act toward others and the public with watchfulness, attention, caution, and prudence that a reasonable person in the circumstances would. If a person's actions do not meet this standard of care, then the acts are considered negligent, and any damages resulting may be claimed in a lawsuit for negligence." [3.7]

This will be your guiding light. Please do this not to avoid litigation. Do it to avoid causing damage and harm to others.

3.6 Disaster Chains

Almost no serious incident or disaster happens due to a single point of failure. Managing a single failure cannot yield sufficient safety or protection.

Chains

You have heard the expression "a chain is only as strong as its weakest link." It suggests that you want something at the end of that chain. You pull on the chain until that desired thing is in hand. You want the chain to hold. You make sure that each link is strong enough. Each link represents a successful component (Figure 3.6.1).

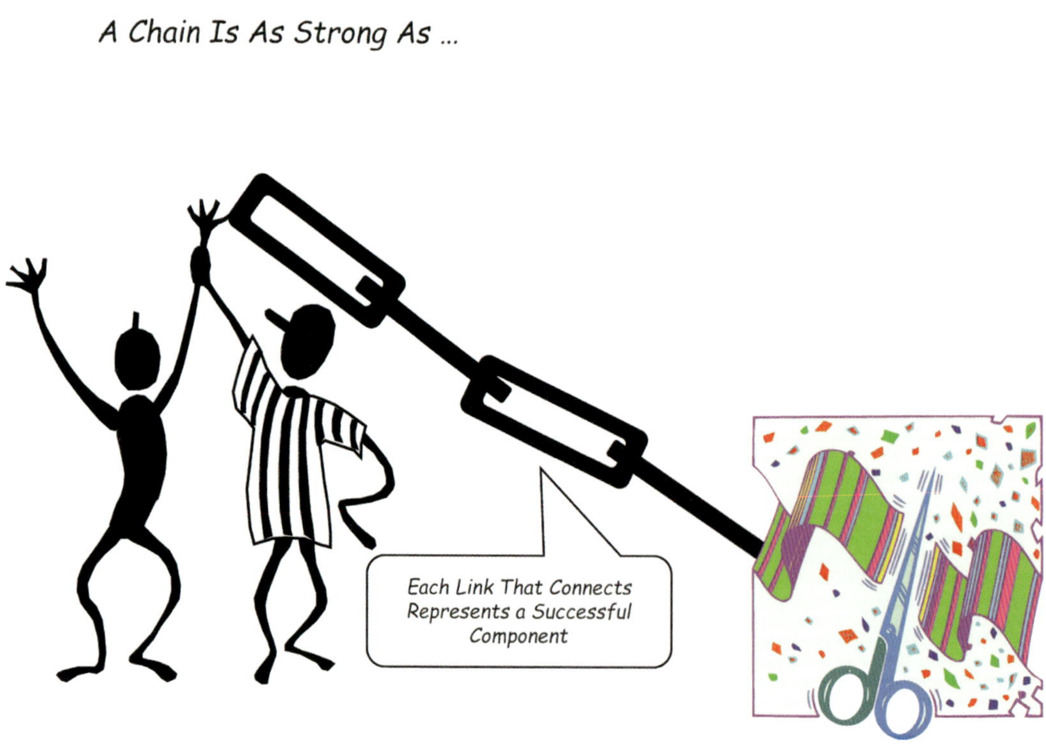

Figure 3.6.1. A chain is only as strong as its weakest link. When it stays intact, you can have whatever it is that is attached. If it breaks, you get nothing.

The "chain is only as strong as its weakest link" wisdom can motivate us to do the right thing: to do our share of the work. If you were the one whose failure caused that chain to break, you would naturally be the bad guy. When the chain breaks, the prize at the end is lost. You do not want this chain to break!

Disaster Chain

But chains can work both ways. If the chain is attached to something that no one wants to have, we now want it to break. We do not want what is at the end (Figure 3.6.2). A single broken or missing link will provide the chance to avoid the problem. An intact link might stand for improper training (for the link to break, training needed to be fully up to the task), another for faulty alarms, and still others for inadequate maintenance, a faulty HazOp policy, a failed lockout/tagout policy, and so on. If any single one of these safeguards were to work, the corresponding link in the disaster chain would break leaving the disaster behind. This is being safe.

Figure 3.6.2. A disaster chain represents an intended safeguard where each link that *does not break* represents a critical safety factor that *did not work*. If none of the links break, you pull the disaster right into your lap. Most serious industrial incidents have disaster chains more than three links long. This means that all three safeguards did not work as designed (and required) to ensure safe operation!

The Swiss cheese model of incident causation presents this message in another graphic way. [3.8]

The Swiss cheese idea is shown in Figure 3.6.3 below. [3.9] Each slice stands for an infrastructure safeguard. Each hole is something important that is missing or done so poorly that it becomes ineffective. Even if there are a lot of slices, the protection is not there if all of the individual ones responsible for the protection against an unwanted production safe operation condition are not fully working so that holes line up. Not working means that there will be a "line of holes" for a

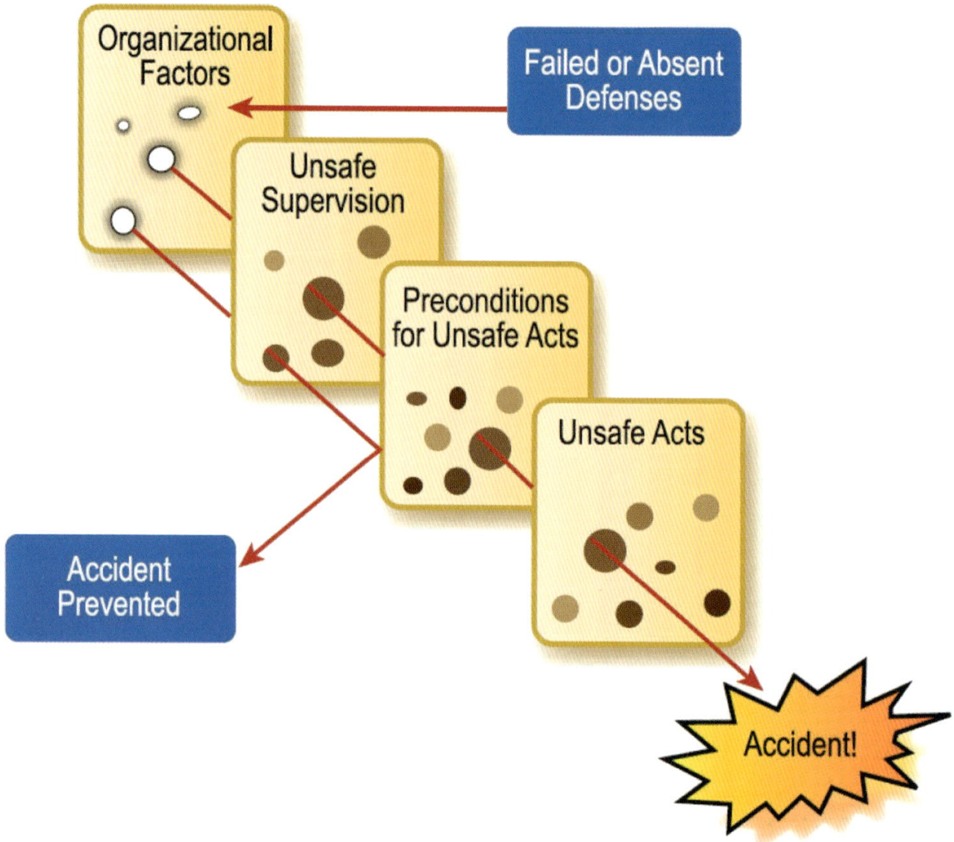

Figure 3.6.3. Swiss cheese model of how an incident (accident?) can occur despite several presumed safeguards in place. If each safeguard has significant weaknesses or faults, it is just possible for a single event to find a fault for each safeguard and therefore cause the sum total of all safeguards to fail. Therefore, it is the quality of the safeguard that is important, not just their number.

problem (a line depicted in the figure) to pass through every one of the individual safeguards. Danger and damage will go through every one of those lined-up holes. You may have heard of this model during the recent coronavirus (COVID-19) pandemic where slices included social distancing, sanitizing hands and surfaces, and wearing face masks.

Safety is not gained through the number of safeguards. It is gained by each safeguard being competent enough to not have a hole that lines up with others.

3.7 Challenges to Safety System

A challenge to a safety system is identified by any notice, warning, automatic action, or any other notification that suggests or announces something is not where it is expected or supposed to be. It could be as simple as a lit physical indicator light, or a display element change to warning or caution. It could be as complex as an automated speed reduction or movement that stops a pump or repositions a positioning actuator. The challenge is the abnormal situation. The notice is the way you are told about it. You are required to explicitly examine the challenge, fully explain and understand all of its ramifications, and then act to fully remedy the condition. The challenge is not managed until it fully returns to normal (no notices are present).

In a properly designed process control and safety operation management system, no routine operation, operator action, or operational change of state may include the activation of any malfunction warning indication whatsoever, unless there is an actual challenge to safety. All actual challenges need explicit action to remedy. Too large a number of incidents and disasters have a direct line to a challenge that was not seen, was ignored or minimized, or was managed perfunctorily. This clarification is explicitly intended to bring a vital safe operating concept to the front of your mind as operator. This means that every challenge must be assumed valid and require explicit action to accommodate the situation announced. The importance of understanding challenges and the actions you need to take because of them, comprise a critical activity in achieving operational safety. If you are having difficulty resolving the validity of any challenge, it is important to get help [**M34**] or invoke *permission to operate* [**M35**].

3.8 Safety at the Boundaries

On the Line

In sports, the "tie goes to the runner" and whatever occurs "on the line" is still in play. You are not an athlete in the control room. Sports rules do not apply here. What do apply are two of the most important principles:

- If any aspect of your control room operations has you operating at the line of maximum excursion (high or low) or very close to it, you should categorically count that as being <u>over the line</u>. When you are that close, even the most carefully managed efforts will not be strong enough to always ensure safe operation. Since you know where you are, use that knowledge to go for the safe harbor.
- If you are approaching a potentially unsafe operating condition or other bad situation, you must assume that you are not going to be successful enough to avoid the bad consequences. A "tie" must always be understood as going against you!

Negotiating with Failure

None of your past success when close to unsafe or bad operation is of any value. None. Never. All of your past experiences that were risky and yet you succeeded provide no lasting skills. Sure, you did learn a single way to skirt disaster. But every disaster has more than one moving part. Even if you think the same one is coming at you again, it most likely will not come in exactly the same way. You have no reliable assurance that you can do as before to resolve it.

Every time you are approaching any problem situation, always begin anew using all of the tools, training, and safeguards you have. Work each out on its own merits with a fresh understanding.

Mindful Deviation

You probably know that it is unreasonable to lay down a rule that is so extensively constructed that it can only be used exactly. No deviation allowed. No assumptions allowed. Follow it to the letter.

When you are in charge, your management expects you to understand everything you do and consider. This also applies to rules and procedures. Of course, there are exceptions for those dire situations that are so impactful that no one should do anything except follow to the letter: a fire (sound the alarm first); a spill or release (go cross-wind and get out of there); a sudden natural event (shutdown and render safe). For the rest, here is a suggested approach for you to consider every time you are working to rule, that is, following a rule or requirement that is expected. We name this activity *mindful deviation*. Emphasis is on the mindful part.

1. Begin with the rule (or requirement). Understand it fully. If need be, refer to appropriate documentation, operations protocol, and the like. Familiarize yourself with all specific conditions that affect how you are to choose it, understand it, and use it.
2. Take your current situation (the reason why you think the rule is applicable and appropriate) and fit it into the rule. Fitting it means that you are able to identify what the rule is asking with what you are facing.
3. Explicitly identify each and every deviation or mismatch between the rule and your situation. What is important here is finding what does not fit. No matter how much fits, it is what does not that will govern your success.
4. One by one, evaluate each mismatch and *prove* to yourself (or get help [**M34**] to do it), that each deviation or mismatch is taken care of by the rule or takes care of itself; hence, you do not need to take it into account.
5. If everything from step 4 is taken care of, work the rule with what you have. If anything from step 4 is not taken care of, the rule cannot be worked properly. Escalate if you know you have enough time; otherwise invoke permission to operate [**M35**].

3.9 Bottom Line of Safety

Lord Cullen, a safety assessor to the Queen's Court in the inquiry into the Piper Alpha disaster in the North Sea where a fire and explosions destroyed an oil production platform on 6 July 1988, provides the following insight: [3.2]

> Safety is not an intellectual exercise to keep us in work. It is a matter of life and death. It is the sum of our contributions to safety management that determines whether the people we work with live or die.

Your enterprise's management defines the characteristics and components of safety for your enterprise. They will determine the standards and guidelines that you, other enterprise personnel, and visitors are expected to follow.

Safety information in this section, and throughout this handbook, is provided to enhance the success and safety of what you do to carry out your duties and responsibilities. Nothing here is intended to replace the safety procedures provided by your enterprise. If there is any discrepancy between information contained in this module (or elsewhere) and your site's practices and requirements, you will want to discuss it with your supervision and management.

PART 2
OPERATIONS

Your important role as control room operator is introduced. Discussion includes your functions and motivations, readiness and overload, what you use, and what supports your many varied activities.

The importance of successful communication is explained. Guidelines are given for safe and effective conversations in your control room setting, including ever-increasing virtual conversations.

The myth of multitasking is explained, and its dangers to operations are discussed.

Operator tools, including policies, procedures, and protocols are defined, and their benefits and uses are examined.

The functional aspects of operator training are covered, with a discussion of the enterprise responsibility to provide and maintain your full measure of qualification.

The benefits of understanding and using models are covered. The importance of interpreting and using data in context is explained.

Part 2 concludes with a brief discussion of the role of control room operator supervision.

PART 2 — OPERATIONS

Module 4: Operators

4.1 The Operator's Function

As the sitting *operator* of an industrial plant, or other operator-managed enterprise, you are entrusted with successful operational activity, second-by-second, minute-by-minute, hour-by-hour, day-in and day-out. You manage the real-time performance of an enterprise easily worth hundreds of millions of dollars. You are asked to accept responsibility for everything that happens during your watch; good, or heaven forbid, bad. You must be able to view and appropriately adjust every control loop, most sensors, all equipment, and supporting utilities within your area of responsibility and authority.

Yours is not an easy job. When it goes wrong, the inability to maintain situational awareness is a major loss. Loss of situational awareness directly contributed to almost every disaster event that was not the result of a complete and spontaneous surprise. No one wants an incident. But incidents and disasters do happen. We now know to a high degree of certainty why. They happen because those in charge of ensuring that they do not are not fully aware that they are happening. They are unaware of what is really going on. They are unaware of what is likely to happen. Or they think that what is not happening, actually is happening.

Successful operator performance centers directly on what is expected. What is expected of you is simple and yet profound: watch to make sure that the operations are properly progressing or whether they are not. When they are not, understand what is going wrong and take corrective action. This requires two skills: the cognitive skill of understanding whether or not something is going wrong, and the manual skill of performing the direct intervention necessary to make proper changes that work. The main thrust of your successful performance includes those cognitive skills as well as the resources for guiding intervention. The specifics of what intervention should be utilized must be left to experts in the enterprise as documented in all of your approved operating procedures and protocols. This handbook works to knit those resources together into an effective whole, supplementing missing structure and tools, where possible.

Definition of an Operator

An operator is defined as follows:

1. The person who operates equipment for its intended purpose. Note: The operator should have received training appropriate for this purpose. [4.1]
2. The person who initiates and monitors the operation of a process. [4.2]

4.2 Professional Operator

Even though intervention may be infrequent, your role is not babysitting. You are expected to fully understand your enterprise. You must keep fully abreast of what is going on through active monitoring and testing. You are asked to initiate and follow through with any and all corrective actions required to keep things properly operating. And you are required to seek outside assistance for all situations that challenge your ability to sufficiently understand and properly manage. From the moment you assume operational control until the moment it is either ended or passed on to another, you are expected to do all in your power to ensure safe, effective, reliable, and responsible operation. You are an essential member of the enterprise team. You are a professional. [3.52]

Operator's Creed

A creed is a deeply held expression of dedication to a belief. It is a sign of professionalism. Among the better-known technical creeds are the Engineers' Creed (USA) and the Obligation (Canada). No operator's creed currently exists. So we propose one. In recognition of the responsibility society places upon the shoulders of *operators* and as an expression of and respect for this level of professionalism, this Operators' Creed is offered for your consideration as a professional operator.

> As the operator, I accept responsibility for the safe, reliable, effective, and responsible operation of the facility under my control. I arrive fit for duty. I take responsible charge. I properly utilize the resources and tools at my disposal. I place safe and responsible operation above all else. When duty is done, I effectively close operations or pass on responsibility to my relief. I leave duty in a responsible manner.

4.3 Qualified Operator

A *qualified operator* possesses the full kit of necessary knowledge and skill required to operate safely and effectively. Operating safely includes preventing injury and environmental exposure. Operating effectively includes meeting production schedules and quality standards. Management has set your qualifications, tailoring the requirements to your specific needs. They expect you to be fully qualified and possess the skills to operate effectively. By meeting these qualifications, you will be able to use the resources available to observe and correct for all abnormal operations including safety-related ones. Only qualified operators should be entrusted with the operation of an enterprise.

The best working definition of what constitutes qualification can be found in the Pipeline and Hazardous Materials Safety Administration (PHMSA) Regulations [4.3] for US pipeline operations. We have substituted the term "owner" in place of the PHMSA term "operator" below.

> "[The PHMSA rule] requires [owners] to document that certain employees have been adequately trained to recognize and react to abnormal operating conditions that may occur while performing specific tasks."

No other regulatory or compliance entity has provided alternate definitions. For our use, it would be the best available. Reliance upon this would be responsible.

> "Note: To be qualified, an individual must be able to properly perform assigned covered task(s) and be able to recognize and react appropriately to any AOC [abnormal operating condition] that may (reasonably be expected to) be encountered while performing the covered task—whether the condition arises as a direct result of his/her work performance (e.g., be specific to the covered task being performed) or not (e.g., be generic in nature, but still observable because the individual is present on site)." [4.3]

Next, we define a few terms also taken from the PHMSA guidance mentioned earlier. We do this because, while the terms each have an intuitive meaning, it is important to agree on exactly what they do mean. Their terminology has been adjusted to follow more general usage. These definitions clarify what we mean when we use the terms throughout this handbook.

The basis of the PHMSA requirement is to ensure pipeline controllers are properly trained to ensure safe and reliable pipeline operation under their watch. (We have applied it to operators in a general context here.) They must possess adequate skills [**M10**]. The reference to specific tasks is understood to encompass all activities and operational states that are safety-related, which are broad and important.

Resources
"...accurate information and the training, tools, procedures, management support, and operating environment where [an operator's] actions can help prevent accidents and minimize commodity losses." [4.3]

Observe
"The act of watching; to watch or perceive. For purposes of conducting qualification evaluations using on-the-job (OTJ) performance, observations must include the interaction of the evaluator and qualification candidate to ensure that the candidate's knowledge of the specific task requirements and procedures (and the reasons for key task steps) is adequate to ensure the continued safe performance of the task." [4.3]

Operate
"Starting, stopping, and/or monitoring a device or system." [4.3]

Safety Related
"... any operational factor that is necessary to maintain [enterprise] integrity or that could lead to the recognition of a condition that could impact the integrity of the [enterprise], or a developing abnormal or emergency situation." [4.4]

Skill
"A demonstrable competency to perform a given task well, arising from talent, training, or practice." [4.3]

Qualified
"An individual has been evaluated and can (a) perform assigned covered tasks and (b) recognize and react to abnormal operating conditions.

"... a written qualification program that includes provisions to identify covered tasks and the intervals at which reevaluation of the individual's qualifications are needed. Requiring additional [operator] qualifications to measure or verify a [operator's] performance, including the prompt detection of, and appropriate response to, abnormal and emergency conditions likely to occur. ...to ensure that [operators] would have the necessary knowledge, skills, abilities, and qualifications to help prevent accidents." [4.3]

Roles and Responsibilities

"Each [enterprise] must define the roles and responsibilities of [an operator] during normal, abnormal, and emergency operating conditions. To provide for [an operator's] prompt and appropriate response to operating conditions, [management] must define each of the following:

"(1) [An operator's] authority and responsibility to make decisions and take actions during normal operations;

"(2) [An operator's] role when an abnormal operating condition is detected, even if the [operator] is not the first to detect the condition, including the [operator's] responsibility to take specific actions and to communicate with others;

"(3) [An operator's] role during an emergency, even if the [operator] is not the first to detect the emergency, including the [operator's] responsibility to take specific actions and to communicate with others; and

"(4) A method of recording [operator] shift-changes and any hand-over of responsibility between [operators]." [4.3]

Message about Operator Qualification

The bottom line message is that only qualified operators should be entrusted to manage control room operations. Unless they themselves have met the requirements for being a qualified operator, no one, not even supervisors or managers, should attempt to require a sitting operator to conduct an action or engage in a specific control or management activity. Otherwise, their guidance is limited to just that, guidance. While the details of what constitutes adequate qualification are left to management to determine, the requirements for an effectively qualified operator are quite clear. This section has laid out those essential needs.

4.4 Operator Locations

The Mobile Operator

A *mobile operator* is an operator that is not tethered to any specific place or position. Movement is not what makes an operator mobile. In order to be mobile, you must have seamless access to the information, communication, and controls as you move. Care must be given to understand what this *mobility* is all about.

Mobility for operators is only important if the ability to move about is beneficial. This type of mobility can be several feet, hundreds of yards, or even over tens or hundreds of miles. In order to be mobile, the individual operator moves about at distances that are meaningful for the job at hand. If mobility is just to avoid extra personnel, that objective must not overshadow the corresponding complexity of tasks and the inevitable competition for resources (time, problem-focus ability, and all of the rest).

There are three general categories of mobility: control room, plant area, and large geographic area. All are based upon distance. Of the three, control room mobility, done well, should offer important refinements to your performance. All of this presumes that your primary location is in the control room. Operators without a control room are a special case not covered here.

Control Room Mobility

Most control room operator stations are designed for a tethered operator. The displays are generally in a fixed location (although the horizontal platform they are mounted on may raise or lower). The interaction devices (keyboards, mice, or other pointing devices, including touch screens) and the communication devices (telephone or station intercom to plant radio system) may or may not be fixed or tethered by cords. Your ability to move about may provide important added benefit. In a single-operator control room, the simple ability to be yards away from the chair location may help. You can attend to personal needs, cook or eat food, access exercise equipment, and engage in light maintenance or other chores. For multi-operator control rooms, the benefits might increase substantially. Now there would be the ability to support each other without the need to be entirely absent from a primary responsibility. We do not for a moment endorse multi-tasking—that does not work [**M7**]. However, if you can move a primary task to the background for a short time, you might be able to competently pick up a new primary task during that period.

This level of "mobility" is relatively easy to achieve, once the requirements and options are understood. The first requirement is the ability to see and manage display content. A simple switch of the current operating screens at the control desk up to the large overhead or wall displays could do that job handily. You could manage that at the touch of a control. Better yet, your location could be sensed, and the screens automatically switch. Such sensing technology is already widely embedded in smart phones. From a human factors perspective, the specifics of how the screen switches progress (duplicating screens vs. switching them) might well be different when you are moving away from the control desk as opposed to moving to the desk. The second requirement is to have all communication devices mobile. Telephone and plant radio systems might be on a headset with appropriate controls. Pointing and clicking actions might be switched to voice activation rather than via a cord or short-distance wireless link.

Plant Area Mobility

Once we consider allowing you to move beyond the vicinity of the control room, technology suddenly becomes important. The voice part is easy. Equipment with adequate range and appropriate noise controls and connection security are readily available. Obtaining appropriate and useful visual information becomes the challenge. Not only must the equipment be mobile, but also it needs to be in a portable format that is easy to transport and use without excessively tying up your hands. And it must be safe to use in hazardous and harsh locations.

Large Geographical Mobility

In this arrangement, you use your mobility to arrive and manage at locations quite distant from each other and any central operating facility. No longer can you wait until returning to the control room to complete tasks. This level of mobility requires new concepts and procedures for collaboration and specialized interaction, visualization, and controls tools.

Requirements to Support Mobility

Now that you are going to be mobile, the enterprise needs safeguards, and you need the right tools to do it properly.

Safeguards

The moment you becomes physically mobile, everyone on the team needs to know that. The board is the baseline. If the other operators or close-support personnel at or in view of the board know where you are (and why), all is fine. If aid and assistance are needed, it can be summoned by a glance, a hand gesture, or a quiet voice. Nothing extra is needed. An operator away from view is a different matter. Everyone else on the team needs to know where he is, why he is away, and how to communicate with him. His location is important for physical protection in the event that his location puts him at risk. This knowledge also helps when the primary means of communication with him does not work. Knowing why he is away aids others in understanding what might be going on as well as in predicting likely next steps or locations he might visit.

Tools

Control room operations are designed with the intent that operators will be physically present. Physical security, environmental comfort, and information access are all covered by the design and operation of the room. If you are control room-based but can be mobile, no control room design considerations can be traded off to facilitate the mobility. Extra must be added, but nothing primary is foregone. This means our focus is on what needs to be added. There are two areas: information and communication, and protocols and processes.

Information and Communication

Mobile operators need access to all required operational information necessary to support control room obligations, as well as the specific information to support any remote tasks. This would include, but is not limited to:

- alarms (if the mobile operator retains alarm response responsibility)
- process conditions and states
- procedures, drawings, and all needed information to support understanding and operation
- communications delivered to the control room (that need to follow the operator)
- controls necessary to change operational conditions (set point changes, start/stop, etc.)
- information and communications to perform/participate in shift handover if it will be during that event, etc.

Proper consideration must be given to ensure effective communication channels are maintained to support all Human Machine Interface [**M18**] devices and voice communications, including in hard-to-receive locations. Special attention may be needed to accommodate low-light or no-light situations, high-glare situations, high-noise areas, and situations requiring no hands or limited hand use (both for equipment holding and interactions). Adequate protection from weather is required. Where interactions with others are needed, the equipment needs to properly support that activity. This not only means shared mobile devices and communications, but the ability to annotate and "point" to displayed information in order to focus others.

Protocols and Processes

The very nature of being remote from others requires careful attention to the procedures and personnel needed for coordination, collaboration, and escalation. The mobile operator requires knowledge of the locations and availability for all appropriate individuals. Explicit procedures need to be in place to support this activity. Where severe operational stress, including weather, is taking place, everything needs to be able to properly function in those conditions. There is more discussion in Module 8.

"Long Arm" of the Operator

Control rooms provide separation from potential dangers from the process. This separation provides a safe and comfortable location for you. You are physically at a distance from the actual equipment you are managing. But being away also eliminates a lot of sensory information. You cannot sense the vibrations of things nor directly hear the sounds of the equipment working (or notice when it is not). The emotional feeling of being there to take in the operation is missing. While advances in video surveillance (with audio) can provide a measure of seeing and hearing, it does not really replace actually being there.

There are situations where it is common that some equipment is unmanned except for occasional checks and maintenance. Examples include pipelines and other transportation networks. No matter the distance, there may be personnel at site for operational purposes acting under your direction. Often it is to check or verify some operational aspect of operations, such as the position of an unmonitored valve. Just as often, it is to make some sort of operational change that you need but cannot control remotely. The command and control of the coordination between you and at-site personnel parallels that of a maintenance operation [**M30.3**]. You retain ownership of equipment and actions at all stages of this activity.

Where other personnel are at the remote site, they should clear their presence with you. The process of "clearing their presence" is determined by appropriate enterprise procedures and is beyond the scope of this discussion.

Long-Arm Operator Checklist

Here is an illustration of the coordination between you and remote personnel [**Q2**]. More details are provided in Module 8.

1. Operator identifies all equipment to be observed or adjusted by remote site personnel.
2. Operator reconfigures and adjusts current operations as needed to enable safe and appropriate observations and adjustments by the remote site personnel.
3. Using independent means, remote site personnel verify that they have properly identified the equipment to be observed and adjusted, and it is ready for their activities.
4. Remote site personnel take the proper steps to observe and adjust the equipment.
5. Operator verifies, step by step, that the equipment can actually be taken back for normal operation and is fully ready for service or is sidelined for other work. This is an active process where the operator and remote site personnel check, test, coordinate, and otherwise ensure that the appropriate adjustments are done and equipment is operationally ready or suitably set aside.

When activities are complete, the remote site personnel check out with you. Once checked out, all adjustments and presence within the remote physical operational equipment perimeter should end.

4.5 Operators and Motivation

It is midway through your shift: do you know what your internal goals are? How are you able to work effectively through the long hours of each shift? What keeps you going day after day? How is it that you keep doing it as a career? What are your personal goals? How do you incorporate them into what the plant needs you to do? How do you manage to work so that each shift was a good thing to have done—or will you relegate it to just another interference between days off?

You want to understand and respect your personal objectives in order to be both comfortable and good at situation management. Personal objectives go a very long way toward preparing you to be a good operator. They assist you to develop acceptable and effective internal goals for job

your shift. To do that, you need to be aware of everything that is happening. When something is not right, you are going to find out what and remediate. Figure 4.6.1 introduces the four-step process [**Q3**]. This will be just the introduction. We go into this in detail later on in "Situation Management" [**M29**].

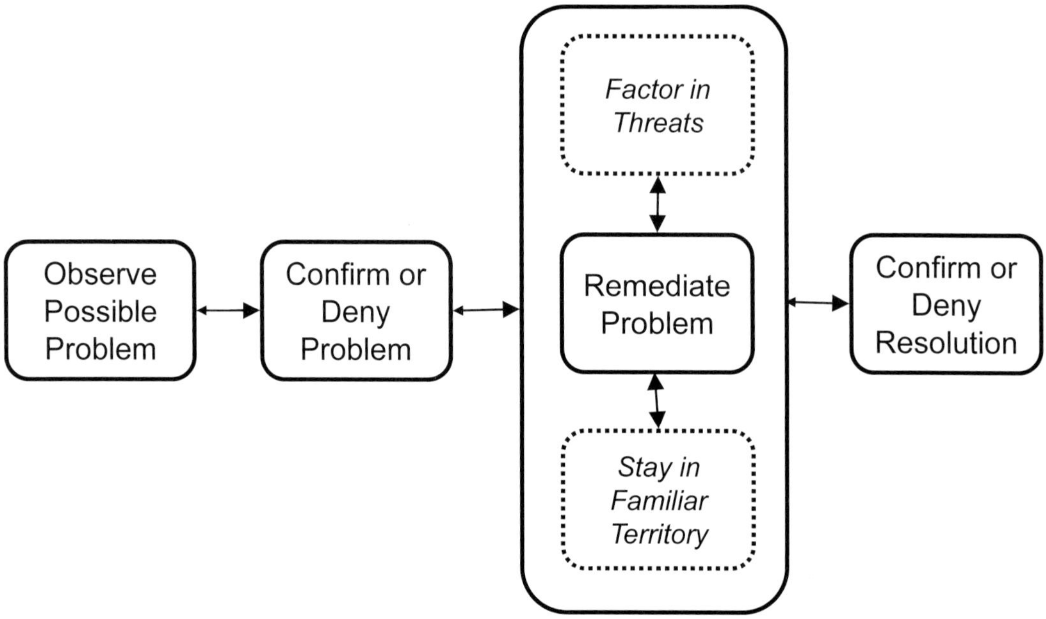

Figure 4.6.1. Basic four-activity situation management process depicting the progression from "observe" to "confirm" to "remediate," with some possibility to move back a step when evidence is not there or not enough to accept. Escalation and collaboration may take place at each step.

However, remediation goals cannot be the same for all situations. It depends upon the operational mode at the specific moment. Figure 4.6.2 lays out the overall picture. Focus on the green (normal) and yellow (abnormal) bands.

In the green band area, the operation is normal and the plant is normal. You are expected to keep it normal by (preventively) observing and testing the validity of what you suspect or see, followed by proper considered action.

All of this changes the minute the operation becomes abnormal (into the yellow band area). If, and only if, the actual plant is just abnormal (meaning just that and nothing worse), your responsibility is to return things back to normal. If it is anywhere else, you are not even going to think about working hard to try to restore normal. This is a fundamental safety net. It is fully covered in *permission to operate* [**M35**].

Operational Modes:	Plant States:	Critical Systems:	Operational Goals:	Plant Activities:
Emergency	Disaster	Area Emergency Response System	Minimize Impact	Fire Fighting First Aid Rescue
	Accident	Site Emergency Response System		
Abnormal	Out of Control	Physical and Mechanical Containment System Safety Shutdown, Protective Systems,	Bring to Safe State	Evacuation
	Abnormal	Hardwired Emergency Alarms DCS Alarm System	Return to Normal	Manual Control & Troubleshooting
Normal	Normal	Decision Support System Process Equipment, DCS, Automatic Controls Plant Management Systems	Keep Normal	Preventative Monitoring & Testing

Figure 4.6.2. Enterprise operational modes, plant states, critical systems, operational goals, and plant activities. Note the clear change in operator goals and activities as the situation changes from normal to abnormal to emergency.

Module 5: Readiness

5.1 What is Readiness?

No plant or enterprise that utilizes operators for management can be complacent unless the human operator is on the job. As operator, you are ready when you are present for duty and fully capable of performing all necessary activities. Readiness needs to be managed from the moment you begin your journey to your shift site until the moment you return to your end-of-shift intended destination. Travel time to and from work is counted towards your total fatigue load. It affects adequate time off for normal recovery activities like personal hygiene, eating, sleeping, and relaxation. The "trip to end-of-shift destination" responsibility is broader. It extends to ensuring that you are sufficiently alert—not otherwise emotionally limited due to shift activities or physically limited—for your safe travel to your end-of-shift destination.

There are three important readiness categories: **impairment**, **fatigue**, and **overload.** Each needs to be understood and managed with assured effectiveness.

5.2 Impairment

> Impairment is "the state of being diminished, weakened, or damaged, especially mentally or physically." [5.1]

Impairment of operators is attributable to a wide variety of causes. Their effects on good operation run the spectrum from reduced benefits by operating too conservatively, all of the way to endangering the enterprise and surrounding public. You will need to ensure that you are far away from any impairment. The following are the primary causes:

- **Fatigue**—overtired to the point of significantly reduced sensory awareness, diminished motor skills, inability to focus and concentrate, and manifestation of situational detachment. (Fatigue is one of the more ubiquitous causes of impairment, but not the only one.)
- **Stress**—a state of being where one's ability to cope with current conditions is overtaxed. It manifests as a burden. It causes one to feel as if current matters are exceeding your available resources, endangering well-being or the safety of self or others.
- **Chemical impairment**—senses are dulled, even if slightly, due to illicit drugs or side effects of prescribed medications; can also be chronic, as in addiction.
- **Emotional or psychological impairment**—pronounced distraction from duties and current happenings due to life events (e.g., very sick child or spouse, loss of a dear one, falling in love, unmanaged anger).

- **Pathologic mental impairment**—succumbing to underlying mental disease, distortion, delusion, extreme fetishes, etc.
- **Non-pharmacological addiction**—gambling, pornography, excessive risk taking.
- **Undisclosed illness or physical limitations**—physical effects of cancer or response to cancer treatments, diminution or loss of senses (hearing, sight, touch, smell), movement impairment, life-cycle events, dementia, etc.
- **Not-yet-known or expressed, yet actual existing disease or physical distraction**—latent, subtle, and early effects of physical disease or injury, including dementia or related personality disorders.

Managing Impairment

Your enterprise should take steps to understand that these impairments can be present, yet masked (intentionally or unintentionally). Therefore, there is a need to monitor and compassionately but effectively address them. Periodic physical and psychological wellbeing testing for early detection is common. As the sitting operator, you personally are on the front line for recognizing impairment, managing it when you can, and reporting it when you cannot. When sufficient sponsored resources are available to assist you to understand and manage it, we find that incidents decrease. Shift handover is a very good time for monitoring. Do not forget that both incoming and outgoing operators have a monitoring role.

Your enterprise will need to evaluate and understand that the above effects may be temporary, based upon a particular situation or point in time, or they may be chronic. While shift monitoring is tasked with checking on the immediate presence and degree of impairment, nothing there is equipped to deal with any chronic aspects. Certainly, the immediacy is evaluated and necessary steps taken to ensure that you are now fit for duty and leaving duty. Recurring aspects must be looked for and managed by the enterprise.

Operator Alertness

Twelve hours on duty is a long time. Even eight hours on shift can tax your ability to remain fully alert and attentive to the myriad of responsibilities and tasks. Therefore, enterprises must find effective ways to ensure you remain alert. To some extent, you will discover and use your own personal "tricks" and task variations. Usually, they are not enough. Shift design will need to incorporate appropriate task schedules and actual physical changes. The ability to work sitting or standing can help. The provision for and ability to prepare your own meals is another. Close-at-hand training aids and modest exercise equipment help. Supervisor visits are helpful for both formal activities and informal coaching or just a brief chat about last night's ball game.

Handling Monotony

No magic bullet here! But there are lots of very effective ways to push monotony to the back burner. Some of them are intrinsic to how your duties are arranged and scheduled. If you have flexibility, use it to tailor activities to fit your way of staying involved and interested. For example, rather than push back on all of the effort to get ready for a shift change, take it as a convenient way to pick up a task that you already know how to do. Jump in and make your "rounds" to find the status and other information you need, think about what you are finding, prepare good reporting, and reach useful, in-the-moment status evaluations.

Pick a single skill that you want to brush up on during your current shift rotation schedule. Maybe some of your shifts overlap with a maintenance activity. Learn more about what they are doing, why, and how to better understand or evaluate the results. If your unit has many control valves, try to find out how they work. Dig into it a bit. Of course, you are not going to fix one. But once you know more about their inner workings, you can think about that when you suspect a problem later on. If a co-worker is working the same shift, you might ask about a resource he used earlier that was new to you, so you can better understand it.

> There is one aspect concerning the practice of practical jokes or teasing or baiting that sometimes finds its way into the control room. It can be very effective for a few, but damaging overall. Please ask management to policy these out and substitute things that work.

One of the best ways to manage is to take frequent breaks. It is amazing how five minutes of doing something completely different can change how you feel. Plan breaks when you know you will need them. Take an unplanned one now when you need it. But remember, breaks do not excuse inattention. If your duties require constant attention, you already have a relief operator protocol in place. It may be as informal as to ask the field operator to watch the board for you. It may be as formal (required in PHMSA control rooms) as doing a mini shift handover, go on break, and do the mini handback when you return [**M22**]. You know how this should work. Use the process purposefully.

The bottom line is that you know when you are losing interest and finding yourself drifting in and out, detaching from what is going on. Use that knowledge to change direction for a while. If you are having this problem and find yourself really distracted, it is time for help. Please do not wait until you have a near hit or wake up from a catnap in the chair! (A near hit is our new term for the familiar near miss.) Management is on your side. No one expects you to be a superman. No one expects you to ignore it either.

5.3 Fatigue

Being too tired is something to which each of us can surely relate. You might be somewhat near that right now—as you are reading this! Yet not all fatigue is the same. Getting away from the immediacy of the moment and getting rest is usually restorative. Yet fatigue is a really complex situation. It can pile up so much that your ability to fully recover is harder to reach, causing a deeper deficit. When you understand its causes, you will begin to recognize the potentials for fatigue. You can then start setting out plans to manage both its onset and its damaging effects.

> According to Moore-Ede of Circadian,[5.2] excessive sleepiness is perhaps the most useful definition for fatigue.

This definition is widely accepted in medical diagnosis, regulatory agency approval of prescription drugs, public policy, regulation of work-rest hours, and criminal law.

> "Thus fatigue and its major symptom excessive sleepiness, is well recognized in law and regulation as a significant body impairment, malfunction and source of ill-health and of accident and injury risk." [5.3]

Fatigue management is quickly moving to the top of the list of enterprise safe-operation activities in the control room. Operator fatigue is a double-edged sword. On the one side, since alertness is significantly dulled, it is likely that subtle changes will be missed. Thoughtful, regular monitoring will be much reduced in effectiveness. This reduces the chances that small problems and issues will be discovered in time to properly manage them. On the other side, when you do take actions, either in response to identified abnormal situations or misperceived situations, it is more likely that that action will be flawed. Procedure steps will be missed or done without due caution. Escalating situations will be missed entirely—or either overstated or understated. Either way, the risk to proper operation is much higher than it need be. If all of this were not enough, chronic fatigue itself is damaging to health.

> Be ready to use your enterprise's fatigue bottom line defense: (a) respectfully accept a supervisor's assessment; and (b) do not hesitate to declare yourself too fatigued to be fully at the job.

Fatigue needs to be carefully understood and effectively managed. You do not always know when you are becoming tired enough so that it affects what you do. You eventually know that you feel tired. However, there is that in-between time—the time when what you are thinking and doing is affected, but you are not ready to see it. To avoid this, vary what you are doing and how you might do it. Do this often. Make that short walk to check on something rather than assume or delay it until you have more than one reason to move about. Turn on that special daylight light (if they work for you). Instead of more coffee, substitute activity and variety using your enterprise's plan for finding what works for you. [5.4]

Circadian Rhythms

The human body uses a daily cycle that greatly assists us to do the things we do. This cycle has been honed into our genetic structure. It is called our circadian rhythm (Figure 5.3.1). When we are in sync with it, our body has a natural way of assisting us as we move through our day. It prepares us to awaken in the morning, it aids our work or other activities during the day, and it gently moves us into the restful state conducive to sleep at night. This cycle is reset each day by sunlight.

Fatigue and Incidents

There is an almost straight-line link between operator fatigue and serious incidents and accidents. Our strongly programmed daily cycle will allow a small bit of tinkering, but in the end is intolerant of most movement efforts. It has been said that the human body never adequately adapts to shift work. Shift workers themselves will admit to that truth.

This internal clock of ours is very difficult to adjust. It adjusts better (but not by much) when we move our bodies to a later time rather than earlier. World travelers have pretty much accepted that it takes about a full day to move their internal clock one hour. Most of us take longer. And it

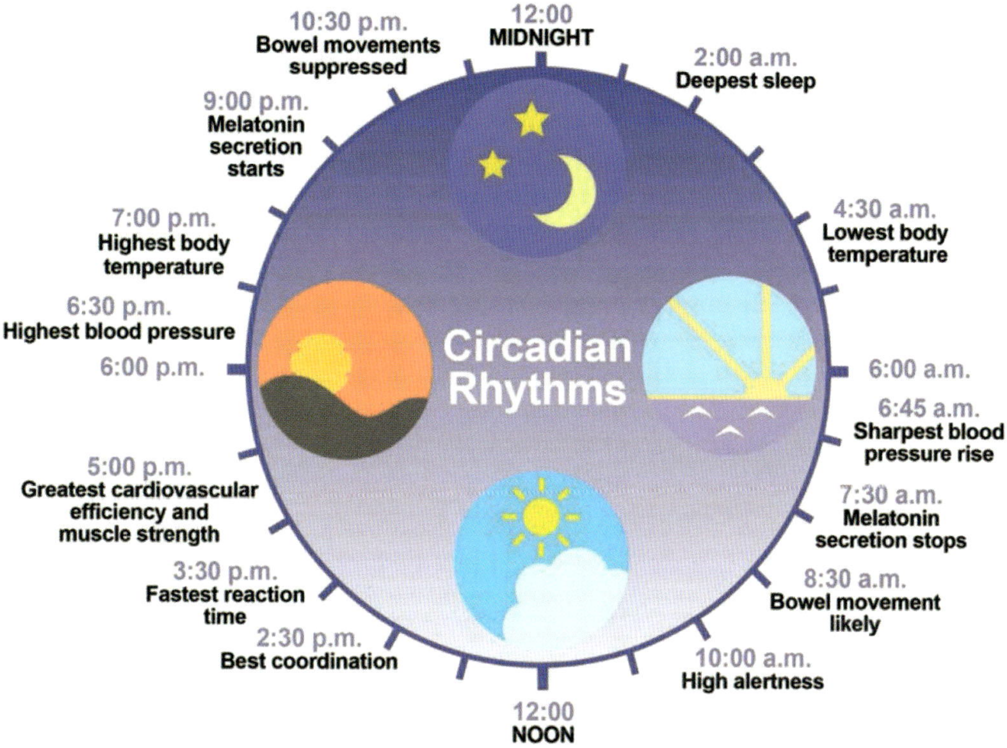

Figure 5.3.1. Our circadian clock [5.5] illustrating how our natural body cycle of alertness and functioning follows a daily rhythm. If our internal clock is much different from the actual time clock, we are not at our best.

is not a smooth adjustment. For an adjustment of six or more hours, there is usually a brick wall at day three. And there is always a lingering sense of being not fully there that lasts for almost as many days as the number of time zones that are crossed. All of this must appear abstract to those of you who change shifts on a weekly basis. Abstract or not, it is real. It is inescapable—we never really get used to it. And our performance and well-being show it.

Causes of Fatigue

The Moore-Ede [5.6] six primary causes of fatigue are:

- **Sleep deprivation**—too little sleep, or sleep of poor quality, including the effects of shift work.
- **Sleep disorders**—effects of sleep apnea, or other related pathology.
- **Illness or disease**—effects of a common cold, other physical conditions inducing pain or discomfort, or because of chronic sleep deprivation.
- **Therapeutic side effect**—side effects from use of prescription or over-the-counter drugs.
- **Heavy, stressful physical or mental exertion**—extended efforts at heavy physical labor (manual road building), intensive mental activity (studying for or taking important test examinations), and posttraumatic stress effects. It should be noted that these effects could be delayed or, on the other hand, recur as rebound phenomena.

- **Stimulant drug use**—rebound or after-effects from stimulating street drugs or legitimate pharmaceutical prescriptions.

Managing Fatigue

It is generally understood that not all fatigue effects are fully reversible. Therefore, an effective fatigue management program will include the following:

- awareness training for operators and supervisors into the nature of fatigue, its causes, and how it is best managed
- setting out expectations for operators to obtain sufficient time away from the job to allow for reasonable outside activity between operating shifts. This is specified by law for some industries (airline pilots, pipeline controllers, truck drivers, and such).
- developing effective protocols for assessing degree of fatigue sufficient to deny work to fatigued operators. A very effective aspect of a program is the no-fault ability of an operator to self-declare being too fatigued to work, either before reporting, during reporting, during a shift, or at the time for leaving work.
- designing careful work schedules that take into account length of each shift, number of consecutive shifts, the provisions of specialized home sleeping accommodations for day sleeping, and careful management of extra-job employment to ensure lack of conflicts
- careful and moderate use of caffeine and other properly consumed alertness enhancers
- providing on-shift work variations and limited recovery options (temporary relief for a power nap, exercise, light therapy, etc.)
- ensuring proper medical and emotional health through encouragement and access to appropriate medical resources (with some form of monitoring to identify problems and issues)

Your enterprise should provide you with good fatigue awareness training. Set up a program that works for you, put it into place, and make sure it is working.

5.4 Overload

Setting all bravado aside, we really do have our limits. And just like most other systems with feedback, performance will not vary proportionally as limits are approached. This is important. You want to pay full attention here. Early warnings of overload will likely to be very hard to recognize and even less understood. Things may be fine until that uncomfortable moment when they are not. And when they are not may not be a graceful event. To be sure, there will be small signs early on. Morale may show it. Absenteeism may be an early result. An increase in near misses or small accidents is almost always present. And there is the ever-present fatigue.

One or more of the following elements causes or contributes to your overload:

- the actual workload really does exceed your available resources
- a mix of situations, which quickly build up to a very high cognitive workload
- a lack of adequate supervision, leading to either operational reluctance or excessive risk taking

Workload

Obviously there is the physical part that comes from the sheer magnitude of the tasks to be done and the time and effort to do them. Even without unusual complexity or extra effort to understand what to do, if the list of things is too long, you will run out of time to do all.

Running out of time automatically adds another weight to the load: what to do about the things that cannot be done? Not only what to do, but there is often an added behavioral burden of feeling both overwhelmed and inadequate. "After all, I was hired because they thought I could do the work."

For these reasons and more, it is important to understand that your management, and not you, has the primary responsibility to prevent overload when it is possible and to manage it when it is not. The culture must be set for you to let management know when things are getting tight, with the expectation that management will affirmatively manage things when they do. This also means that in addition to assessing your workload, effort needs to be directed to uncovering and managing individual stress to a responsible level.

Work Complexity

When operations become operationally or situationally complex, your understanding and capabilities may move outside your ability. This is one of the primary reasons for the concepts of *permission to operate* [**M35**] and *crew resource management* [**M34.4**]. These assist you in managing complex and demanding situations. Both are intended to keep your performance matched to situations you are facing now.

It is up to first-line supervisors and trainers to assess the full effects of complexity upon operators and take steps to ensure appropriate operation. Fixing it may mean that procedures need to be re-examined for clarity and directness. It may mean that the present process design has inherent inadequacies or excessive operational difficulties. In any event, situations that lead to excessive operator attention or an increasing number of abnormal operating events should be reviewed for overcomplexity, inadequate instrumentation and controls, ineffective training, and inadequate procedures and protocols. Your job is to recognize any difficulties and report them up.

Managing Operator Load

Operator loading needs to be explicitly and adequately assessed. It is also a formal requirement of proper alarm management. [5.7] There is a common misconception that operator loading is set to be a balance between normal operations and abnormal operations. Neither is the actual case. During normal operations, most operators are loaded around thirty percent doing general monitoring, preparing reports, preparing and conducting shift handovers, coordinating maintenance, and such. This allows the remaining seventy percent to be used for a bit of training, responding to alarms [**M19**], following through on weak signals [**M23-M26**], and other operational needs.

When a significant abnormal situation comes along, you are immediately transformed to having way too little time to understand what might be happening, deciding what to do about it, and acting to remediate. Studies from significant planned abnormal events (major startups or shutdowns) have repeatedly shown the need for one to two extra pairs of expert hands. See Figure 5.4.1 for a comparative illustration.

At this point, finding any sort of balance in operator loading between these two situations could not possibly work. Rather, the approach is best done by ensuring adequate time is available to handle the manageable upsets that come along. When the extra reserve is not needed at the moment, use the time for training, special reporting, writing work orders for optional improvements, and the like. We function best over the long haul when our load is variable and reasonably far from full steam ahead. For all larger upsets, some other mechanism must be in place.

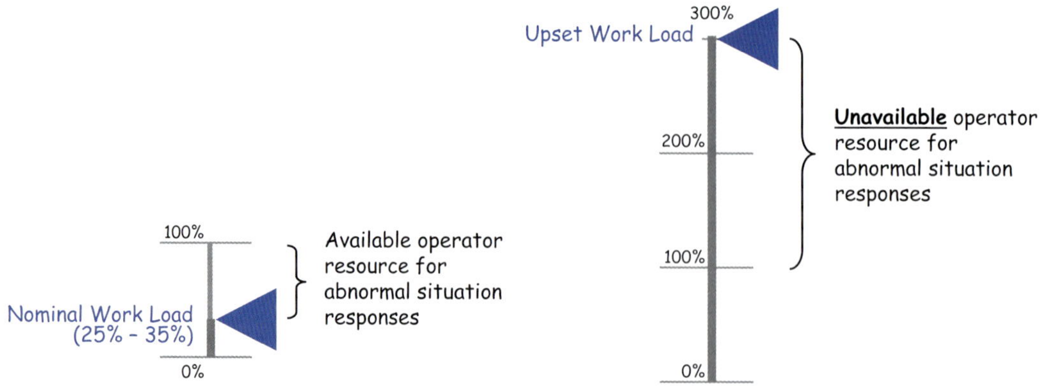

Figure 5.4.1. As operator loading changes from normal to upset situations, the amount of available time to attend to the problem not only disappears but, without aid, will overload.

Supervision

You should be more than adequately prepared and able to manage things when they go reasonably as planned. Yet few operators have the wealth of experience, full set of tools, and the wisdom to enable them to handle everything that comes their way. And even for those few, if the situation threatens to exceed their boundary of operation or authority, others are required for guidance, assistance, and permissions. In all situations where the requirements for operation may be adversely affected, ready access to supervision is required. It must be made clear that even where supervision is directly in the loop, the requirements of *permission to operate* should never be bypassed.

5.5 Handling Bias

Your job as an operator is to make sure that you use the tools at hand to do the operating job without getting off track because of biases [**M16**]. Understanding and managing biases are part of that job. Biases in the control room push or nudge you to think something is more important or less important than it really is. They shove your thinking so that you shy away from one conclusion in undue favor of another. And in a lot of other "below our own personal radar" ways, they cause our thinking to fail to be as clear as it must be. More often than not, if we were vigilant, observant enough, and questioning enough, the information would be there. Later on in the handbook, you will visit Module 16 "Biases and Pitfalls" to better understand them and work out ways to avoid their sticky effects.

Module 6: Communications

All of us need to share information, concerns, and more with others with whom we work together. Communication is an important way to do that.

6.1 Communication

Few places have more at stake about what happens than an industrial control room. At all times when there are others present, everyone must know what they need from each other. You must communicate. Effective conversation is the basis of how communication is understood, accepted, and managed. This module explores what it takes to ensure that what is communicated is what is intended—**and** what is intended is what is needed. But sometimes, due to improper communication, what is intended is not what is heard.

"Communication (from Latin *commūnicāre*, meaning 'to share') is the activity of conveying information through the exchange of ideas, feelings, intentions, attitudes, expectations, perceptions or commands, as by speech, non-verbal gestures, writings, behavior and possibly by other means such as electromagnetic, chemical or physical phenomena and smell. It is the meaningful exchange of information between two or more participants (machines, organisms or their parts).

"Communication requires a sender, a message, a medium, and a recipient, although the receiver does not have to be present or aware of the sender's intent to communicate at the time of communication; thus, communication can occur across vast distances in time and space. Communication requires that the communicating parties share an area of communicative commonality. The communication process is complete once the receiver understands the sender's message and the sender knows it." [6.1]

6.2 Communication Attributes

A *communication* has the following attributes:

1. Content must have enough significance (or potential importance) and need to merit sharing it with others
2. A sender (for our purposes, usually an individual) and a recipient (usually another individual)
3. A referenced and understandable message (including to what or where in the enterprise does this concern relate, its content, its need, and importance)
    a. Specific purpose of the message
    b. Specific thing (part of enterprise, situation, etc.) that the message relates to
    c. Specific message content, clearly expressed
    d. Message content delivered via a reliable vehicle with reasonable traceability

   e. Message content is understood by the recipient
   f. Message content acknowledgment is given to the sender
4. If a communication is required (say by procedure, standing operations requirements, or any other), then it must be formally archived as to content, sender, recipient(s), and timing.

The following are a few types of communications:

- Verbal messages from one operator to another, or one operator to any other member of the operations team, including field, maintenance, and supervision
- Virtual messages such as text messages, email, video conferencing, etc.
- Entries in reports, logs, or other operational annotations (recipients implied) [**M8**]
  - Shift handover reports
  - Operator logs, etc.

6.3 Conversations

Conversation is one of the most important activities that happens in the control room. Conversations occur between board and field operators, between one board operator and another, with supervisors, during shift handovers, and more. Three key communication types are communication, collaboration, and escalation. These are your conversations. A troubling number of terrible incidents and disasters could have been prevented or reduced in severity had the participants understood the proper form, purpose, and steps to use for accurate and reliable communication and used them effectively. Please take the time and effort to fully understand these recommendations and guidance.

Communication may be actual or virtual. Virtual conversations are addressed later in Section 6.4.

Mindful Conversations

Mindful conversation is an important communication behavior to understand and practice. At a minimum, all control room conversations should be mindful. According to Rosa Antonia Carrillo [6.2]:

> While effective safety management systems are part of every successful organization, the most successful ones also pay attention to creating superior communication and collaboration that support the free-flow of information. Absent a trusting relationship among team members and between different teams, there is little chance that information will be shared.
>
> Mindful conversations are a powerful tool for incident prevention, but they require a productive safety culture to take hold and a strong leader/sponsor. A socially reinforced approach using conversation and relationships cannot succeed within a managerial climate of excessive or harsh command and control. People will not freely contribute their ideas and observations if they fear retribution or do not feel valued. Thus, the leader as creator of conversations is an important role in creating a climate for open communication.

Mindful conversation deals with behaviors, not content. When each party uses these suggested behaviors, they are able to form a trusting and honest connection. Then real communication takes place without undue distraction, bias, or excessive discomfort. These are Hernandez's [6.3] suggestions for mindful conversations [**Q4a**]:

1. Turn off or away from all other devices and individuals not necessary to conduct the conversation. You might need them or refer to them later, but for now, get them out of the picture.
2. Listen using your eyes to make contact and pick up on all needed clues about how the other person is reacting to the need for communication or is feeling about communicating with you.
3. Listen to the tone of voice to pick up on what might be below the surface or being kept hidden, or is hard to say. Hear the words said, the way they are spoken, and the care and feeling, or lack of, that they convey.
4. Say "no" to thoughts that have nothing to do with the conversation. These thoughts are yours, not the other person's. Get them out of the way.
5. Attempt to truly understand the communication so you can hear without your personal interference due to your judgment, your critique, and your own discomfort.
6. Watch out for external conditions unrelated to the conversation where you are distracted by clothing, gender, mood, etc. All of these can lead you to presuppose a message without paying full attention.
7. Respond instead of reacting. Give yourself the time to process what you hear and plan what you need to respond. You are in the listening pathway now; listen so you can actually hear. **Do not plan your response before its time.**

Safe Conversations

Safe conversation is at the heart of how communication in the control room is understood, accepted, and managed by individuals. Safe conversations ensure that what is said in the control room is responsible and effective. A safe conversation is not about safety directly. It is about making sure that the conversation itself is safe in the light of being effective. Almost all conversations in the control room have a safety component. So all except the most casual, such as how strong might the coffee be, all should be approached with an understood structure. When something is going wrong or is important to share, place that issue at the forefront. Even if nothing about safety seems involved, all must be sure to communicate clearly. The need is for both parties to fully know what the conversation is about, why the conversation is happening, and what each expects of the other.

> Safe conversations are a must-follow protocol for every functional conversation in the control room including shift handover, collaboration, delegation, escalation, and the rest.

Let us assume all is mindful conversation. We now move to what is said and how it is best managed.

Guidance for Safe Conversation

1. Explicitly state what is **known** about the situation and how it is known.
   - State precisely what is intended to be communicated and how that is known.
   - **Read back** each point to the other party similar to the way that an aircraft pilot and the air traffic controller confirm voice communications. Each party must also

specifically ask if there is anything else that needs to be known that has not yet been mentioned.

- **Confirm** what is known by both parties and anything that must be actively observed and challenged where there is any doubt or discrepancy.

2. Explicitly state what is of **concern**.

   This sets the *reason* for the conversation, but not the *outcome*.

3. Explicitly state what is **expected** of each party.

   This sets the working relationship between the parties.

   - Is confirmation expected or is another opinion needed?
   - Is actual operational assistance being requested, including escalation, hands-on assistance, handoff, or something else?

[**Q4b**]

Safe Conversation Examples

Safe conversations are best explained by examples. Note the clear construction and hierarchy of confirmation for each. This is essential to ensure careful and accurate communication. All participants are in the same communication space during the activity. They are either present in the same place or virtually linked with the ability to see, hear, and respond in a shared information and communication environment.

Example 1

Known (sender): *"The scrubber outlet flow has been decreasing for the past hour. Here are the trends."*

Read back_(recipient): "You're telling me that the scrubber outlet flow has been decreasing for the past hour and that the trends appear to suggest that this is happening. Did I miss anything?"

Confirm (sender): *"You didn't miss anything. The trends show the scrubber outlet flow has been decreasing; and it appears as if the sump flow is decreasing as well, which usually does not happen when the scrubber outlet flow is dropping."*

Concern (sender): *"I don't ever remember the sump outlet flow dropping off like this for no reason."*

Expected (sender): *"We need to know if there is a piping leak. Please find the lead pipe fitter and ask him to come here immediately to look for leaks. If you cannot find him right away, can you do it yourself?"*

(recipient): "Understood. Either way, I'll find the fitter or take care of it myself and get back to you in 30 minutes with a status."

(sender): *"Ok. I'll expect your report in 30 minutes."*

Example 2

Known (sender): *"I've received no deliveries of additives this shift, and the last delivery was a week ago. Here is the delivery log for the past 30 days."*

Read back (recipient): "You are telling me that you are about to run out of 0.4% copper sulfate for the first-stage treatment plant effluent. And, to your knowledge, there were no recent deliveries and the last shipment of record was over a week ago and nothing since is in the delivery log. Did I miss anything?"

Confirm (sender): *"Correct. I am not sure how much we will need this week. I can also see that the copper sulfate tank level is within an inch of the bottom of the outtake pipe. According to the current use, that gives us about 6 hours of operation left."*

Concern (sender): *"We seem to be running a bit low in the tanks. If we run out, the water treatment plant will have to be shut down, which will require the entire west processing facility to shut down."*

Expected (sender): *"Could you check on the coordinated delivery board in the field office for the current delivery status when you go on break? If nothing is scheduled, can you place an emergency order for a 2-hour delivery of enough to take us to the next regularly scheduled delivery?"*

(recipient): "I'm not going on break for 3 hours. I'll call now and make sure a 2-hour emergency delivery is arranged and report back before I go on break."

(sender): *"Ok. I'll expect your report within 3 hours."*

Example 3

Known (sender): *"All of my logged production values but one are within normal operating ranges. Here is the shift handover report; let's view the current values on the screen."*

Read back (recipient): "You are concerned that the fuel flow to burner #4 is low by over 45%; but all of the rest of the readings are pretty much normal. You'd like me to review the shift handover report and check the current operating values on the HMI to see what they are. Did I miss anything?"

Confirm (sender): *"That's the full problem. It does seem like the propane flow to burner #4 is much lower than all of the other 3 burners. The temperature controls for all 4 have the same setpoint and are in automatic. I see that the total fuel flow does match the combined flows for all 4 burners, with burner #4 reading low as it reads now. It looks like that flow is really too low."*

Concern (sender): *"I can find nothing to explain that this is okay or that I might have an unseen problem that needs to be fixed."*

Expected (sender): "*What am I missing that could explain what we see here?*"

(recipient): "Let's look at this together and find out."

Recapping Lessons from the Examples

We recap how the communications worked by focusing on the fullness of the back and forth communications:

1. Sender has a message (request, information, or any other communication) to convey to a recipient.
2. Sender transmits the message to the recipient by direct conversation, over a voice communication medium, or via some other secure way (fax, text message, etc.).
3. Recipient transmits acknowledgement and specifics of the message back to the sender via an appropriate method (not necessarily the same as the sender uses, but most often the same).
4. Sender transmits to the recipient the exact concern (reason for the communication and the selection of that particular recipient).
5. Sender and recipient both agree on exactly what each expects of the other and how the desired activity results will be confirmed.

Later on, the recipient will need to respond to the sender in the way of a report or other appropriate feedback resulting from the initial communication. Effective communication is also essential for this "response back." The same protocols are used for it, but the current sender was the recipient in the earlier exchange, and the current recipient was the sender before.

6.4 Virtual Conversations

Virtual conversations have become routine in the control room. You have been using text and email for many years—and there are new modes being added constantly as technology expands. These might include blog-like summaries, Zoom©-type video conferencing, and others. When using virtual conversations, it becomes even more important that the message be understood. Because, unlike face-to-face conversations, certain cues are missing, and new ones are present that confound the situation. It becomes even more important that virtual conversations be mindful ones.

There is no real eye contact in a virtual conversation. Yes, you can sometimes see eyes, but that is all. As a result, you cannot easily pick up on body language cues as to how the other person is feeling about this conversation—or how they are feeling in general. There may or may not be verbal cues. When there are not, such as in a text, there is no tone of voice to help you determine things that may be affecting the comfort level of the other person.

"We've all sent an email or a text where we meant one thing and the recipient took it in a completely different way. They never saw the smile on our face, the touch of our hand, or the look of concern. There were none of the emotions that we take for granted in the real-world to guide them in the virtual world to a proper conclusion." [6.4]

Without body language and tone of voice, it has been shown that memory is affected. This is because our subconscious mind is addressing the effects of these upon the content being presented. It then ties these social clues to the content. Without this connection, our ability to remember is detrimentally affected. Our conscious mind tries to make the interpretation, sometimes in error. Morgan gives the following example: "Does the silence in response to what you've just said mean everyone's in rapt agreement, or everyone's tuned out – or people are on mute so that they can have a party?" [6.5]

The rules for mindful conversations still apply to virtual conversations and may be missed without face-to-face contact. Safe-conversation protocols should still be followed, but it may be more difficult to be explicit. What follows are a few protocols for virtual communication:

Email

Email has been around long enough that its use is expected and understood. One concern with emails is the possibility that the reader might easily misconstrue the writer's intent because the text lacks sufficient social cues. Remember that using all caps is the same as yelling in a verbal conversation. Emojis and abbreviations should be avoided. You should refrain from using email to say something that should be said in person. Always, always keep a full awareness that everything written in an email has the potential of being misdirected, forwarded to unwanted recipients, or simply never read (ignored, sent to the wrong party, or ending up in spam). And, of course, remember to proofread and spell check.

Texting

Texting is becoming accepted in the workplace. It can provide a means for very quick remote communication. Responding to a text takes the average person ninety seconds compared to the nine minutes it takes to answer an email. [6.6] The general consensus is that texting should only be used for brief communications. Abbreviations should be avoided and emojis should never be used. Always proofread. Yes, even your texts. Auto correct can change your text from concise and pertinent to something very unprofessional and sometimes even awkward. When you proofread, please do it before sending (we all do it afterwards, and see much more than before). Position yourself at the screen of your recipient reader. Ask yourself what part of your message could be unclear or misunderstood. Fix it before sending.

Virtual Meetings

Protocols should be developed. Is the camera on or off? Without a visual appearance, there will be a lack of social cues, and the effects discussed above could lead to unclear messages. Background noise can be distracting, particularly when in a control room.

6.5 Logs and Reports

The effective use of logs and reports is discussed in Module 8, Section 3. Here, we address how to effectively communicate using them. Almost all of the content, form, and schedules are fully determined by the enterprise. Nothing here is meant to change any of that. However, everything here can be relevant and quite useful as you go about the task of doing them. Rather than repeat those concepts, we will just list them with a few helpful comments.

As you write (or keyboard, or touch, or dictate), ask yourself:

- Why am I adding this content into the report or log?
- What might be missing from my content that could change the way it is viewed, understood, or acted upon? Should I add it in?
- Does my content depend on other content in this report or log? In another report or log? How do I let everyone know about those dependencies?
- What does the reader or viewer expect to see—not the actual data or values or words, but what is being looked for?
- What does the reader or viewer need to do in response to the content I am providing?
- What is intentionally not in the content that needs to be flagged as not present, with reason(s) for its not being there?

Check over your work to ensure that it is complete, in the right place, and properly entered.

6.6 Control of Work

Everyone understands that control room activities are important, and doing them correctly is essential. As with everything we do, some things are more important to get right than others. Here are a few of the activity types to become familiar with.

Control of Work

Control of Work (CoW) activities are defined as follows:

"In essence, a CoW System is a structured communication mechanism, to reliably communicate information about hazards, control measures, and so on. During these activities, good communication between management, supervisors, operators, and maintenance staff and contractors is essential." [6.7]

CoW activities include permit-to-work (PtW), hot work permits, confined space entry permits, isolations, and more. What is being done for the work is important, and getting it right is essential for safety and effectiveness. Thus, interpersonal communications are critical. Not everything that is needed for proper conduct is apparent, written clearly, and otherwise complete. Hence, added discussion or points are employed to assist; there are always verbal clarifying comments and questions to be asked and answered. Moreover, the way something is communicated—say, by emphasis or dismissive phrases, or not said, where expected—can carry importance. All of this needs proper communication, that is, mindful communication.

Shift Handovers

Shift handover communications are generally agreed upon as being the most extensive and important communication event in the control room. Module 22 covers them. When you get there, as you work your way through the handbook, please remember what you have seen here and put it to good use.

6.7 Site Protocols

This is a reminder that the concepts, lists, and examples in this module are for illustration purposes only. All of your site protocols specifying or prohibiting particular communications actions take precedence. Where none exist, or the existing ones are open to your adjustment, then the ideas and materials here would be useful and could be used. Run all of this by your supervision.

6.8 One Last Communication Point

There are situations where communication will be ineffective simply because it does not happen. The first is when questions are not asked. Sometimes, this occurs when a junior operator is fearful of being judged negatively by a superior for not already knowing the answer. Other times, someone simply may not know what to ask. There is a saying that "you don't know what you don't know." And there are the times when the *sender* holds back information. Sometimes it may be to maintain a feeling of superiority, or due to insecurity. So asking for more information that the speaker has not yet shared, as demonstrated in the safe conversations above, is important for successful communication.

As a closing note, the vital resource of effective Crew Resource Management (CRM) [**M34.4**] critically depends upon open, full, and trusting communication.

Module 7: Multitasking

7.1 Myth of Multitasking

Multitasking is the intentional act of trying to work on more than one engaging activity simultaneously.[7.1] [7.2] Or, simply put, it is trying to do two or more tasks at the same time. "At the same time" means that we try to interweave and overlap those activities so that it appears as if they are concurrent. These activities can be highly related or significantly different. No matter, it still will not work. So why would you want multitasking in your toolbox?

> Multitasking will slow you down and increase your mistakes. Humans are just not "wired" to concentrate on two things at once. Take multitasking completely off of the table in the control room.

Overload happens. Operators sometimes try to use multitasking as a resource management tool when they find themselves overloaded. Think you do it effectively? Maybe you do, but chances are very good that you do not. In reality, we are not capable of doing multitasking well enough to rely upon it as a useful skill or tool.

Less than 3% of us are acceptable at it. Not even good; just acceptable. This means that when we try to multitask, we are really switch tasking. When we switch tasks, we stop paying attention to the task we left and focus more on the one we just picked up. The task we left alone moves along without any attention, as if on autopilot. But in reality, there is no auto piloting whatsoever. This shift in attention is not so obvious. We feel we are doing all tasks at the same time. That feeling comes at a price. Every switch takes time. There is the pick-up time it takes to drop the earlier task and focus on the new one. We spend this time even after leaving just for a second. Then there is the recovery time to switch back to the task we earlier shifted away from. This time is to reacquaint us, to remember where we were and what we need to think about to get it going again.

Recovery times are five to twenty times longer than the time it takes to shift activities. This occurs because different parts (lobes) of our brain are used for different types of activities. We coordinate these parts of our brain very well for consistent tasks but not as well for different tasks, such as mathematics/reasoning, memory tasks, and visual/spatial processing. A fifteen-second quick look at something could cost nearly a minute to fully refocus on the original task, although it might seem much quicker.

Research estimates that individuals who **routinely** engage in multitasking suffer a 20% to 25% overall loss in productivity. [7.3] When we are doing a consistent task such as looking for a lost pair of glasses, we are really good at coordinating the individual parts. We do not do so well with different tasks such as looking for a lost pair of glasses and dressing for a very important meeting.

Either we forget a wallet or we take way too long looking. While a missing wallet is inconvenient, if, after a switch, an operator missed seeing an abnormal operational reading, it can spell real trouble.

Real-Life Illustration

Since driving a vehicle is such a common situation, we use it for a brief illustration. You are driving (even without a fancy navigation app), trying to find the right road to turn on next. Simultaneous activities would be driving and talking on a mobile phone—or driving and trying to decide whether you got the job after that interview you just left. Driving while engaging in almost any phone call creates the same level of visual and situational awareness impairment (to safe driving) as a blood alcohol level of 0.08%. [7.4]That is sufficient for a legal finding of driving while under the influence (DUI). This multitasking distraction thing really matters.

7.2 Multitasking in the Control Room

Multitasking should never be a permitted behavior in the control room. Control room management protocols need to be designed and followed to ensure all work is managed as a series of stepwise related tasks and activities. Start one. Work it. Finish it, or park it for a while. Go on to the next. In this way, a general schedule is followed without paying a distraction price.

Your schedule needs to be managed enough to provide you with the confidence that the alarm system, combined with weak signal processing [**M23-M26**], will keep operational risk under control. You are then freed to pick up and follow other related tasks without undue urgency or expectation of quickly jumping back and forth. At times of reflection, say, between threaded tasks or upon returning to the board after picking up a coffee, a visual review of the human-machine interface [**M18**] and alarm system [**M19**] should provide sufficient awareness orientation. This all depends, of course, upon the alarm system being proper and the HMI design being effective. Take a close look at countdown timers in the next module [**M8.4**]. Everything depends upon your infrastructure being appropriately used.

7.3 Alarms and Multitasking

You are busy working through your regular activities. Your mind is occupied with the tasks at hand. Along comes an alarm. Alarms insinuate directly into your real-time activities. Now your whole attention is turned to the new alarm activation. This is as it should be. You must quickly assess the alarm and just as quickly cease all current, non-alarm activities. What if the things you were in the middle of before the alarm activated still needed to be addressed? Now you have to work on the alarm and at the same time wind up other activities. Suppose those other activities are critical enough to need completion or movement to a pre-determined proper status? What do you do? Doing both is not a proper option.

Normally, the built-in alarm response time is determined to sufficiently process the alarm. There is a bit extra, but not enough to do much else. Now you have a time resource problem. You need a good way out. If the alarm is active, all else must be dropped to work it. If the other stuff cannot be quickly handled (in a few seconds) and mishandling it will open up other issues, it must be passed on to another individual. Or the operation must be rendered safe by shutting down or safe-parking it. Simple. Powerful. Uncomfortable? Sure, but since we really cannot do both, we need to do the right things. Running out of time to properly handle the alarm is not one of them.

7.4 Takeaway

Multitasking is going to slow you down and increase the chances for mistakes. This is really about distraction. No matter the cause, when you need to concentrate and really think through something, anything that diverts you will get in the way. You are going to bounce around in your mind between what you want to focus on and whatever is waving a "see me, too" flag from somewhere else. Even if you are pleased by the chance to take a breath and see something else, you must pay for it. The price is having to spend extra time getting back, the chance of missing something you already saw, and the missed opportunity to seamlessly move to that just-imagined next step. You are distracted. You lose that thread or it unravels before you can catch it.

How does distraction work? Every distraction is really a pull to change your focus or attention at the moment. Each disrupts and interrupts your current attention. "Disruptions and interruptions are a bad deal from the standpoint of your ability to process information." [7.5] This applies to all tasks requiring even a modicum of concentration. Understanding will need all of your conscious attention. When you do not concentrate and understand, your performance slows down and declines in efficacy. [7.6] Focus on a single task. Stay there. Only switch when important or necessary. Multitasking is just not an acceptable working activity in your control room (or anywhere else, for that matter). Work with your supervision to settle on a workable way to flow through your shift without distraction and any need or reliance on multitasking.

Module 8: Operator Tool Set

Your basic operator tool set goes hand in hand with training and all the rest of your basic skills.

8.1 Basic Operator Tool Set

In this module, we lay out a few of the basic tools. These "tools" direct or assist you in day-to-day operation. We cover how each tool supports situation awareness and situation assessment, and how it all fits into situation management. Most of these tools should be quite familiar to you. Here, their basic functionality is clarified to give you a better understanding of their underlying use and purpose. This way, you should be better able to both use the tools and accommodate for issues and problems that may arise during their use. The benefits and limitations of the tools are also explained to help you evaluate when one might work better than another and how they work together. It should also aid you in shaping them to better suit your intended purposes. We are not suggesting modifications, just sharpening your way of using the tools approved by your enterprise management.

These tools should be used only if your enterprise has formally designed and approved them. Some of these tools have a name here that is different from the way your enterprise refers to them. But be sure you know that they are the same. For those that you have, this module provides guidance and tips for use. Additionally, if you remember something that worked great at a different job, it is not okay to use it now unless you check it out with your current enterprise and get approval.

The discussion covers:

- what the tool is all about.
- why the tool is useful/important.
- how to utilize the information in the tool.
- benefits and limitations of the tool.

There are more tools than the ones shown here.

8.2 Directive Tools

Policies and Procedures

> Procedures and policies must specifically take into account all of the naturally occurring intuitive foibles and limitations of the operator.

"A well-defined set of policies and procedures provide the roadmap your employees need to be independent and effective, so they understand their role in the organization and senior management doesn't have to closely watch every process." [8.1]

To use policies and procedures effectively, they must be available, and you must be fully trained in their use. You must know where and how to access them, and thoroughly understand everything. When your entire enterprise follows policies and procedures, things run smoothly with consistency. Everyone knows what they should be doing and is doing it efficiently. You can do your job with confidence. And your enterprise will be a safer place. [8.2]

Policies

Policies are the enterprise's plans for how everything should fit together for predictable outcomes. They identify a strategy for you to carry out your daily duties as an operator to achieve the goals of the enterprise. All activities required to carry out the goals of the enterprise are defined within your existing set of policies. [8.3] There are policies for how to purchase raw materials, from whom, and how to package and deliver them. There are policies for how to handle normal as well as abnormal events and situations. There are more. They are used by everyone to enable a predictable, approved, and documented way of using all of the enterprise's individual parts, including what goes on in the control room.

For example, there would be a policy for determining how shift rotations are to be arranged and managed. That policy could be something like this:

- All shifts will have a formal handover.
- All shift schedules will follow a plan for ensuring adequate time off, as well as including training, severe weather, and the like.
- Special shift schedules will be designed and followed for extended illnesses, significant damage to the operational equipment, and the like.

Specific activities to be carried out during the actual shift handover would be covered by procedures.

Procedures

A procedure is fully explained guidance for how to conduct a specific operation or task. Procedures define your responsibilities and how those responsibilities affect your team. They also cover emergency situations. You do not have to guess or invent proper actions. They can also protect the enterprise in the event of a legal issue. They are generally written in a narrative form. This means a procedure would be read and then followed. It may be designed for a step-by-step use, either fully or partially. Different from checklists (later in this module), each procedure may reference the applicable protocol (defined below) rather than include it within the procedure document itself.

Procedures need to address the following:

Applicability—under what conditions or situations is the procedure to be used.

Intent—exactly what is the purpose of the procedure.

When employees know why they have a procedure and understand its value, they are more likely to think of it as important and follow it diligently. [8.4]

Explicitness—exactly what to do, how to do it, when to do it, and who should do what. Documentation and reporting requirements are listed and explained or referenced.

Expected Results—what to look for when things are correctly done; what to look for as signs of problems; any special confirmations needed to move to the next step or pause (indicate why and how long) and then resume; what to look for as indicators that the step should be skipped or the entire procedure stopped or ended. If stopped or ended, specific instructions on how to stop or end it.

Limits of Operation—what are the safe limits of operation that must be respected, regardless of the process or activity under way; what are the preferred limits of operation that should be used as indications of normal.

Complete—the procedure should be designed so that faithfully following it should produce the expected results (unless obtaining the expected results is not possible due to existing problems or outside effects). It must include what to do in the event of problems carrying out the full purpose of the procedure.

Do not try to rely on your memory. In order to use procedures effectively, they must be documented and available. If a procedure is verbally provided to you, you should request a written or electronic copy. Be sure to understand the purpose of the procedure. [8.5]

Protocols

A protocol is a description of a general process for doing a specific class of operations. For example, there might be a general protocol for starting up equipment. It would apply to all equipment. It would contain items like locating the controls, checking with lockout/tagout to see if it is present, checking work orders to ensure that the piece of equipment is not on one, calling for additional assistance if such is usually required, and other aspects common to all or nearly all equipment startups. Protocols are different from standards, policies, and guidelines. Although the words are sometimes used interchangeably, there are differences and advantages.

Here are a few useful thoughts. A protocol: [8.6]

- is a framework for a complex, specialized sequence of activities.
- provides increased autonomy with a focus to shape future work.
- ensures consensus within the primary care team.
- provides a measure of legal protection.
- identifies training needs.
- may facilitate responsible change.

When using procedures, the protocol will indicate what specific procedures are to be used and where they might be found and verified to be current. The protocol will specify whether there are any global operating situations that affect the covered task. The usefulness of protocols is that there are fewer of them, and they are broadly applicable. Consider that they provide the upper-level framework so that procedures are properly used.

8.3 Supportive Tools

These additional tools are used to assist you in day-to-day shift activities.

Checklists

A checklist is usually a part of a larger procedure or protocol. It consists of a specific list of required items to check, the order of the checking, and the condition to be observed or confirmed against. Each item on the list has a description and a box to check off. The order of the list is the order in which it needs to be performed. Explanatory instructions are given to clarify what a check mark (√) means or under what circumstances a check mark needs to be skipped. The main purpose is to ensure that attention is given to every item that needs it. It is useful universally for routine tasks that might be constantly performed or for tasks that might only be performed once. It is extremely useful for high-stress situations where the likelihood of something being missed or done out of order is important to control. Checklists are a compact tool that combines in one place, what to do, any specifics for how it must be done, and confirmation that it was done. The immediate ticks (√) and any accompanying annotation document the progress to ensure tracking of what has been done and what has not.

One of the dangers of using checklists is that they can be done perfunctorily without actually reviewing the full item and without doing a full check to see what is what. Training and insistence on proper use helps. For one or two vital steps or items, an additional user entry (not a simple check) can be requested on the checklist item line. Missing entries there indicate an incomplete or perfunctory check.

Each checklist needs to meet these structural requirements:

Identified—Where a checklist is used, it must be documented as to when it is being used, on what is it being used, by whom, and for what purpose.

Operationalized—Where certain conditions must be present in order for the list to be used, those conditions must either be built into the list (say, by having a different form of the list for each condition) or be able to be added during the check-off process (say, by tracking date and time where the checking process might be paused or interrupted).

Explained—If special instructions are needed in order to check any line item on the checklist, those instructions must be present, close at hand, or fully referenced. In addition, a note must be added to document that the instructions were referred to as this line item was evaluated or checked.

Clear—Every item must be written with conventional terminology, phrasing, and expressions so that they are unambiguous and understandable to the expected user. Any bilingual aspects are to be handled by enterprise policy and according to applicable statute.

Complete—Every item that is required to be checked or performed must be explicitly on the list. Even if an item or task is minor or is included as a part of many other items, it must be listed explicitly. If order is important, a strict order matching the requirements must be followed.

Fully Optioned—There must be an area on the checklist to mark the exact situation to check or document and to match the found conditions (for example, if an item's status cannot be verified or is normally important but today is not). If an explanation is required to appropriately denote the meaning of the specific "check," it must be present and adequate.

Contemporaneous—The checklist must be immediately at hand so each item evaluated or checked can be entered on the checklist without remembering or jotting down notes or other reminders for later entry on the list.

Archival—Every aspect of the checklist must be sufficiently expressed, documented, and annotated so that it would be clear to anyone at any later time to be able to fully understand what was checked and the condition for the check to be entered.

One method of using checklists is with a *challenge* and a *response*. This is where one person would read aloud the item (the *challenge*) and the other would verbally give a *response*. This method is used in the airline industry. When completing checklists, you might want to develop your own discipline to help remind you to perform the checklist, and to keep track of your place while doing so. [8.7]

Interruptions should be avoided. These cause errors and failures. If you are interrupted, the responsible activity is to begin the entire checklist from the top.

Simulators

Simulators are predominantly thought of as training aids. If their quality and fidelity are good, they are excellent for that task. Simulators can also be extraordinarily useful during actual operations without disturbing that process or operation. One example is when an operator is planning to make an operational change that might have broader implications than usual. A simulator can test various ways to make these changes without risk and can provide feedback as to which alternatives might be more appropriate.

Simulators can be useful for understanding and possibly handling abnormal situations, though it would be unlikely that you would have enough time to make use of them for most on-shift support. There are exceptions. If time permits, they could be useful to search for causes. You might use it by "guessing" a problem and then using the simulator to see if the expected abnormal situation develops for it. Also, if a problem is strongly suspected or even known, various remedies might be simulated to determine which might work and give better results.

Simulators are also effective for evaluating the interactions of a team during a situation. They are particularly effective in helping team members understand their roles. Without clear knowledge of what your role is when you are part of a team, particularly in an unfolding abnormal situation, the problem could escalate. Behavioral traits may be noticed during simulations and incorporated or corrected if necessary. If other personnel are also available and have access to and are trained how to use them, simulators might be used in parallel as the operator works the problem so long as the expected available time for operator success permits. In order to be effective, feedback should be given after simulator use.

Lateef offers tips for successful use of simulators during simulation activity: [8.8]

- **Avoid premature closure of ideas:** Do not stop searching for ideas after the first workable one appears. Often the best idea comes second, third ... or even tenth. Think of ideas as stepping stones to other ideas rather than as destinations in themselves.
- **Get outside a problem and look at it from different angles.** For example, try approaching a problem in a production optimization simulation from the points of view of a compliance officer, an engineer, or an equipment vendor. What would be their top-line interests and concerns for the exact situation you are examining?

Tip for successful use of simulators also offered by Shirts: [8.9]

- **Give your subconscious a chance to work on the problem.** The solution to an especially troublesome problem will often pop into your head when you least expect it—on the freeway, in the shower, at the beach. Give it the opportunity.

Logs

A log is a record of important events. It can be virtual (electronic) or handwritten. The date and time is noted along with a description of the significance of the event. [8.10]

In some cases—the military, for example—logbooks can be official or even legally binding. "Proper use of a log book provides solid proof to regulatory agencies, auditors, legal entities, and courts." [8.11]

Well-maintained logbooks can be useful in the event of failure of process control system recording devices and the HMI.

Bitsize suggests [8.12]:

Often when going back to review a log you find that you neglected or forgot to enter an important piece of data. It might be a temperature, or something else. Reasons for these oversights are many, but may be avoided by using a predetermined procedure when using a log. It is a good idea to develop a consistent plan that works for you. Your enterprise may have precise methods or entries they are looking to see. You should, of course, follow those. In addition, here are some suggestions to help you fill out your logs successfully.

Keep notes throughout your shift. Do not save everything to shift's end when you may be rushed and anxious to be on your way. Notes will help you to document things as they occur. Notes taken in the moment will be more accurate than trying to remember.

Try to be thorough and consistent. Do not rush through entries. Do not be cryptic. Include all units of measure. For example, if you are documenting temperature, include Fahrenheit or Celsius to avoid questions later.

Reports

Reports capture information in an organized way (the report form and format) and make it available to others. The operable part of that is (a) capture, and (b) available to others. Properly designed and appropriately used, reports ensure that what was important during production or operation was captured and is available later. Now everyone can rely less on remembering (we rarely remember what really happened). It can be shared later on with others. It can be searched, if in electronic format. Reports are a special form of communications. They are discussed here because they are structured enough to be a tool for operators.

> When you write a report, you present material in an orderly manner. A report should have a purpose, present relevant information and draw conclusions, and possibly make recommendations. When a report is well-written, the reader will clearly understand its purpose. [8.13]

The type of report you as the operator may prepare is sometimes called *operational reporting.* This type of report produces a detailed account of day-to-day operations. This type of reporting is useful to the enterprise because different areas do not operate independently of each other. By comparing operations in the production area, for example, with resource (raw materials) costs, management can make intelligent operating decisions based on current data and trends. [8.14]

When preparing an end-of-shift report, excessive language should be avoided. Concise is better, because the reader will not have to sift through extraneous information. If the reader is not interested in reading the entire report, they may skim, and miss something important. [8.15] Exercise care not to skip items that may seem trivial. This often leads to problems and possibly even a near hit.

8.4 Timers

This last tool in your kit is a complete change of direction! A countdown timer is just that: a device that you can set for a certain time duration (say, twenty-three minutes, two hours, or ten minutes). After you start the timer, it will count down to zero and then make a noise for you to hear. The easiest ones to get are kitchen timers. The ones you will be looking for are direct set (you enter exactly the time duration by pushing numbered buttons); start, pause, and stop at will; and several together will conveniently fit at your operator station, each in a separate numbered bin. Figure 8.4.1 shows what this looks like.

Figure 8.4.1. Timers for operators. A timer will be used to remind you to check something at the end of the time period. This will free you to forget about watching and pick up other tasks. From the left: {L} timer (with #2 on the back), a bin (labeled #2, to match the timer), your operator station on the bottom (in case you need to pass the watching on to someone else), and the note identifying what is what; {M} note in the bin; and {R} everything together.

By using a timer to remind you to recall or look at something at a fixed time later on, you will free yourself of keeping track of each timed activity. Until the timer goes off, nothing about that situation needs your attention. Until the timer goes off, you do not have to watch out for it. You do not have to multi-task between keeping an eye out for that task while you work on something else. You can devote your complete attention to the present. Generally, how you will use these timers is up to you. Specifically, you will use them to help you keep track of whether or not certain activities or processes unfold as they should [**M27.3**].

We suggest that you have three timers available. For each numbered timer, you will have an identically numbered bin or cup that is large enough to fit it and a note. Each set will be used to remind you of whatever you decide needs to be reminded. Procedurally, you would set a timer for the time lapse you desire, write a post-it note with the item you are watching and the time the timer is started, and put it and the timer in the same numbered bin. If you are done with watching, at the end of the elapsed time, you clear the timer and trash the note. Bin and timer are now ready for the next use.

What if a timer goes off and you cannot get to it? Not good. The timer was for a real reason. Either pass the timer to an informed colleague to handle or shift your current job to a qualified colleague. Don't try an end run around multitasking [**M7**] and think you can do it just this once.

If you find that you are almost always using all three timers, maybe you are overdoing things. If you want more than three, rethink the whole thing. More may be necessary, but it likely means that you are under too heavy a load, or your job responsibilities are way too scattered. Get advice soon.

Virtual Timers

Alternatively, you can use Siri® or Alexa® to set virtual timers. It is recommended that timer names and purposes still be used rather than simply stating:

"Alexa® set a timer for ten minutes to remind me to stop the effluent rundown pump."

A better setup would be:

> "Alexa® set timer #1 for ten minutes to remind me to stop the effluent rundown pump so that the reaction rate can be maintained above 88% conversion."

Because relying exclusively on the virtual timer(s) without a tracking mechanism may not completely remove the timer(s) from your thinking, it is still recommended to use the bin and note method instead. You can see the bins. You can see the timers and notes at a glance. Remember that the purpose of this tool is to prevent multitasking.

Module 9: Models

9.1 About Models

A model on a basic level is the rationale each of us uses to think about what we are doing. It affects what we do and how we do it. Sometimes our individual ways of thinking may make it more likely that we might fail instead of succeed [**M16**]. Sometimes this can be embarrassing, or simply humorous. But in the control room, failure is a problem. A model is an invented representation of a part of or all of the enterprise. Each represents a mental idea based upon knowledge and understanding of one or more piece(s) of equipment. Models are used as a substitute for the real thing. In this module, a few useful models upon which operators rely are examined. There are three general categories of models: replica, explanatory, and descriptive.

Replica models are scale models that architects build to depict a new building or village or chemical plant. We will not use these.

Explanatory models, which we will use, are ones that engineers and process experts build, using mathematics, to provide operational data to appear close to the real thing. In the control room, these are used to simulate the actual process for information or are used inside sophisticated process control algorithms. They are built to behave closely enough to the real thing for you to use them to predict outcomes.

Descriptive models are used to classify and track where process units or elements appear in the plant; why they are where they are; and to annotate control documentation, including on the HMI.

9.2 Explanatory Models

Enterprises and their associated major pieces of equipment can be complicated. The complication can arise due to size; the complex nature of its pieces; or simply because what happens inside them is hard to see or hard to understand, even if it could be seen. Suppose someone could build a model of that enterprise or piece of equipment that sees and understands everything. The model would be a substitute, produced using knowledge and detailed understanding. While there are a large number of ways to build models, let us suppose this one is built of mathematical equations using chemistry, physics, and other sciences. And suppose this model is "tuned" so that it behaves closely enough to the real thing. Because someone constructed this model for us, it is built of pieces we understand. It is easy to look "inside" the model to see what could not be seen in the original real thing. We can also use this model and make it do all sorts of things, even things that might damage actual equipment. Test runs using the model are called simulations. They permit us to see what might be going on inside of the actual thing. Knowing what is going on inside allows us to draw conclusions and make decisions. This is called model-based reasoning. Figure 9.2.1 illustrates it.

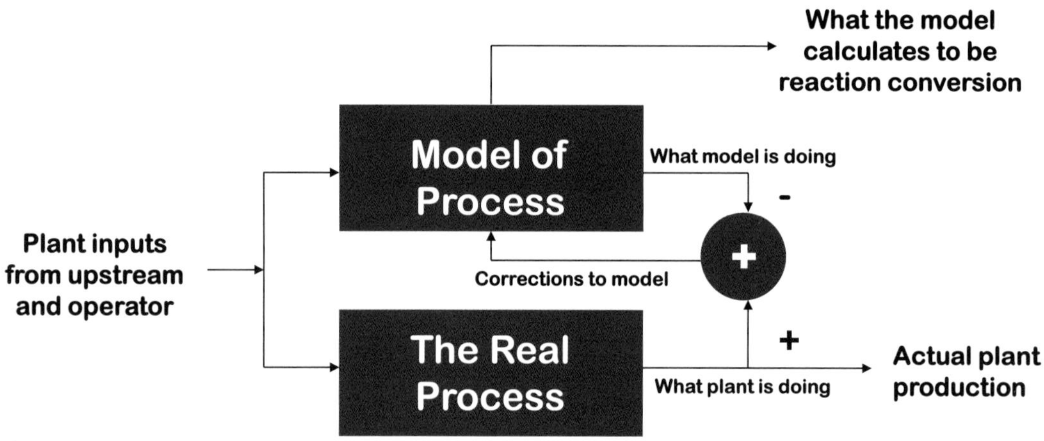

Figure 9.2.1. Model reference design diagram illustrating how information from the real process is used to "build" a model, which then permits a "look inside" to see what could not be seen in the real process: in this example, the reaction conversion.

The lower box depicts the actual process. The model is the upper box. For the example in this figure, the model calculates its results as numerical data to tell us reaction conversion. When the operator knows the specific conversion rate, he uses his process knowledge to adjust the proper operation to move that rate closer to its desired value. Without knowing the conversion rate, he must wait until the actual products change. Usually that is only after the product quality has changed. Knowing beforehand and being able to make proper adjustment reduces this delay.

9.3 Enterprise Area Descriptive Model

An enterprise area model is descriptive. It is all about geography and labels. Using this type of model, the labels in a Process Control System (PCS) can be constructed in a way that incorporates information as to exactly where in the enterprise they are located and with which equipment they are associated. This gives you the ability to quickly navigate to and clearly manage the intended parts of the enterprise.

Refer to Figure 9.3.1 for an illustration of the concept. The figure depicts a single enterprise. In this typical enterprise, there are four to six operator areas. Within each typical operator area, there may be three to five individual, significant process units. Each process unit might contain ten to even fifty individual pieces of equipment (e.g., pumps, heat exchangers, towers, etc.). Within each piece of equipment, there might be between one to several hundred individual sensors, valves, and so on.

The enterprise area model is used to associate related equipment entities (equipment, sensors, etc.) to each other. A tag is a label given to each unique entity appearing in the process control system. The tag clearly defines the entity by giving it locational and relationship information. This information is later used for control, computation, or view. For example, if the tag incorporates locational and relationship area information, it will be easy to manage alarms. Consider the following example flow controller tag identified as FC-1-3-REACT-MIX-03452, using descriptive information in the label. Enterprises need not use letters to identify geography. Tags can also be constructed of only numbers, perhaps after a few letters for a prefix. For situations with a fixed format, the dashes or spaces are not needed. Hence, our example tag for the flow control loop shown in Table 9.3.1 might just as well read FC131241503452. But do not do that; it is way too hard to read properly.

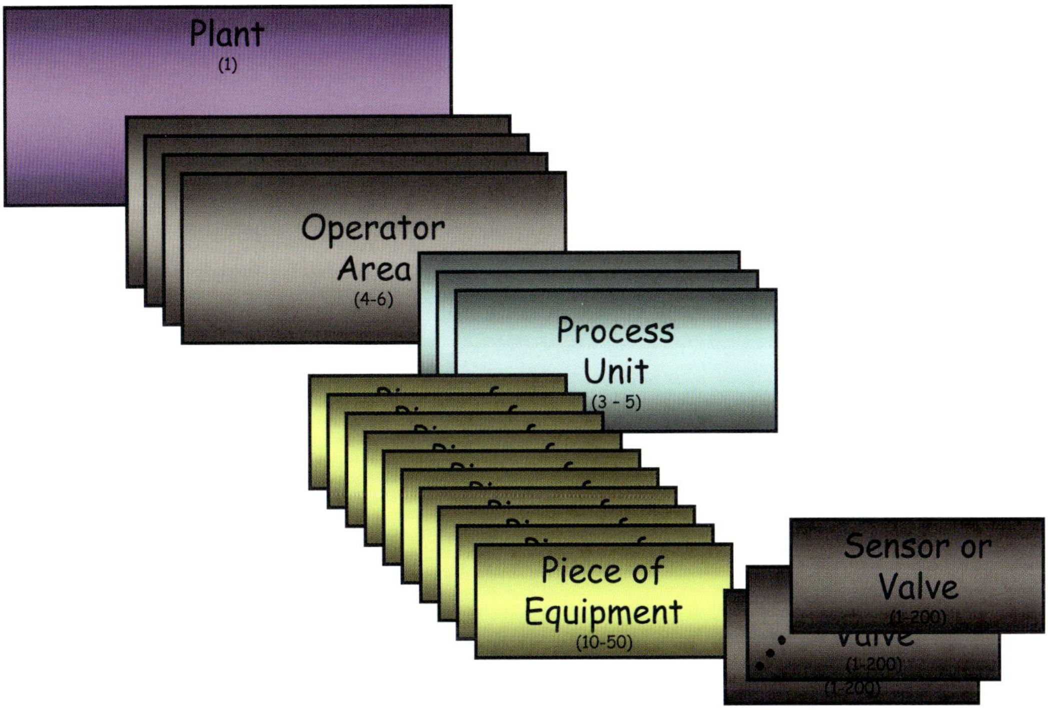

Figure 9.3.1. Conceptual enterprise area model showing the various levels of equipment and how each level is "attached" diagrammatically to the levels above and below. When such a structure is used for naming conventions, it is much easier to use automated or semi-automated procedures to modify operating conditions or alarm settings and the like for the intended equipment.

Plant ID	*Operator ID*	*Process Unit ID*	Piece of Equipment ID	Sensor, Valve, or Element ID
1	3	REACT	MIX	03452
1	3	124	15	03452

Table 9.3.1. FC-1-3-REACT-MIX-03452, showing two possible naming conventions used to construct a tag that incorporates a structured descriptive model.

Table 9.3.1 shows the information used for the construction of a tag incorporating locational information. There are many tags for this reactor mixer, all with a unique label and number. All of them have information in part of the tag construction that tells us where in the enterprise it is located. We are now ready to make use of this capability. Suppose that the mixer in the reactor was shut down. It is a simple matter to identify related equipment. Related equipment would have a tag that looks like this: ??1312415 X X X X X, where the five Xs refer to all such tags in the database with the same prefix: ??1312415 (?? stands for any symbol). Using this prefix as an automated search filter (say in a script), we find all related equipment and, for example, shut it down. In order to do the same thing without this ability, you would either need a list of items that are part of ??1312415 and use some sort of automated way of using the list, or read the list and individually shut down the required parts manually.

9.4 Understanding Surrogate Models

A *surrogate model* is a simpler or more familiar stand-in for a real thing that appears too complex and/or too hard to visualize or understand directly. It is a substitution—something that works much like the original, but is better known and more familiar. However, we must be extremely careful that what we pick is a good surrogate. A bad surrogate acts as a filter or "red herring" that hinders, rather than being useful to aid understanding. For example, the last shift experienced a lot of problems with a certain electric motor driving a large, important compressor. On your shift, you see laboratory values deviating. One reflex may be to focus on the compressor as the likely culprit for what might be causing the abnormal laboratory values. Yet, a moment of thought about what the compressor does and what might go wrong would have quickly ruled it out for causing your particular suspect lab values. So that "big problem" on the last shift had moved to the front of your mind. It got in your way.

The message here is twofold. First, ask why you are considering each surrogate? Rather than settle on one, or working to decide which of several possibilities to consider, quickly list and test all for reasonableness. Here, the burden of proof is on ruling in, not ruling out [**M27**]. Each is out until demonstrated in. If any (or even all) are not clearly appropriate, do not use. Second, as any surrogate is used, ensure that the ever-watchful eye remains on the actual process, not on the surrogate. You use the surrogate to plan; you use the actual to check. Watch the lab values carefully and investigate the broader nature of those implications (bad samples, bad laboratory analyses, piping leak, etc.). Give up the surrogate as soon as you begin to understand what might be the real problem or issue. Remember that the only purpose of the surrogate is to jumpstart problem-solving.

The Surrogate Model Test

Surrogate models can be helpful, but they can stymie progress in a very big way if they prematurely or unduly narrow your thinking [**M16.4**]. There are two simple test questions that can be helpful. Look at both the surrogate (in your mind's eye) and what you can actually observe from the real situation. Then ask the following questions to make sure things match up well enough to use the model:

- What are the actual (real) specific similarities between what I see and what my surrogate predicts?
- What are the actual (real) specific differences between what I see and what my surrogate shows?

Remember that surrogates are always a supposition made in the absence of something more tangible. No matter how many similarities are found, if any **real differences are present, it is quite likely that the surrogate is inappropriate** and should be immediately questioned. If insufficient, then discard. Don't forget that getting help is always a good idea [**M34**].

9.5 Using Models

Almost everything you do as an operator is guided by your understanding of how your enterprise works, its goals, what the others in operations expect you to perform, and many other things. For example, as you just begin to open a very large valve, you expect the flow to start in a way quite different from when you open a valve on a small line. You expect a difference based upon your experience with large and small valves. Your experience is the way you develop a particular mental model of how a valve adjusts flow rates. If somehow, a new line was installed and it had an extremely large valve, your expectation of how the flow begins would be imagined based on what you expect of other large valves, not the small ones.

As your shift progresses, and you see something unusual, say, with a heater, maybe a bit different from what you saw before, you are going to rely on how you think your heater works. You rely on your understanding of that heater—your mental model of it. When you are looking around for any problems in your enterprise, you might pay more attention to the places where something earlier went wrong and caused trouble. Sometimes, you remember specific problems you saw earlier. Most times, you are thinking about how it should work and what you see now that might not be good operation.

All of your alarms will have had their alarm response information developed with a designer's knowledge and understanding of how a piece of equipment works (their models). When you use weak signals to understand whether small indications of things getting a bit out of the ordinary might be important, you do forward extrapolations and backwards projections using your models [**M24**]. The better the models are, and the more everyone shares the important aspects, the more effective everything works. Whenever you and another individual do not seem to be seeing eye to eye about what is going on, always check to ensure that both of you are thinking about it with the same understanding—the same model. Models are tools, nothing more.

Module 10: Operator Training

10.1 Training and Skills

Training—your training, to be clear—is vitally important. This module does not discuss your specific enterprise training. All of that is up to your enterprise to design, provide, and monitor. Here, we discuss the various responsibilities of your enterprise to provide it, how it should work, and important functional aspects of what different types of training are designed to do for you.

Many operating positions are hourly. They are often supported by union representation. They are recruited largely from the general labor pool in the community. At times, other enterprise members transfer in to staff these positions. Personnel receive hiring qualification testing, some level of individualized and/or group formal or semi-formal foundation training, and a sizeable amount of on-the-job situational-response training. This training is to make sure you have the needed skills so that the production facility produces the required amount and quality of product at an acceptable material and energy consumption for those aspects under its control.

Most simple tasks can be easily managed by exercising ordinary skills, using ordinary care. Things that are not simple, especially the highly complex, are not so readily understood. To fully understand them often requires proper skills and capabilities. There are two general forms of training: skills and competency.

Skills Training

Skills training is the "dexterity or coordination especially in the execution of learned ... tasks." A skill is "The ability to use one's knowledge effectively and readily in execution of performance." [10.1] Skills are acquired by review and practice of specific tasks in order to master them. For example, when network routing centers experience an overload, the existing procedures are referenced and carried out as designed, with care to accommodate the current state of equipment and available alternate routes, you followed your earlier training. This reinforced that skill. Another example would be the process of shifting a major product flow from a main pump to a backup one without unduly disturbing the flow characteristics. The advantage of acquiring skills is the knowledge that the specific task or tasks trained for should be able to be executed competently. And so long as the procedures are accurate and robust, and nothing is encountered which was not trained for, this would likely be the case.

Skills training is the "dexterity or coordination especially in the execution of learned... tasks." A skill is "the ability to use one's own knowledge effectively and readily in the execution of performance." [10.2]

Safeguarding against the use of personal tools is one of the challenges during training. Using personal tools is a large contributing factor to the failure of on-the-job training (OJT). Management should insist that no OJT be used unless done by the most experienced and qualified trainers. When the operator who has been sitting in the job for a year or so attempts OJT, he passes on all of the errors and misunderstandings from his earlier training as well as some new mistakes (and personal tools) of his own.

A good illustration for how OJT can begin with good intentions and eventually end up in a very unacceptable place is the children's game of Telephone.

> In Telephone, everyone sits in a large circle. The first child begins with a "telephone call" (whispering in the ear) to a neighbor with a clear and explicit message. The neighbor "calls" the next neighbor and repeats the message as accurately as possible. This continues until all in the circle have been called. By the time the original caller is called by the last caller, the message received is almost never recognizable as the original message.

The reason for the change is entirely understandable: people hear messages not exactly as they are said, but through the "filter" of their experience and expectations.

> Suppose in this telephone game a child heard the phrase "in hot water." He thought he understood it, but maybe missed a bit of the part that went before. His family always used that phrase for "in trouble." He might repeat the more direct phrase "in trouble" because it was much easier to remember than "in hot water." But in this case, "in hot water" was meant actually to be in water that was too hot.

The children's game of Telephone is far afield from real control rooms—but the lesson is not. Control room management must be comprehensive. Small aspects have considerably more importance than may seem at first glance.

While private and personal tools developed by operators may seem useful and appear important, they must be demonstrated and approved to be in concert with the established operations technology, procedures, and good protocols for utilizing them. This is not to say that all must be disallowed. Management must first properly evaluate them. If a personal tool is determined to be beneficial, it may be added to the infrastructure through established management of change process. The rest of the personal tools are to be quickly removed and proper ones explained and re-established.

10.4 Specialized Behavioral Training

The "behavior" here refers to practicing the important people interaction procedures and tools covered in your handbook. While all these skills are rather straightforward and easy to understand, you do need to develop them and make them your own. Practicing will provide a basis for you to be comfortable and competent when you need to invoke *permission to operate* [**M35**]; *collaboration, delegation, escalation,* and other forms of getting help [**M34**]; control room *conditions* and *codes* [**M31**]; and *operator redeployment* [**M36**]. An important behavior is how you, as part of the team, are able to receive and give information and advice during critical operating situations. Crew resource management (CRM) is the way to ensure that everyone "says something" when they "see something." [**M34.4**] This does not come naturally. You learn and become comfortable by knowing how and practicing. Look for ways to make training for these competencies readily available for you. Keep at it until using all of them becomes second nature. Remember, when the need jumps out in front of you might not be a good time to learn that you are a bit "rusty" or mistake one competency for another. Either way, you are not going to be at your best.

Notice that training also comes up in the next section on process understanding. You work to master the interaction parts. In the next section, you are ensuring that when you do the interacting part, you are doing it with a full understanding of your plant operations. They work together. You operate better.

10.5 Process Understanding

Operators need to have a proficient and clear understanding of all parts of the enterprise they are in charge of operating. A passing understanding is definitely not good enough. From a process design point of view, it might appear that the chemistry or mechanics behind the plant or production activity are too complicated to fully master. This design-level understanding is not needed for operators. However, you do need to know how the part of the enterprise you are in charge of managing reacts to the full range of operations, both normal and abnormal. You need to understand the full range of operational faults that may result from an inability to maintain safe and responsible operational conditions.

There are several specific places where good understanding is particularly important:

- *Collaboration, escalation,* and *delegation* [**M34**]
- Alarm response [**M19**]
- Weak signal analysis [**M23-M26**]

The bottom line of training is that it is your responsibility to recognize every situation that you are having difficulty understanding, difficulty managing, or difficulty finding the right resources. You will need to bring this up with your management. It is management's job to provide what you need. It is your job to make sure they know what you do and do not know, that you need to know! And how to use all properly and competently.

Module 11: Using Data

11.1 Data and Information

One of the best capabilities of a modern operator station is its ability to provide a great deal of data to you. In addition to control variables, measurements, and control loops, it can provide elaborate temperature profiles, imbedded analyzer values, valve positions, and a host of other important and relevant data. As an operator, it is probably not unusual for you to have way too much data and troublingly too little information. Operators who were unable to monitor, detect, or conclude information from all of the data available, helped a significant number of serious incidents along. This speaks strongly against more data; it speaks directly to the need for effective tools to understand the data that comes your way—or does not come and that you need.

Two useful tools for gaining information from data are *weak signals* [**M23**] and the ability to manage the untoward effects of *bias* [**M16**]. To say it another way: look for what might be really going on, but look without deciding until you know enough. All of this is geared to help you answer the situation management question:

> The Situation Management Question: Determine what are all of the reasonable and plausible explanations from the "data" observed, via all of your situation awareness tools and activities, and only after that, decide what can be really going on.

Data into Information

You will never use data directly! Data needs to be interpreted to discover and understand what it actually means. Proper data in context becomes information. Every piece of data, whether an instrument reading, lab value, or observed value must always be tested against what is reasonable and evaluated for its meaning. Nothing reasonable is automatically true. Nothing unreasonable should be discarded or disbelieved without being fully investigated. There are two general classes of information: (1) objective (direct observations and recorded data); and (2) implied or inferred (what you conclude or suspect from other sources).

Objective information comes directly from readings and other material that are visible (on screens, reports, instrument readings, etc.). Data that differs from what actually exists in the plant or operation may be due to error such as: (a) instrument malfunction, (b) report writing error (noting a lab value in a wrong location of the report), (c) transcribing errors, (d) intentional deception, or (e) something else just as troubling.

Implied information is what you deduce from the data using your personal supposition, conclusion, or conception. You are working to determine the data's relevance. The knowledge you need must be anchored by, or suggested from, objective information.

Knowledge Fork

Sorting out the proper information from all of the data is vital. When you bring information, opinions, and observations into play, you want to make sure everything is clear, accurate, and verified. Using the *knowledge fork* (Figure 11.1.1) ensures that each bit and piece brought to the table is dependable. Every observation, every bit of data, and every other item must be clear. As each bit of information comes in, including observations as well as procedures that appear to be relevant, use the tines of the fork to classify it as known, unclear, or assumed. Make sure that everything is classified. Use written notes, annotation, and post-it notes on documents or screen shots to track. This can be done as each bit comes in, or after a few are in. Remember that the later it is done, the easier it is for assumptions and unclear parts to lose their proper category and just fall into the one of knowns. So do it early. Only known information can be relied upon. The rest must be ignored if and until properly verified.

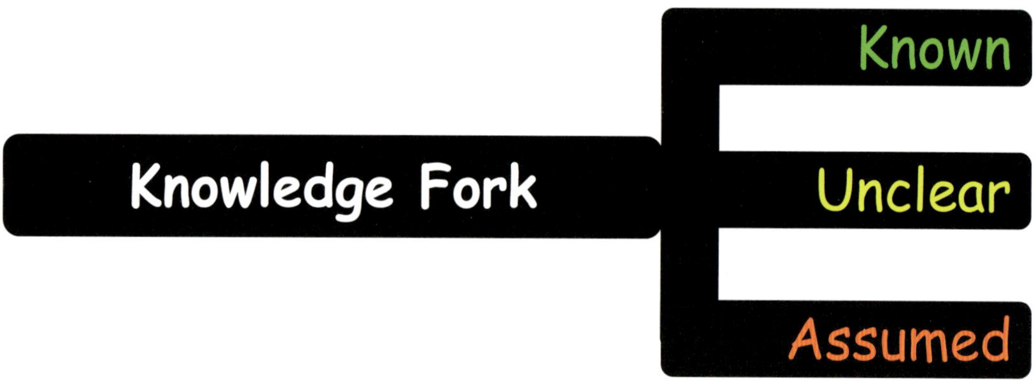

Figure 11.1.1. Knowledge fork with three tines describing how to qualify every bit of information available and intended to be used for problem identification and remediation (if needed). These classifications ensure that decisions are made from only known information. Anything unclear or assumed must be verified to the level of known to be used.

The knowledge fork categorization needs to be used to clarify all information used by operators. It is of course, a *thought* tool. Consider it to be a prominent part of your tool kit! "Pick up" and evaluate everything with it.

11.2 Using Data

Data is the currency of exchange between you and everyone else you interact with. Every time you coordinate with someone in the field, both of you rely on data to tell you what is going on. What is going on depends upon where it is. This is indicated by readings, settings, operating objectives, and operational difficulties. The data is the status. Only known information can be relied upon. The rest must either be verified or ignored.

- Every observation, each bit of data, and every other item that you bring to the table during collaboration, delegation, or escalation [**M34**] must be clear.
- Written reports, orders for testing, or repairs all depend upon the information (from the data) that you relied upon to decide what to ask for.
- Taking over a shift requires you to absorb and contextualize all of the handover information so that you are ready to take ownership [**M22**].

11.3 Partial Information

We humans are really, really good at working with limited information to try and understand a situation. As soon as we can grasp a nice bit or piece, what was initially very vague and incomplete quickly snaps into focus. Even without a framework, the signal can jump out of the background. The most often used example of this is hearing your name in a very noisy, crowded room. Immediately, without any attention directed at listening, your name catches your attention. This can be a useful trait. Most often it is not.

Boroditsky [11.1] illustrates this concept with a very simple word phrase test.

Figure 11.3.1. Partially obscured phrase that appears to be quite readable due to our ability to infer meaning from incomplete information. Often, we are wrong and then we act inappropriately.

Looking at Figure 11.3.1, even though the black bar obscures more than half of the message, there really is not much of a problem reading it and seeing the words "jumping to conclusions." We are adept at making sense out of what might at first appear too indistinct or fragmented. We make sense out of fragments by inferring from our experience and expectations. But all of this jumping is a problem. We often conclude along a biased pathway. See Module 16 to identify these and how to avoid them. Many of our individual biases are shared widely with others. Fortunately or unfortunately, this shared bias will foster a shared induction. Just because others decide to see what you think you see, it is proof of only a consistency of people seeing. It has nothing to do with the correctness of it all.

Look at the letters in the figure above again. But this time without the black bar. Now see the actual phrase:

IUMRING TQ GQNGIUSIQNS.

The actual phrase makes no sense. But what if it did make sense, and it was a different sense than before? While observing obscured or partial information, it is human nature to jump to a conclusion that tries to explain it. But that is only what we think it might be. It is not necessarily what it is. We must be very careful to train decision makers to carefully confirm their assessments. Looking ahead to Module 23, there are very clear steps for collecting the clues, considering what more serious problems they could suggest, and then looking for proper evidence. This will not work if we decide what's wrong at the beginning. Looking for potential problems from sketchy clues works only if looking is not prejudged by what you want to see. No jumping to conclusions here.

11.4 Missing Data

"The wise man is one who knows what he does not know." [11.2] Just hearing this sounds like a paradox. But for you in the operator's chair, what you do not know cannot be factored into your thinking. If it is important, you are going to miss out and likely not operate where you need to be. The four quadrants in the matrix of knowledge by Goodman [11.3] (Figure 11.4.1) may be used to help define known and unknown knowledge and how and when to use it.

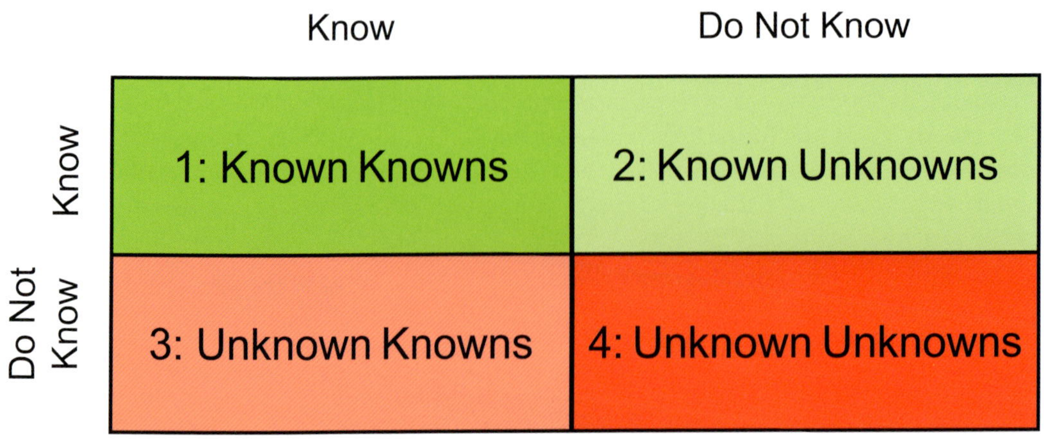

Figure 11.4.1. Matrix of Knowledge with the four quadrants. The upper two are what is known that you either know (upper left) or know that you do not know (upper right). The problem lies with the lower row: what you do not know that you really do know; and what you do not know that you actually do not know that you do not know.

> The upper left "known knowns" is clear. Everything that you know, and you know that you know, belongs there. You can count on it. Using it for decisions, where it applies, is proper.
>
> You also know the upper right "known unknowns." For example, if you are informed that there will be a cybersecurity drill in the control room today at 1400 hours, and you have never done one before and have received no training, you know that you do not know how to do the drill. Another case is when you attempt to perform an operational activity that is not working out well, and you have done everything you were trained and experienced to do, you will know that you really do not know how to do it for this circumstance.

The problem comes with the lower row. Let us take a look.

> The lower left "unknown knowns." Here you actually do know information or procedures, but you cannot recognize that you do. There is a mismatch between the actual situation and what you think you are doing, or how you are thinking about what you are doing. If you are able to make a proper association of what you see now going on to your background of experience, you could draw on that knowledge. Most times, you cannot see. To resolve this, you need ways to find out that you actually do know, given that you think you do not. You need outside help [**M34**].
>
> The lower right "unknown unknowns." Here you do not know what you do not know. This means that you actually do not know anything relevant. At the same time, you do not know that you truly do not know anything. If you are not finding anything, you might waste precious time that needs to be used for another remediation approach or understanding. On the other hand, anything you do find would be something that is not a proper fit. You will get it wrong. Or you will not look at all.

Again, a useful remedy to the problems in the lower row is collaboration. A key question asked is "what do you know that I do not?" There, you rely on the collective knowledge of team members. But this still leaves a big gap. The best-working way to cover this gap is to always remain vigilant for anything and everything that does not seem to fit in. All are important reminders that there is more to know. Weak signals [**M23**] is made to order for exactly this situation.

Module 12: Operator Supervision

Experience is a hard teacher because she gives the test first, the lesson afterwards.

—Vernon Law (star baseball player)

12.1 Understanding the Role of Supervision

This module covers the direct responsibilities of supervisors as they supervise operators. It is intended to help you understand and appreciate what to expect from your direct-line supervisors. It will allow you to see what your supervisors do. It will also show you why it is okay to ask for assistance [**M34**].

12.2 Front-Line Supervisor's Role

In situations where control room operations either caused the issue or failed to lead to appropriate resolution, studies have been done to determine contributing causes for serious operational incidents and disasters. When resolution should have been possible, we find a clear link to supervision deficiencies. These failures fall generally into five categories.

Supervisors failed to:

1. adequately ensure operator training effectiveness.
2. appropriately evaluate shift operators' fitness for duty.
3. be readily and consistently available for consultation, collaboration, and proper escalation decisions.
4. call for additional assistance in a timely manner.

   Poor behavior usually gets in the way. This is because senior management fails to recognize that it is better to provide assistance (even if eventually determined to be unnecessary), than not to have the proper resources ready to use when needed.
5. be effective in attaining a broader level of operator situation awareness.

   With situation awareness, you will delay jumping to conclusions and avoid continuing an inappropriate or inadequate solution approach.

The enterprise must not place any burden upon you, the operator, beyond your reasonable and appropriate capabilities to manage. This is why you have control room supervision. Proper supervision protocols for your control room depend on the enterprise understanding what supervision is necessary. This impacts whether or not you, as the sitting operator, should be able to rely on supervision being readily at hand and fully up to the task. Control room operators and supervising managers are a team. They work hand in hand.

We explore supervision in three categories: administration, development, and operations.

Administrative Supervision

Administration ensures that the functional structure of an enterprise is followed and maintained. Here is where performance standards are set. Personnel schedules are developed and fulfilled. Administrative supervisors maintain training needs and keep proficiency records. They ensure that proper equipment, tools, and documentation are available and used as needed and intended. This includes keeping and utilizing appropriate production records, setting product/production standards, and making certain those standards are in place and followed effectively.

Development Supervision

Development is the life-cycle process for operations personnel. It delivers proper and sufficient safety awareness and training as well as operations skills assessment and coaching. To provide development, supervisors employ mentoring, a programmed sequence of critical skill building, and performance feedback. All of this would be done in a coaching and team-building atmosphere—an atmosphere that is absent of blame and understandingly tolerant of honest missteps.

Operational Supervision

Operational supervision is when the active supervisor participates with the control room operations team. This ensures that suitable experience and collaboration is available. It will ensure appropriate personnel resources are present and functional during operational times that require it. Such resources may include additional hands-on personnel, decision guidance, escalation management, and management coordination for policy implications. These resources may be physically present or available via effective tele-collaboration.

12.3 Experience Mismatch Caution

Experience and competence will vary between individuals in the control room. Hence, the operational responses of inexperienced operators and less-experienced supervisors would naturally be limited. Supervisors usually rise from the ranks of operators after extensive experience. Or they are new to their post. Operators similarly vary from the just-hired or in early training, to fully seasoned and savvy. During abnormal situations, it is essential that competent operation and adequate supervision be present. But this natural safeguard can be compromised for a shift when both operator and supervisor have limited experience. The same concern would exist if first-level assistance were from other operators or on-call resources that have limited experience.

> Inexperienced operators and inexperienced supervisors must not be paired on the same shift at the same time. At least one must be experienced.

To address this situation, it is strongly recommended that inexperienced operators and inexperienced supervisors not be paired on the same shift. It is left up to your enterprise to determine the level of experience involved for those decisions.

12.4 Takeaway Message

No reasonable amount of personnel selection, training, enforcement of performance requirements, or anything else like it can make up for the lack of an operator's ability to perform. Management must provide appropriate hands-on supervision. Gaps are recognized, accommodated for, and remediated by management. By doing so, the result will be a clear and continuous partnership between operators and supervisors. Supervisors must be ready. Most operations failures can be traced to supervision failures. Neither is necessary. Leaders almost never fail a test of leadership during a crisis that they did not already fail before. Responsibility is developed before it is needed. It is practiced when it is.

PART 3
KEY CONSIDERATIONS

Part 3 covers key considerations that frame important concepts of control room operations—concepts that affect success or contribute to failure. It will help you to better understand what might get in your way as you work.

It begins by examining a critical resource: time. When you fully understand that time is your most valuable resource, it will guide your every activity. You must have enough of it available to manage problems. When it runs out, you cannot manage any more.

Different types of abnormal situations are defined. We show you how to see, understand, decide, and either act or defer action.

In the discussion on doubt, you are encouraged to embrace doubt and make it work for you.

Part 3 concludes with a module to recognize and understand the different types of biases and pitfalls we all confront. It then explains how the many ever-present but often unrecognized biases influence decision making. It touches on differing norms and conventions of the many different cultures that may share the same control room.

PART 3 — KEY CONSIDERATIONS

Module 13: Power of Time

Time is the most valuable resource you have with you in the control room.

13.1 Time

Success or failure is always, every time, limited by how much time you have and how you are able to use it to manage a problem. If you ever run out of it, you have run out of options. So, while time is still available, your available options are to effectively and efficiently:

- identify potential problems early enough.
- work them properly.
- verify the effectiveness of your actions.
- modify those actions if needed.

Situation management is time management. So time management is everything. Time is *the* most important thing for successful control room activities. **No one and no action can intervene if there is not enough time for it to work!**

> If somehow, magically, one could stop time during an upset, all upsets could be successfully managed! If every disaster in the making could be paused, it would be possible to spend as much time as needed to understand what is going on, to assess potential damage if it continues, and to fix it if we wanted. Even if we could not reverse time, the mere ability to stop time or even slow it down would be the wonder tool. Alas, such daydreams cannot come true. Yet exploring this fantasy helps focus on your most valuable commodity during an upset: time.

You are in charge of how you use time. Learn and use tools that work for you. Avoid multitasking [**M7**]. Use those countdown timers [**M8.4**]. Never run out of time before you stop the clock using *permission to operate* [**M35**].

13.2 See, Understand, Decide, and Act (SUDA)

How you *See, Understand, Decide,* and *Act (SUDA)* is the important operational reality of successful production management (Figure 13.2.1). There must be sufficient time to complete all remedial actions before damage occurs. The following definitions are useful:

fits the current situation, you have possible doubt. It is important that *possible doubt* concerns are not dismissed out of hand. One of the ways to keep this under control is to realize that weak signals [**M23**] usually fall into this category.

Probable Doubt

Probable doubt is 51% certain. That is, taking everything into consideration, it is more likely true than not. Start with the list of possible doubts. Pick all that are most likely to be true. They remain as the most likely. In the control room, an operator with probable doubt leaning toward something being amiss means that something must be followed up. In general, these situations would be abnormal ones. You are beyond weak signals. Now something is more likely amiss and needs attention.

Reasonable Doubt

Start with that earlier mental list of probable doubts. Remove all the other doubts that a reasonable person would find above but close to the 51% dividing line. We are looking for what now seems to be significant. But so far, we cannot yet fully believe it. What we have left is a list of significant doubts. If any fit the current situation, we need to evaluate them for proper understanding or action. Reasonable is the benchmark that captures both intelligence and street smarts. It is used in the United States as the legal test of criminal guilt beyond a reasonable doubt. In technical terms, reasonable doubt is 95% certain. There is a bit of room for being uncertain, but one feels that anything there would not be reasonable enough to consider.

> Asking operators to only act (or to actively reject action) based on needing confirmation **beyond a reasonable doubt** is much too high a requirement! There are no responsible mechanisms to enable timely operational decisions to be made that confidently.

As soon as probable doubts are evaluated and found to be well beyond the 51% point, they are reasonable enough. Each must be fully evaluated to decide whether or not action is warranted. The result only needs to be credible. Any remaining uncertainty needs to be understood. This is a good place for collaboration and escalation [**M34**]. The constant and reliable safeguard is to always rely on permission-to-operate protocol [**M35**]. This is the rule that keeps everything within the safe-operational boundary.

Shadow of a Doubt

Shadow of a doubt is anything perceived as 99.9% or more certain. It is unreasonable to even think that everything important must be known beyond a shadow of any doubt. Actions that require knowledge beyond a shadow of a doubt are not useful. Try to take this extreme degree of certainty off the table, please. No operational situation should be required to even come close to meeting this degree of certainty. None. Never!

15.2 Managing Uncertainty

Operators' decisions in the face of uncertainty or during unexpected situations are rarely crystal clear. There is always uncertainty that must be accounted for and acknowledged as part of the operating culture. This suggests the following consideration: design and practice open communications during operating events to encourage participants to fully express what they see and understand. These considerations are [**Q6**]:

Know

- the symptoms of the issue at hand.
- what is known and the quality of that knowledge (probable, certainty—take note of the knowledge fork) [**M11.1**].
- what is not known but is essential to know—and why.
- what are the clearly perceived operational or other risks being faced.
- what are the reasonable choices and why.
- what must be avoided.
- what plan is the safest and most likely to be successfully executed.
- what is necessary to execute that plan.
- is what is necessary to execute that plan available.
- what is the backup plan in case that plan cannot be selected or executed.

The tools and principles in this handbook are designed to help you do the job of safe and reliable operating. Keeping the focus clear on these points can go a long way toward minimizing guesses about relative importance. It will discourage jumping to conclusions caused by filling in missing items with conjecture. By the way, any time that you have a plan but no backup plan, it is cause for worry. This is exactly the type of over-committal that demonstrates loss of perspective.

> A plan without a back-up plan is not a plan.

Lingering Doubt

We have all had them. A lingering doubt is one way our subconscious is trying to tell us that something needs attention. Sometimes we think of such doubts as being in the back of our minds. They might go away for a while, but they keep coming back. We have trouble saying exactly what they are. They appear real, yet they appear vague and uncertain. They just keep hanging around.

Lingering doubts are our internal self's way of being exposed to a small indication of a weak signal. Something needs to be decided, or understood, or examined more. In the control room, light must be shined on any lingering doubts to see if they need to be worked using your situation awareness tools. If nothing useful seems to be there and the doubt goes away, okay. But if the doubt still seems to be lingering, it must be escalated. Get others involved.

Managing "Truths"

Looking ahead to the discussion of alarm management [**M19**], you need to clearly see how you can investigate all of the possible causes first and then test each separate cause for validity and significance. A clear benefit is to be able to test truths in a structured and unbiased way. Many of the methods are laid out in the alarm management module. This is preparation for why they are so important.

Every time you are in a situation where truths are conflicting, either one or the other truth needs more examination. Or neither is true. Look carefully. Decide only when you have evidence and are sure.

Module 16: Biases and Pitfalls

16.1 Advisory Perspective

All of us are profoundly influenced and guided by our personal beliefs. These beliefs often come from cultural heritage, geographic locale, and other influences. We visit this important topic in order to advance your understanding of how these beliefs impact control room situations. This module discusses insights rooted in anthropology, psychology, and sociology. But do not think for a second that, because it is not engineering, this topic is not powerful and important.

This module speaks to the heart of what you face as operator. It talks about and respects the very essence of human nature. Operators are individuals with distinct personalities, histories, aspirations, preferences, dislikes, outlooks on the world, and a host of other traits, including situations where they have no opinion whatsoever. You perceive the world and who you are in your own personal ways. But what you perceive is not the only thing going on. You also bring along the deep-seated effects of your rich, individual, human nature past.

Over vast millennia, we have developed important traits that have stood by us, protected us, and allowed humanity to evolve to today. One of these traits is expectation. When faced with a situation, we instinctually work out how it matches what we expect and what we have had experiences with understanding and handling. It is why we immediately move out of the way of a falling object. No time is lost by considering exactly how far away it may be or how heavy or dangerous it might be—we just get out of there fast. It is how we are. But just because it is part of our survival nature to always be comparing what we see and feel to that vast, yet mostly hidden, instinctual database does not mean it works everywhere. Where it does not work at all is in the control room.

You read earlier about mental models and doubt. We now look into how we think and the biases we bring along as we do. You may ask why such material belongs in a practical handbook for sitting control room operators? It belongs because these very insights can be just as important to your success as traditional design and engineering. In fact, social scientists have long been involved in investigating the large number of psychosocial contributors to disasters. Doing so, they gain important insights into their understanding and management. Drawing upon their analyses and applying their lessons help us avoid embarrassment, increase job satisfaction, and dodge or lessen disaster. All these worthy goals give an advance warning of what gets in the way of successful control room management [**M31**]. Some are going to pop up exactly at the time and place where you least need to get it wrong.

16.2 Why We Make Mistakes

One of the biggest mistakes we make is to think that we do not make mistakes. An even bigger mistake is to avoid understanding why. To introduce this, observe what we will call "traveler's immunity." People often do more daring things when they are out of their familiar element, like on a trip away from home. Why is that? Part of the reason is that certain sense of freedom that comes from getting out of regular habits and away from the daily grind. But dig a little deeper, and something else comes up. We somehow think that being away from the familiar seems to grant us special powers over physics. The ground there cannot be as hard as back home; the water cannot be as treacherous as at the local beach; the roads are not as dangerous. It is as if being away can convey a special immunity from the ordinary dangers back home. But physics is the same everywhere.

Let us examine why we make mistakes. One of the best lists comes from Hallinan: [16.1]

We look but do not always see
We all search for meaning [before understanding]
We connect the dots [even when they aren't]
We wear rose-colored glasses
We can walk and chew gum—but not much else
We are in the wrong frame of mind
We skim
We like things tidy
Men shoot first
We all think we're above average
We would rather wing it
We do not constrain ourselves
The grass *does* look greener

This module is about understanding and working to keep items in this list from getting too much in your way.

Looking without Seeing

As operator, you pretty much use seeing (with your own eyes) as a big source of acquiring information and keeping track of things. When you see things that look normal, you expect that they are. If you are able to see things going awry early enough, most alarmed abnormal situations could be worked on before the alarm activates. You also depend on seeing problems on the HMI screens [**M18**] and on laboratory reports and the like. But it is really hard to look and see everything that needs seeing. Our eyes and brain are not set up very well for this. Missing things can be problematic. If nothing else in this handbook seems useful to you, this should be. Understanding and managing what impacts your ability to see things as they really are is why we spend time on HMI design and all of what goes into carefully building them.

Let us review a few of these important earlier design aspects:

- Information displayed to the operator is for a purpose—everything is needed; observe it all.
- Displayed information must be placed in context—to facilitate needed associations.
- Information must be in a form that is useable—jobs that need doing are supported by where they are placed on the screen and how the information is displayed.

Before leaving this part on the importance of looking and seeing, prepare for the important ways of looking to work better. You need to be able to find situations before they become problems, yet human nature gets in the way. So you need to find ways of looking that really can make the seeing work. What does not work very well at all is the old fashioned "looking around." Keeping an eye out is not going to be a good work plan. Best operator practices should help turn this around.

16.3 Creating Our Own Problems

Failure Avoidance

Every enterprise sets their own standards for performance. The culture of the enterprise should encourage compliance with standards and manage situations of noncompliance. This leads to an important message: history and human nature tell us that where the balance is not fair between meeting standards and not meeting them, disaster thrives. To clarify, when fear of failing is more important than the full consequences of actual failure, operational failures will almost always result. They are often tragic.

> When fear of failing is perceived to be more important than the full consequences of an actual failure, operational failures will almost always result.

The problem is not the nature or fairness. The problem is that punishment fear, or even peer pressure fear, can distract operators from their duties to the point that they are unable to focus on the job at hand. Operations are at risk for any eventuality that circumstance presents. The concerns are both a failure to act when action is required, and the taking of inappropriate action for the situation at hand. Fear of punishment is an extremely powerful force. Consider the Amagasaki Rail Disaster. [16.2] It is a powerful lesson for us all.

Amagasaki Rail Disaster

At 9:19 a.m. on the 25th of April, 2005, a seven-car commuter train traveling at high speed around a pronounced curve derailed and struck an adjacent apartment building in the Amagasaki area of Osaka, Japan. The driver and 106 passengers were killed. The cause of the crash was excessive speed for the track configuration.

The driver was experienced with this route and schedule. The maximum speed allowed in the curve was 70 km/h (43 mph). Engineering calculations determine that derailment would likely occur at any speed above 106 km/h (66 mph). The actual train speed entering the curve was 116 km/h (72 mph). The driver had an eleven-month unblemished record at the time of the crash. However, the young driver had previously faced harsh punishment—financial penalties and months of intentional demeaning punishment in a retraining program called "Nikkin Kyoiku." [16.3] The initiating event for the punishment had been his overshooting a station by 100 meters during the third week of employment.

The formal Amagasaki investigation noted that it was likely the driver, fearing a return to harsh punishment, became so focused upon meeting the rigid schedule that he failed to properly monitor track conditions, thereby operating the train at too high a speed in an attempt to make up lost time. When the speed and curve were eventually noticed, the driver avoided using the emergency brake. It would constitute another reportable breach. He had accumulated two breaches already on this single run.

Logic-Tight Compartments

A *logic-tight compartment* [16.4] is "the act of accepting beliefs that are illogical in spite of insufficient justification to support them." We do not always think consistently. In some parts of our lives, we carefully measure and deeply ponder: What is the right decision to make? We use logic and consideration to weigh all of the necessary factors. Yet in the stroke of a single moment, we are prepared to accept beliefs that logically we should have no earthly business believing. Believing means we deny compelling contradictory evidence and distort logic. This is the basis of *confirmation bias*, discussed later in this module.

Your enterprise's engineering design and management protocols should have included anticipation of such risks. First of all, your operating culture must allow you to reject misinformation. Critical judgments and effective action cannot result from attempts to resolve the wrong problem. Procedures must be designed from proven protocols, not anecdotes of a few odd anomalies of practices gone amiss. Near-miss reports must be objective and carefully studied to identify the cause as misstep, honest error, or logic-tight misconceptions. Action plans must be in place to guide proper response to correct the situation. A logic-tight compartment cannot be broken open by force. When a compartment is confronted, carefully and respectfully find ways to gently but consistently pry it open. The most difficult type of learning is unlearning and replacing it with a new behavior or even a new paradigm.

Logical thinking is powerful. It is generally useful and beneficial. Trouble comes when some of the logical or compelling evidence of the truth is inconsistent. Perhaps you see evidence that appears to support different positions. Both positions cannot be true at the same time. You will need a good way to decide. Faced with this dilemma, you are likely to pick the more agreeable truth. Most will pick the position that represents preconceived notions and what is wanted to be true.

Avoiding this is a manageable problem. You need to find the actual truth. We next explore ways to do so.

Loss of Scale

Losing mental scale is like taking the first step into sand and sinking just a bit. Take another step and sink a bit more than you expect, but not too much. Take a third step, expecting the same proportional response as before, only to discover too late that you are in quicksand. This is not proportional! It is off the scale. Any significant mismatch between the real thing and what we think it might be, is damaging. This scale concern works both ways. Thinking that a problem or issue is much larger than it really is can be just as concerning because it causes us to overreact. When we mistakenly approach a problem that is actually much smaller than we imagine, it results in unnecessary delay or waste; we think we need more planning and resources. Even after its true size is understood, the earlier thinking has a biasing effect that often results in more time and effort spent at resolving it—or using a completely inappropriate process or methodology than what the true issue requires.

The Two-Reasons Trap

The *two-reasons trap* is explained as follows:

> If there is only a single reason for something, our natural instinct is to ponder more. Yet when we see or find a second reason, any second reason, to either explain something or to base a decision upon, we quickly lean toward accepting the decision.

This "trap" is subtle and pervasive. It can seem quite logical, even responsible. Two or more things pointing to the same conclusion seem convincing. Without being aware, we use it all too often. It even appears to be subtly implanted in places we actually depend on to be reasonable. When there is a reasonable explanation for something coming from one direction, and another reasonable explanation that appears to be coming from a different direction, and both align, most of us tend to be satisfied that we have sufficient grounds. We would, except for an important problem. The two (or even more) reasons need not be anywhere close to the same weight or importance. Whereas the primary reason may be well-founded and convincing, the second one might be trivial or insubstantial. Making matters worse, the second reason is often sought out knowing that one needs to find a second one to ensure that what we want to believe is well-founded. Sometimes, even the first reason is shaky.

Either way, this trap needs to be fully recognized and managed. The best way to manage it is to:

- ensure that all reasons are independent and of sufficient strength to be believed in their own right.
- look for disconfirming information as diligently as for confirming.
- always look for more information about any reasons that are even the slightest bit weak or uncertain.

It is the quality of each reason, not their number or order of discovery, that matters. You can manage this.

Remembering

We all are good at remembering. At least we think we are. We see something interesting and intend to mention it at the first opportunity. We make a mental note to remember a task to do when we finish what we are doing now. We decide to remember. Yet almost always we cannot do it well enough. This is not to suggest that our memory is deficient or flawed. Our inability to remember is inherent in the way our minds work. When you make a mental list, you are making a conscious effort to remember important things. You review your new list in your mind a few times to let it sink in. But if another brief but absorbing task interrupts you, most of the previous remembering will be lost. We quickly revert to preconceived ideas and methods. We stop looking and start concluding.

Bottom line: remembering is not a skill you can rely upon. Rather, when one piece of information is to be compared to another, all of that information needs to be kept readily available for viewing at the same time in the same view. Thus, the text for the written procedure needs to be in view at the same time as you take the action step. It means that all tasks that remain to be done need to be readily available to view and select. It could be a task return point needed for follow-up monitoring of an activity started earlier. It could be an alarm that has not yet been handled. It could be a series of tasks to be completed before shift end. Regardless of the nature of the tasks, all should be clearly but not invasively available for pickup by you without the need to remember what they are or where to find them. This exact issue is the reason for the important emphasis on the "alarm acknowledge" formal procedure for *alarm management* [**M19**].

Illusion of Skill

No competent airline pilot, not even one who has flown the exact same airplane for years, will do preflight checks without fastidiously using a checklist and referring to the procedures. There is an old saying that "the more you know, the less certain you get." With most wisdom comes caution. Unfortunately, caution does not always prevail. People, even experts, often confuse the power of their own skill and experience with an illusion of skill. They give themselves permission to cut corners or allow downright recklessness. This illusion can blind an individual from seeing or taking into account clear clues that were obvious to the non-expert. And it may be difficult even for experts not to over-claim—acting like they know something when actually they do not. Using an expert requires that you must ensure that what they say and what they do must make sense. Ask for explanation. Seek clarity. If you do not fully understand, it is the expert's fault, not yours!

> If you do not fully understand an expert's advice, it is the expert's fault, not yours! Work toward a proper clarification.

Good Is Not Good Enough

The "80:20 Rule" [16.5] states: for any task, getting 80% of the reward takes only 20% of the effort. As if that were not enticing enough, earning the final 20% of the reward requires 80% more effort. No wonder everyone would like good to be "good enough." The reality in the control room is that everything needs to be done properly. In the control room, good is not good enough. Much needs to be done as close to exactly right as reasonably possible. There is too much at stake. Control room management is finding what is needed to do a proper job and doing all of it, short of perfection, but well above "just good enough." Your enterprise will have decided exactly how far it needs to go and will have built that into their infrastructure. You should have the necessary tools and will be expected to do it that way all of the time.

16.4 Biases

A *bias* is an undue influence or predisposition. Biases are real. They are important. They can be hard to recognize. We all have them, and they can get in the way of good operation in a big way. (Social or personality aspects are not part of this discussion.) Biases can be recognized, described, and managed enough to make sense. They are subtle affects and for the most part are largely unknown by the individual. One of the telling ways to observe where biases are present is to look for "highly trained, experienced people breaking their own rules…." [16.6]

Confirmation Bias

Confirmation bias [16.7] is the tendency to take a group of facts and believe mostly the ones that agree with what we already believe. Anything that is not what we expected might be noted but is then often set aside. Almost everything consistent with or that advances our original belief is embraced and categorized as, "See, what did I tell you, it's true." We use this bias (unintentionally, for the most part) when information is gathered, remembered, and interpreted. It is strongest for emotionally charged issues and where deeply entrenched beliefs are involved. We all have confirmation bias to some extent.

With confirmation bias, once you think you have found the problem and start to work it through, you rarely question that decision. New information is almost always dismissed because it cannot be right. After all, you already know the situation. But, likely, you do not. It is not clear why people act this way. Part of the explanation is related to loss aversion. Other explanations for the observed biases include wishful thinking and the limited human capacity to process new information. "Another explanation is that people show confirmation bias because they are weighing the costs of being wrong, rather than investigating in a neutral, logical way." [16.8] Clearly, this leads to adverse and often disastrous operating consequences.

Unfortunately, confirmation bias (also called confirmatory bias or myside bias) is very busy working all of the time, especially under stress. It is a permanent guest in the control room. Some important indicators are:

- accepting and proceeding with the first thing that seems right.
- looking for small things that "confirm" that the present pathway is the right one.
- ignoring big things and warning signs or cautions that suggest rethinking the situation might be a good idea.
- failing to ensure that the current solution is actually working the way it needs to.

You want to take clear and effective steps to make sure you do not bring confirmation bias into your control room. Avoid any reward for simply acting fast. Act right, not fast. A good working safeguard consists of four steps [**Q3b**]:

1. Effective operating protocols for situation awareness need to be provided for you to use. Use them.
2. Ensure that there will always be enough time for situation assessment and remediation activities; otherwise *permission to operate* [**M35**] will be in effect.
3. Locate and use sufficient operational and directional information to permit successful management of the abnormal situation.
4. Utilize all safeguards and support mechanisms [**M34**] [**M35**] in case your intervention is not working as needed, or is not going to be fully successful in time.

There is another important message. Confirmation bias poses a significant danger when it blinds us to the magnitude or true extent of a real problem. Once we mistakenly think that that problem is Problem A when it actually is Problem B, we proceed as if it were exactly Problem A. Even if the two problems might appear to be alike, Problem B will not be addressed properly. Remember to wait to form a belief until all reasonable information has been viewed and weighted. Act right, not fast.

Continuation Bias

Continuation bias is the natural tendency to continue with a chosen action once started. The further along the progress, the more likely it will be continued, even in the face of concerns indicating otherwise. We get caught up in continuation bias for fear of losing our investment (e.g., time, emotion, reputation). Stopping or changing course usually requires recognition of error and the decision to start over. All of the effort and resources already spent may end up being called "wasted." Managing continuation bias is important. The practices and technology for doing it are the same as with confirmation bias. Act right, not fast.

Anchoring Bias

Anchoring bias is what happens with a first exposure to specific information. It turns out that we are unconsciously heavily influenced by the first piece of knowledge or information we come across. This is why sales personnel will try to anchor you by up-selling, that is, by starting first with a much higher-priced product than the one you are interested in. They really do not expect you to take that product and price. But presenting it means that you are now anchored. Later, when quoted a lower price (but still too high for the product you actually are interested in), the tendency is to compare that price to the very high anchor price and then go along with it. Without the anchor price, that too-high price for what you were really after would have been immediately questioned because you were able to see it as too high.

Anchoring is based upon a natural human nature desire to please others. In the control room, you build in others' wishes, almost without conscious knowledge. You defer making a needed repair request because you know that the budget is running low. When analyzing a problem, you use the list started by your supervisor rather than beginning fresh and maybe finding something unpopular or just different. This is part of *situation awareness* [**M27**]. Explore, review, and understand. Work on solutions only after that understanding is verified. Act right, not fast.

16.5 Expected Roles (Get in the Way)

Control rooms do not operate in isolation, even when remote or manned by a single individual who is rarely supervised or questioned. Managers conceive and provide guidance. Engineers and technologists design and maintain. Suppliers and customers keep it real. Operations team members keep a watchful eye and provide a guiding hand. You are part of that team.

Within a single control room, at any point in time, there may be several individuals present. Each has responsibilities and authorities and does what he thinks he is expected to do, counting on others to do the same. But what happens when individuals shape their actions based upon what others expect of them? In your control room, this invites problems.

Halo Effect

The *halo effect* is when we make conclusions about an individual's abilities or characteristics based upon observed performance in one area. We then automatically assume that it carries over to other things that person can or cannot do. There is nothing wrong with that if we are talking about respect. But much can go wrong if we cede our own thinking and understanding to that person and act on their belief or wish—or what we anticipate is their wish.

After-incident knowledge illustrates this idea. "I thought something was wrong, but Charlie, who has been on the board for twelve years, said it was okay." "I started to shut down the off-ramp due to excessive traffic backup, per a revised operating protocol option, but Jake said he never did it that way before and it worked out fine for him. So I didn't act." Jake had received performance awards for the last three years.

Bandwagon Effect

The *bandwagon effect* is also referred to as herd mentality or safety in numbers. The bandwagon effect acts sort of like a club membership. It works like this:

> If the rest of the guys think this is cool, I should too. I'm part of it. I don't need to rethink this, because I belong.

This group identification is not necessarily formal. It can be just most of the people on the same shift. It differs from anchoring bias in that little original thought or effort is expended—what the group decided or thought is all that I need. It differs from the halo effect in that I do not spend any time worrying about how good the group is. I am a part and that is enough for me.

Diffusion of Responsibility

When a problem arises in a group situation and all observe it, there is a strong tendency for each individual to believe that someone else is going to take responsibility for it. The group need not be large. However, the larger the group, the more likely it is that no one will act. There need not exist any organizational structure within the group that might suggest a natural responsible person. The fact that no one appears to step up changes nothing. The *diffusion of responsibility* [16.9] effect, also called the bystander effect, can actually prevent anyone from initiating action, with the tragic result that no one does. This is one reason for carefully defining roles and duties for responsibility, escalation, and delegation [**M34**]. It is the reason for the "in-charge rule" [**M34.2**]. When no one acts, you may use "step-up leadership" to maintain operational responsibility [**M2.1**].

The "What Then" Question Remedy

The *"what then"* question is a good way to manage your first thoughts and early conclusions. It is a check to see if there is more to the situation that you need to know. *What then* means if something is actually true, ask what are all logical or reasonable things it might mean? Here is how it works.

> You see something on a laboratory report. For example, you see that a certain fluid density reading is a bit high. You compare it to the normal range, and yours is right near the edge, very close to being over. Not to worry, a bit away from the edge is "far enough away," right? Being comfortably close to a normal boundary is usually not a problem. But before you decide to move on, pause and ask the *what then* question.

What if being very close to the edge actually might mean something? Find out what being on the edge might suggest. Ask or look it up. Unless some other thing(s) are abnormal or suspicious, all can be just fine. But unless you look for those other things, you will not know. Always ask the *what then* question for situations that are: (a) hard to believe; (b) too good to be true; (c) too near the edge of normal, or maybe a bit over; or (d) appear okay now, but if they were not, it would mean that something really problematic is happening or is about to.

> *What then:* If something is actually true, what are all logical or reasonable things it might mean?

Asking and answering the *what then* question serves to avoid jumping to conclusions too quickly. That is always a good idea.

16.6 Inattention Blindness

Inattention blindness deals with "looking without seeing." It is also called inattention bias. It is part of the more general term "selective attention." [16.10] This can be a much bigger and more important issue than first meets the eye. We all have this blind spot. And it really is a form of blindness. Inattention blindness is a failure to notice something unexpected when your attention is focused tightly on something else. That something else need not necessarily be important. It is the reason that multitasking is not a good working tool [**M7**]. We often fail to see something that is fully obvious right in front of us, if we are not looking for it or distracted. It is the side effect of focusing, something we do quite well. When we focus our attention, it acts to filter out all of the surrounding seemingly irrelevant and distracting content outside of the focus. It effectively shuts out everything else. When our thoughts are working on something in the visual field, that focus will actually prevent us from seeing other even more obvious things elsewhere. Things you may very well need to see. You must look in order to see.

16.7 Geography of Thought

An individual's personal way of thinking can be quite different in different cultures around the world. People in different places actually do have a different view of what is logical. What if they had a different view of how one approaches everything? [16.11] They do. We have all heard the expression "a clash of cultures." This is exactly that. The clash is not right versus wrong. We actually do see facts somewhat differently. We often conclude differently. Both are correct. "Understanding the thought processes of other cultures may very well turn out to be critical to the survival of all civilization." [16.12] To begin, think of the world simplistically divided into two spheres of thought: East and West. These are just labels. And the demarcation is approximate, of course.

Figure 16.7.1 suggests very generally how East and West each might appear on a world map. There is the region in between which is a combination. Call this a blended region where we see an actual blend of East and West cultures and norms.

Figure 16.7.1 World map delineating the East, West, and Blended cultural frameworks. [16.13] These regions (marked by the purple and green dotted lines) exhibit different norms with regard to their approaches to life and thought. The East is to the right of the purple line. The West is to the left and below the green line. The blended is between the two.

Norms and Conventions

What is normal or conventional differs between regions. Let us visit what those differences are.

Normal Visual Flow Directions

Start with the basics. Figure 16.7.2 depicts an example display page consisting of eight windows scattered about [**M18**]. The windows are empty, and there is no guidance or suggestion for how to move between them. We will see how the various geographical regions might naturally affect how to view the page.

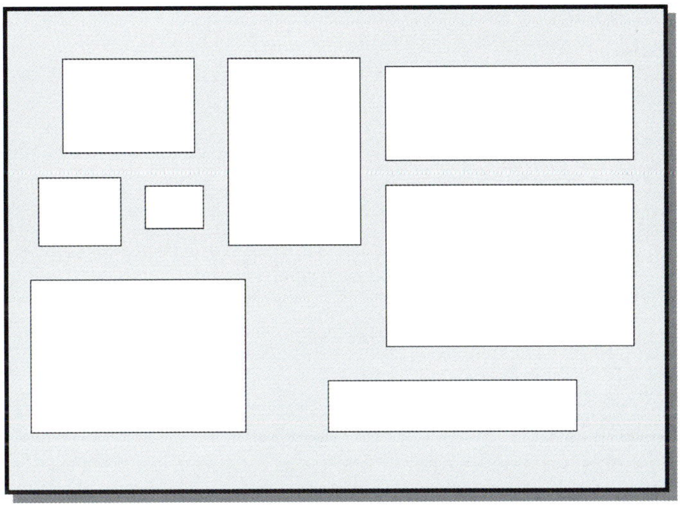

Figure 16.7.2. Illustrative page from an HMI screen showing the usual component layout structures of windows, formats, and elements. No order is visually keyed. The viewer will examine them according to his expectations. Depending on where in the world he is, this can be different.

The three frames of Figure 16.7.3 illustrate the three natural ways of viewing. The way things are viewed must match the intended way they were designed to be viewed. There are assumptions that are integral to our ways of designing screens. Take them one at a time. At the left is the West. The West expects the causal flow of information to be left to right and then back to the left, moving downward on an eventual pathway to the bottom. The top is considered to be the cause or the beginning (from a causal perspective). If the visual frame (*display* in HMI nomenclature) is also divided into sub-frames (think of each as a *window* in an HMI display), each sub-frame is viewed with the same visual flow order.

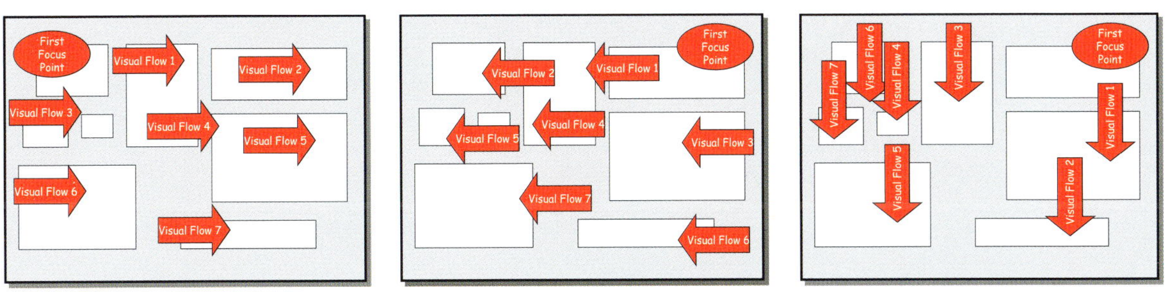

Figure 16.7.3. Eye flow on an HMI page (depicted in three windows, from left to right), illustrating different starting points and flow of view based upon cultural convention. This suggests that layouts will need to guide the viewer's view. In the West, it is left-to-right, top-to-bottom. In the Middle East, it is from right-to-left, top-to-bottom. In the East, it is top-to-bottom, right-to-left.

Move to the East (rightmost frame in Figure 16.7.3), and this natural flow is different. It starts at the top of the frame but the natural flow begins at the right-hand side. The expectation is for the causal flow to begin at the right and move immediately down to the bottom. What is to the left is subordinate. Upon reaching the bottom, the viewer flow returns to the top and moves to the next position to the left. This is all well and good, until the downward direction requires choices. Here, conflict begins. The view process wants to continue downward, but the next frame wants the downward part to end so that the view returns to the top and left.

In the blended region or Middle East (center frame in Figure 16.7.3), the causal flow is expected to go from right to left and then back to the right, moving down for the next causal pathway. These are not rules. It is the way different cultures view things naturally. This means that screens designed for one culture do not necessarily flow in a causal way for other cultures. When this happens, viewers can be confused or misled.

Why is this important? The importance lies in what the information shown is meant to convey. There are always two parties to this activity: the designer and the viewer. The designer has an objective in building the view: to present needed information for the viewer. The viewer has the need to view this information, to reach (sometimes vital) conclusions about what it all means and what, if anything, must be done about it. If the designer and the viewer do not share what is customary, expected, and normal, how can we expect things to have been designed the way the viewer expects? The short answer is that we must do the designing in such a way that the viewer will extract the necessary results in their customary manner. The work must reach a consensus of expectations and the correct methodology to achieve that. The whole HMI consensus about screen design just got much more important.

Illustration

Consider the illustration in Figure 16.7.4. [16.14]

> An enterprising soft drink marketer selling a branded cola decided to develop a succinct advertising message. Because the target market area was multilingual, a graphic ad was designed. The cartoon opens with a man lying on the sand in near utter exhaustion. The next frame shows the same man happily drinking from a can of cola. The final frame shows the man running off into the desert appearing vibrant and active. In the West, we are conditioned to see exactly that story progression.
>
> When the same message was shown to people in the Middle East, the advertisement was a complete flop. How could anyone think that people would want to drink cola that took a vibrant individual and then left him at death's door in the desert? Well, their flow of logic is right-to-left, not the other way round. Same pictures, same placement, vastly different message.

Figure 16.7.4. Nonverbal marketing message where entire meaning is dramatically different when the individual frames are viewed in order from left to right (the intended flow by a Western designer) or right to left (typical for the culture where it was targeted)

Visual Perception

Sometimes the whole may not be a sum of its parts. Take a look at how we might view a nicely designed and stocked aquarium fish tank, as depicted in Figure 16.7.5.

Figure 16.7.5. Fish tank (courtesy of Wake) showing fish and the environment of the tank consisting of the pebbled and planted bottom, added natural features of the bottom, and a rear tank picture to complete the view

The Eastern eye finds comfort in the harmony of the nature scene in the aquarium landscape. The fish, while not essential, provide just another element. The view would still be enjoyed immensely without any fish at all. Simply the diversity of Nature and the movement of water flow would be a cherished moment. Form and content are one. A Western eye focuses immediately on the moving fish. To a significant extent, the grasses and background seem to distract from the view at first. It is all about who are the viewers.

Language Construction

Language construction is the way we place words into thoughts and sentences. Differences can confuse as well. One larger class of language with a limited vocabulary builds sentences as a constructive process using extra words to convey purpose. Another has a different word for (almost) everything. An example of the former would be the phrase "cry that is small" instead of "whimper" because that earlier language does not include whimper as a word. In thought construction (sentence word order), the entire sentence must be read to grasp the meaning. Until completely read, the reader will not know it ends with something good or useful, or something damaging or dangerous. Last words can make this difference. In situations requiring brevity or abbreviation, extraordinary care must be taken to understand these differences. Intent must lead to understanding in order to be effective.

Logic and Reason

General

We think something is logical when it is generally considered to be "reasonable and expected," [16.15] because it is based upon experience. It is easy to classify it as logical or not. It does not have to be fully explained. Just the gist is enough. In a discussion, everyone is "on the same page." But what if your logical is not everyone's logical? Then nothing can be taken for granted. Culturally different people may have far different expectations. Sometimes so different that what makes sense to one person makes absolute nonsense to everyone else. While uncomfortable and possibly embarrassing socially, this confusion is frightening and dangerous in a life or death situation. Make no mistake. The effective management of a hazardous operation can be life or death if done without proper regard to the situation and the requirements. Logic is not the same throughout the world.

In the West, logic is used to prove something that is not obvious or was not experienced before. Oversimplifying, if a certain truth implies a certain conclusion, and if a second conclusion based upon the first is also true, then the first conclusion implies the second truth. For example, if Charlie is a horse and all horses have tails, then Charlie has a tail. We expect reasonable people to follow logic.

In the East, logic is far different. What governs is that which has been experienced by self or conveyed by others down through the ages. Anything that flies in the face of that experience cannot be true. The East also has a radically different view of consistency and contradiction. There, one must find ways to live in harmony in a world full of contradictions and continuing change. In this light, harmony is the goal. Contradiction merely helps reinforce its realness, whereas in the West, it creates discomfort.

Simplicity as Truth

As you saw earlier in *Getting Started* [**M1**], if there are two competing explanations for something useful, we tend to believe the simpler one is correct. Of course, the explanation must not require suspension of intellect. If a basic simple explanation does not work, then the next simplest one that does is the one to use.

Logic vs. Fate

Some cultures believe that events will transpire the way they do, no matter what. If a pipeline wants to rupture, it will do so despite any efforts at preventing it or managing it by people. If a chemical reaction wants to go exothermic, it will do so at a time of its own choosing. Therefore, it would not be logical for anyone to try and prevent these events from happening. The thinking might go:

> As an operator I am only the observer of fate. It is my lot to observe the good; it is also my lot to observe the not good. All is not up to me. It is out of my hands if it happens.

There is no place in any control room and responsible enterprise for fatalism or predetermination. As the operator, it is your responsibility for effective *situation management* [**M29**] to ensure that any beliefs in fate by the operations team do not cause improper operations protocols and behavior. You will want to find culturally appropriate ways to manage all of this.

Nailing Down the Issue

Situation management builds on the expectation that truth will be consistent, that facts matter, and that proper management has a reasonable chance to affect outcomes. Inconsistencies require additional concern until fully resolved. These expectations are core to *awareness* [**Part 6**]. Even before you opened this handbook, as an operator, you were always looking for something out of the ordinary; you were looking for things that did not add up. Real plants and operations act according to their physical reality. Stress corrosion does its action the same everywhere in the world. Heat imbalances portend a problem everywhere. In the control room, it is essential that successful operating principles and beliefs be developed that are fully competent, understood, consistent, accepted, and practiced by operators, supervisors, and managers alike. Good operation is not against respect for culture. It simply asks for practices that can be understood and employed successfully.

These differences do not have to be a dozen time zones away. As societies become more and more multicultural, this issue becomes closer to us all. Take rural compared to urban environments. More than accents and dialects differ. Understanding and respect is necessary worldwide. Personal culture with all of its amazing diversity must be properly understood and safely managed in your control room.

Individuality

A window into deciphering cultural differences hinges on the different views about individuality. In the West, everything centers on an individual's rights and goals. In the East, everything relies on getting along. Higher good is realized by interdependence rather than independence. Cultural differences are real.

Neither view is better. Be it East or West, particular care should be taken to ensure that what you are asked to do respects your underlying view of responsibility. Successful operations must include effective and respectful incorporation of cultural norms into expectations and protocols without compromising safe and effective operations. Success or failure should not elevate or depress an individual's view of and respect for self.

Handling and Reporting Problems

One cultural norm that is universal and important is the concern about avoiding embarrassment of self and others. We are all concerned about this. In the Middle East, a very old yet valuable understanding is that the surest sign of the onset of embarrassment is the mere appearance of any reddening to a person's face. Anyone who causes another's face to fully redden in public has committed a most serious affront. It is that important. In the West, avoiding embarrassment is referred to as "saving face." We are not that much different after all. So it is important to ensure that the handling and reporting of necessary things be done in a way that respects both the individual and the needs of the reporting.

When it comes to learning from mistakes or near-mistakes, it is clear that the first thing we must know is that an actual mistake happened or almost happened. Before the United States Federal Aviation Agency (FAA) instituted no-fault and anonymous reporting of airline issues in 1979, a large percentage was not reported.

"Without proper accounting, experts say, airlines can't learn from mistakes and regulators can't properly assess safety risks. Even smaller incidents—like component failures and near misses on the runway—are key bellwethers for major crashes. Left undocumented, botched procedures are left to grow endemic." [16.16]

Since then, there have been over a million anonymous tips with 4,565 actual incident reports being written to share. However, a large part of the industrial community is slow to adopt similar reporting standards and mechanisms. A much larger number of potential lessons to be learned go unnoticed. This is especially prominent in jurisdictions that impose extremely harsh penalties for errors. Without safe reporting methods, the entire operation of an enterprise is at unacceptable risk.

16.8 Close

The West and East distinctions are used to crisply and cleanly illustrate real geographical differences. They are so real that a merging of cultural norms is seen between those two anchor points. This suggests paying even more care to cultural expectations and roles. Recognize that cultural differences require understanding, respect, and accommodation. The accommodation referred to here will ensure operators are able to adequately and properly perform the needed activities with both personal dignity and enterprise operational certainty. It should contribute in a positive way for you to feel needed, comfortable, and effective.

As we end your module on biases, it is important to draw a broad line between common everyday biases and real psychological issues. We are not discussing individuals' personality disorders. This discussion is only about human foibles and tendencies.

PART 4
THE CONTROL ROOM

The control room is where it all happens. Part 4 examines the broad aspects of the control room as your major center of activity. It covers what a control room should look and feel like and what it should provide for you.

The key tools, activities, and components upon which you constantly rely are covered. Included is how a properly designed human-machine interface (HMI) and the alarm system assist you to monitor situations, to understand and successfully manage your enterprise's process automation features and installed technology.

Covered is the variety of components for managing the shift itself during everyday situations, plus additional activities necessary for maintaining sustained good operation.

Shift handovers are the most important and vulnerable communication events that you perform in daily shift operation. The entire spectrum of shift handover is explained along with recommendations on proper handover management, including coordination and interaction with maintenance and shift supervisory personnel.

PART 4 — THE CONTROL ROOM

Module 17: In the Control Room

17.1 Purpose of Control Rooms

Your control room is the setting for successful cooperation between you as the operator, the interface equipment, and everything else that you are responsible for. The control room itself includes the spatial arrangements of equipment, the environmental management of the workspace, who goes in and why, and the designs of the working equipment and other related tools you use inside. Later on in Part 7 *Managing Situations*, you will explore how good design helps you better manage the job of operating.

This module provides a basic understanding so that you can effectively participate in the evaluations of your control room and assist in any future plans for upgrades or replacements. We also cover the key systems you will be using. Proper design will enhance your working experience and provide the necessary confidence that your space is going to be helpful and trustworthy. When we understand and follow best design practices, operator workspaces can be built to markedly enhance your well-being and success.

17.2 Control Room

The control room is the physical place in which you work. Control rooms come in a wide variety of shapes and sizes. They cover the complete operational control of an enterprise or just a small portion. They are staffed by one operator or more. They can be multi-functional, to include both a board operations team and a field operations team. They can be multi-purpose, having a combined responsibility for production, quality, and management interactions. Sometimes control rooms are local to the production process area. Other times, they are hundreds of miles away or anywhere in between. Some are elaborate, complete stand-alone structures. Others are little more than rented space in a university administration building.

Architectural Aspects

The shape of the control room has evolved from small buildings with a pneumatic instrumentation panel and room for a chair or two, into specialized computer control centers that take full advantage of technology. As technology continues to develop and prices become more affordable, large theater-type video walls have become popular. To better accommodate these, the room shape has evolved, with operators facing large screen displays (LSDs). More discussion of display types, arrangement, and use for operation and coordination may be found in *Using the HMI* [**M18**]. No one size or one design fits all. What is important is that the design of your control room fits your needs.

The Concept of Space

Function alone will not encourage you to stay and work. Proper design of space makes being inside useful and effective; it feels good to be there. This opens up the occupant to greater empowerment and energy. The following section is an introduction into special spatial concepts and why they are important. We are just beginning to see the power of architecture and how it can and could be applied to control rooms.

Spatial Introduction

Spatial introduction is the specific way that the space invites you to enter. Introduction is all about creating a promise. As you come on shift, entering the building prepares you to transition into the role of responsible operator. Spatial introduction follows a carefully defined process. On the one hand, it wants to effectively invite you to come on shift. It should guide you to make the transition from someone else to the role of responsible operator. Entering the control room should perfect that transition. However, we want that invitation to extend only to operators. Not in an overt or confrontational way, but in a functional and respectful way.

One of the more successful methods for accommodating visitors is to have them "visit without entering." Through the use of an observation wall or gallery, the "visitor," be he a co-worker in a support role or a visitor to the company, is invited to visit in a way that is open and inviting for that role, but should not disturb control room operators.

Geometry

Geometry describes the design properties of the space so that it will be able to deliver its intended feel and purpose. Geometry is the tool kit, the art, and the sizeable craft of the designer. A successful geometry will produce the desired expanse, facilitate the transition into introduction, foster collaboration, and encourage retention. It is all about how the areas are designed, how they interact, and how you are afforded the ability to use the space to do your work. It involves using room shapes, designs, and measurement ratios so that everything contributes to an inviting and productive space. It suggests discovering and avoiding those aspects that detract.

Expanse

Expanse is the emotional feel we experience by being inside the space. Expanse is all about feel, but feel is intimately connected to the purpose for the space. There are four degrees of expanse. It starts from the most basic:

1. **Inside**—a neutral space. You do not gain anything by entering and you do not leave anything behind after you leave.
2. **Inviting**—our space, is a place that you and I go to work together. We can work on the same task or different tasks, but we want to share the energy of working. (Remember the time when you went into a study area, found your favorite nook, and got right to work? Work wanted to be done; energy and focus and a need to work came at you from everywhere.)
3. **Intimate**—my space, it is for me. I can invite others to enter, but I do it for me or for them in a way that I want them to experience.
4. **Reverential**—It was everyone's at the same time. Entering was a transition. Being there was an experience. Maybe the place told you something new or let you teach yourself something you were missing. Being there caused you to be in a different place, both in your mind and with your feet. Leaving was hard. Returning again was desired.

All in all, the experience of being inside the space was an elevation of self with the experience of being a part of something larger.

Collaboration

Collaboration is the way a space affects its occupants' interactions [**M34**]. Being in this space together encourages individual operators and supervisors to think of themselves as a team. It facilitates their working together in a way that is progressive, proportional, and respectful. Look for the following important attributes:

- Progressive—allow for an easy start without either a strong commitment to continue or a penalty for discontinuing; it encourages what is needed
- Proportional—allow for the right amount of collaboration that appears appropriate for all
- Respectful—allows for participants to decline to collaborate without interfering with others, without suggesting to others that collaboration might be wrong; yet permits the non-collaborator to decide to join in or to be asked to join in later without bias.

Retention

Retention is the ability of a space to retain its occupants for the purposes intended. The design reassures occupants that they are welcome for as long as they remain, and that they are also wanted. Good control rooms encourage you to be there, on task, and effective. Recall those other places where you happened to be and could not wait to leave? Getting out of there was the only thing you wanted to do as soon as you could! Imagine how important it is to have the reverse. Your control room should have this feel.

17.3 Control Room Needs

Here are the requirements. You should be able to recognize and find them in your control room. When they are not there, ask for ways to make it happen.

Physical Protection and Security

The enterprise must provide you with:

- physical protection from the errant effects of Nature; both for direct protection and well-being of the personnel, as well as the ability to maintain critical operations.
- physical protection from danger and harm resulting from production or operations gone wrong.
- physical protection from violence, intimidation, and related nefarious activity.
- cybersecurity to prevent unauthorized personnel from operating, modifying, or otherwise affecting the operation of the enterprise or the underlying support activities.

Environmental Controls

The control room must provide all necessary infrastructure to ensure proper health, comfort, and absence of distraction for you. This includes:

- properly managed facility temperature, humidity, clean fresh air.
- proper management of noise and vibration.

- effective design and operation to ensure effective visual, voice, and other means of communication for person-to-person, person-to-equipment, and equipment-to-person interactions.
- effective design and management of illumination to support tasks, well-being, and alertness during the entire operation event (coming on duty, during duty, for collaboration, and going off duty).

Ancillaries

The enterprise must provide the following extras:

- Toilets, showers, dedicated personal lockers, and related
- Snack, coffee, kitchen and related eating and storage areas
- Exercise, stress relieving, special light or related treating equipment, and short-rest areas
- Training and refresher tools and options readily available for incorporation to vary the shift process and increase competencies

Information Support Tools and Technology

The enterprise must provide sufficient, fit-for-purpose support to ensure the following ready, accessible, and current information:

- The approved production management targets and any operational constraints or imperatives to be met during production and operation
- The true nature of the production or operation being managed (primary information such as sensor values and control parameters and settings; secondary information including information integrity, support information including weather effects, maintenance, etc.)
- The approved procedures and protocols to be followed or consulted during the entire scope of operation that includes start-up, normal and abnormal operation, and shut-down

Where permitted, the ability to print, transfer to other at-hand devices, and share with others needs to be available. It includes the ability to track its distribution and location.

Sufficient Process and Operational Controls

The enterprise must provide ready access to sufficient controls (the handles) for proper and effective production:

- Reliable and effective direct tools and controls required to properly and successfully perform the full range of operational duties expected (e.g., start/stop; open/close; parametric adjustments, such as flow rates, level settings, temperatures, etc.) and re-ordering or re-sequencing production operations, production steps, or material pathways, etc.
- Reliable and effective indirect tools and capabilities needed or chosen to support production or operation management. Examples include proper personnel operating under your authority or direction, designed to extend your physical reach of influence to remote or separate physical or functional areas located in the control room or operations center and not directly accessible by you.

Operational Support

The enterprise must provide you with ready access to appropriate support personnel (technical, managerial, or operational) to assist in understanding and/or control and effective operation:

- Ability to share information via both informal communications as well as reliable primary mechanisms (e.g., control room display replication, procedure databases, etc.)
- Ability and mechanisms to reliably and effectively support all aspects and forms of collaboration in real time (including communications technology and physical work areas, such as crisis management facilities)

Special Operating Situations

While the control center is for the specifically designated operations, there are special uses that must be factored into its design—specifically, significant plant or equipment startup, cutover, or shutdown, requiring added personnel be present and have access to certain controls equipment. One-of-a-kind commissioning activities need to be accommodated. Times of operational stress such as significant abnormal operation or emergency operation also require added personnel to be present. Other uses involve the operators engaged in additional activities that support their operational role such as training and certification. For facilities without separate training areas, this activity must be done in the control room or control center. All of these added uses are important and need to be factored into the control center and control room design and operations protocols.

17.4 Control Room Access Management

The control room or operations center is not a social center or any form of clubhouse. It really should not be a convenient location to others for restrooms, coffee breaks, guest meals, cold-weather warm-ups, or anything else peripheral to operations. Think of it as a "members only" facility to the extent that the only personnel who should be present are those on the direct operating team (operators, supervisors, and the occasional manager). All other personnel (maintenance, quality control, safety, support services) are invited guests: present and welcome for specific contact and purpose, during the course of that purpose. Past invitations should not be considered a free pass for future admission.

This might sound a bit harsh and unsociable. It is not intentional. Proper access controls reduce distraction and enable the needed level of focus for those inside. Operating can be a very demanding task. Few will have difficulty understanding the need for focus when challenging situations present. Yet most of us underestimate the need for focus when nothing appears to be going amiss. It is precisely at these times where a good watchful eye is needed all the more. As you have seen elsewhere in this handbook [**M7**], our ability to multitask is a misconception. People are really, really poor at doing it. A visitor in the control room who is not directly supporting a current task is a distraction at best.

17.5 Permits, Personnel, and Visitors

Current activities within and affecting the operating area are the first responsibility of the shift operator. These same activities are also the responsibility of shift supervision for continuity between shifts and reasonableness over time. This means that there must be ready communication of all activities with you. This includes all deliveries and withdrawals (removal of material, non-incidental equipment, etc., excluding normal production output), all maintenance, and all other activity that places personnel within the battery limits of the operating area or their resulting effects on the equipment or operation.

Permits

The control room should provide for effective communication and exchanges in a way that does not disturb others or unduly disturb you. This is normally handled by setting specific times for these activities to occur and ensuring that any face-to-face interaction be done without actually entering the main areas of the control room. A permit window or an electronic conferencing facility might work well for this.

Visitors and Other Personnel

Visitors are a distraction, no matter whether their presence is for public relations or enterprise improvement. Where public viewing of the control room is an important part of community relations, the control room is designed for such viewing without involving physical entry to ensure that distraction is low and interaction easily managed.

17.6 It is All About Design

Proper control room design is all about design. And that design is all about you, the operator. Control room design is one of the main influencers on people performance, managing fatigue, and reducing human error. It is about getting the lighting levels to a value that stimulates people and acts as a fatigue countermeasure. It is also about how people work and how design features influence their behavior and performance. There are many options to exploit but none as important as getting the basics correct. This design activity is termed variously as *ergonomics, human factors,* and *user-centered design.* All are intended to match the equipment you use to the task at hand, to your capabilities, and to the best way to help you do what you want to. From this module, you have a good idea of what this means for you in your job as operator.

Module 18: Using the HMI

The HMI consists of hardware and software that provide information about your enterprise and allow your inputs to be translated as signals for machines. The HMI provides the required results to you as content on visual display screens.

18.1 How the HMI Assists Operators

You are already well familiar with your *human-machine interface.* Using it is second nature. It is your most basic of tools. Here, we are asking you to revisit it, but from a perspective of "why it was designed the way it was." We want to equip you to better understand how it assists your work in the control room, thus enabling you to be even more effective and comfortable. While some of the features and usage described here may not appear to correspond to your exact situation, all of these concepts should be useful. Where you find more use for existing features, work them into your activities. Where you discover problems with yours, consider working with the engineers and designers to help make things better.

HMI technology is commonplace in most industries like electronics, entertainment, military, medical, etc. The HMI is also known as a user interface (UI), the man-machine interface (MMI), the computer-human interface, and a human-computer interface (HCI). [18.1]

> It is the designer's responsibility to cast the presentation to always show in-context information relevant to the task and situation. Showing only data means that the importance and significance of those data values are blatant, obvious, or trivial.

The HMI shows you:

- where things are.
- what the current situations might be.
- what might be going wrong and how wrong.
- where and how to intervene to make things better.

You can see, understand, and then manage what you can, using your expertise and appropriate tools. Information is depicted on any number of *display screens* (the individual views that appear on *displays*). Each display can show only one display screen at a time. But there can be any number of display screens designed and stored in the equipment and then called up by you for view. They are also the vehicle for providing operational documentation and task guidance.

18.2 Nomenclature

We will use the BSI Standard ISO 11064 [18.2] for nomenclature. Figure 18.2.1 shows a diagram that will be used in this handbook. This framework is pretty much a universal standard. Take a moment to work out exactly what your own enterprise's terms mean in this framework. The physical hardware device used to show all of the content is termed *display*. All of that content would be conveyed to the viewer on *display screens*. The content builder would be free to design and provide at will so long as it was within the capabilities and limitations of the display hardware and the display screen software.

Any (as many as needed) individual display screen would show content utilizing the following terminology:

Display—the physical hardware (on which content display screens are shown)

Display Screen (or Screen)—the actual content (information, real-time data, text, etc.) that is available for operator selection and showing (on the physical display)

Page—the totality of viewable content, in the moment, that the display screen contains

Window—any of a number of (usually) individually managed visual sub-sets of content on a page. A page may contain a single window (occupying the entire viewable area); a part of the viewable area, with the rest not having any content; or many separate windows (with proper regard to visibility).

Format—any of a number of content clumps (we will leave that undefined for now, but think charts, pumps and equipment, diagrams, status text, etc.) that occupy any given window.

Element—any needed basic content item used to build up a format.

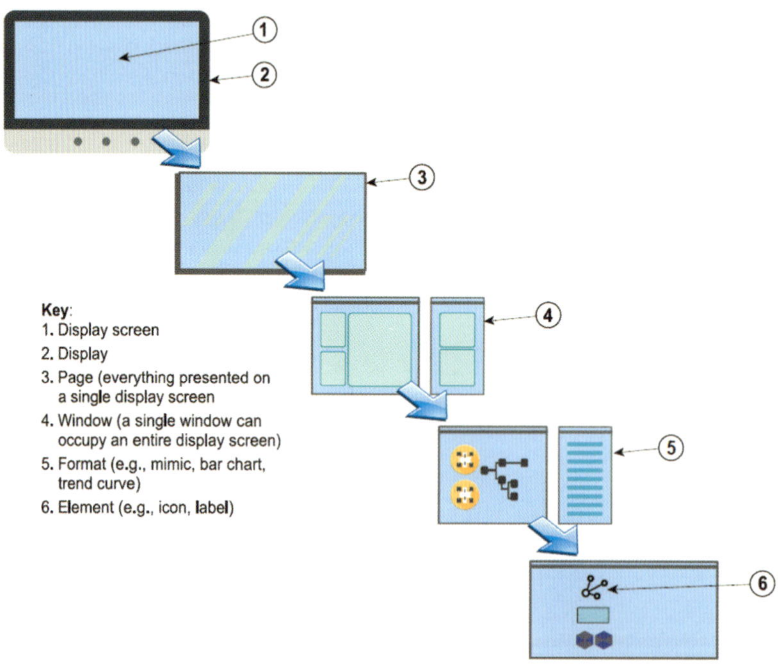

Figure 18.2.1. ISO 11064-5 nomenclature diagram for graphical displays. This level of detail permits specific reference and discussion. Starting from the upper left, each successive component will be found as a part of the component above it in the hierarchical order. [18.2]

Depending on what was selected to show and how it was designed to be shown, the viewer could find, visualize, and understand the content—or be completely frustrated, or worse, dangerously misled.

18.3 Guidelines and Principles

Operator display screens need to be carefully and properly designed to be effective and used successfully. You really do need to be able to find things where you expect them to be and in a place where your practices and procedures will lead you. We will also discuss navigation later in this module. It should help you appreciate whether or not (a) your display screens are designed effectively; (b) you may need to brush up on the proper procedures and navigation practices; and (c) what you are looking for is not there, which means that maybe you are possibly looking for the wrong thing or looking in the wrong place.

Basics for Operator Display Screens

Everything you as an operator need to know or find out about the status of all enterprise equipment under your responsibility and management must either be available for view on the HMI or be obtained by some other specific and dependable way. This might be via external documentation, checking with someone else, etc.

The HMI has four basic requirements for good design: *purpose, understanding, use,* and *completeness.*

Purpose

All information displayed has an explicit purpose. Both the designer and the user (you) must know the specific uses and intentions of all information for every screen. This means that all display screens must be consistent with, and purposefully designed for, all of your expected tasks and other needs: startup, equipment operation, shift handover, troubleshooting, and everything else.

Understanding

All information must be displayed with enough context to enable you to fully understand meaning and intended use. While you are expected to generally sense things like normal, abnormal, unusual, dangerous, etc., screens must provide enough guidance for you to do it without being cluttered up with other information.

Use

Display screens must enable you to do what needs doing. There is proper navigation and coordination between screens and supporting materials (procedures, checklists, drawings, and such). Everything the HMI depicts to you needs to support why it is there and to make sure you can do what you need to do.

Completeness

There are no gaps between design and execution. Everything you need to see is there, and you should be able to readily find it. All actions that you intend to take using the HMI must be there and trustworthy.

18.4 Display Screen Design

Structure

On each display screen, designers use *hierarchy* and *segmentation* to appropriately provide information and needed process management access for you.

Hierarchy

Hierarchy groups display screen information, usually using three levels of view: overview, secondary, and tertiary. A three-level display screen hierarchy is a robust structure that encourages ready access to information while at the same time keeping important situation context. It promotes efficient navigation when delving deeper. The first two levels progress from the general to the more detailed. If these levels are not capable of providing additional deeper detail, the third level provides it. But now expect the third level to switch from detail to providing guidance; think procedures or diagnostic assistance. See Figure 18.4.1. At any level of view, all of the display screens at that level contain data and information consistent for that level.

> Hierarchal display design organizes your views to readily keep an overview situation in mind while you follow a pathway to more detail and proper actions (if needed). It is a powerful navigation aid.

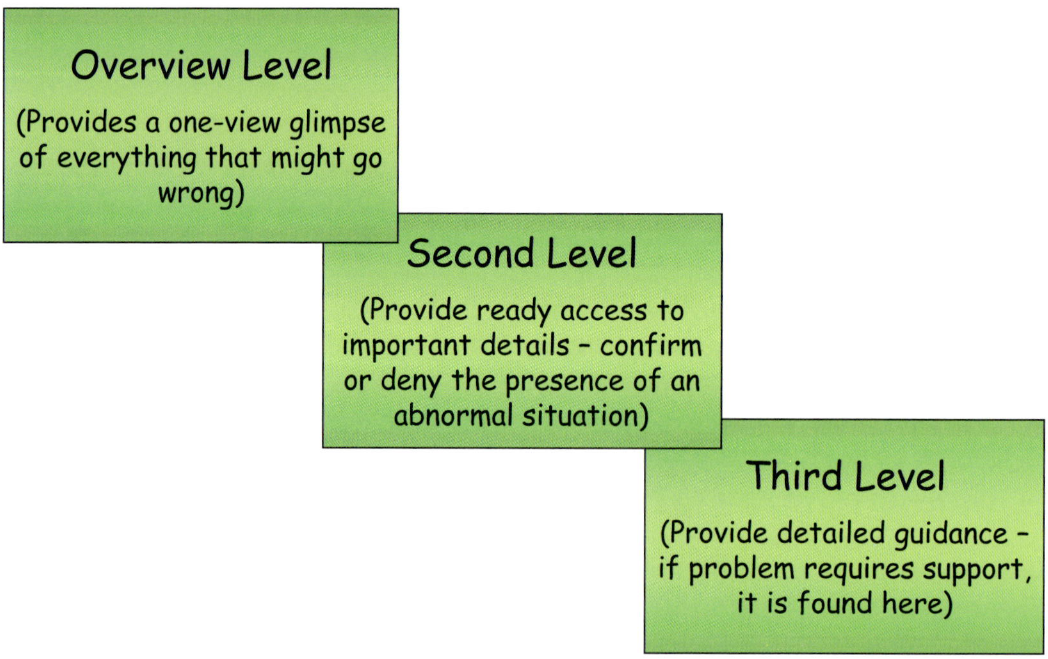

Figure 18.4.1. Three hierarchy levels for display screens. This structure permits all information to be located directly and then accessed without excessive clicking. This structure also encourages uniformity and appropriate co-location of related information.

Example Hierarchical Screens

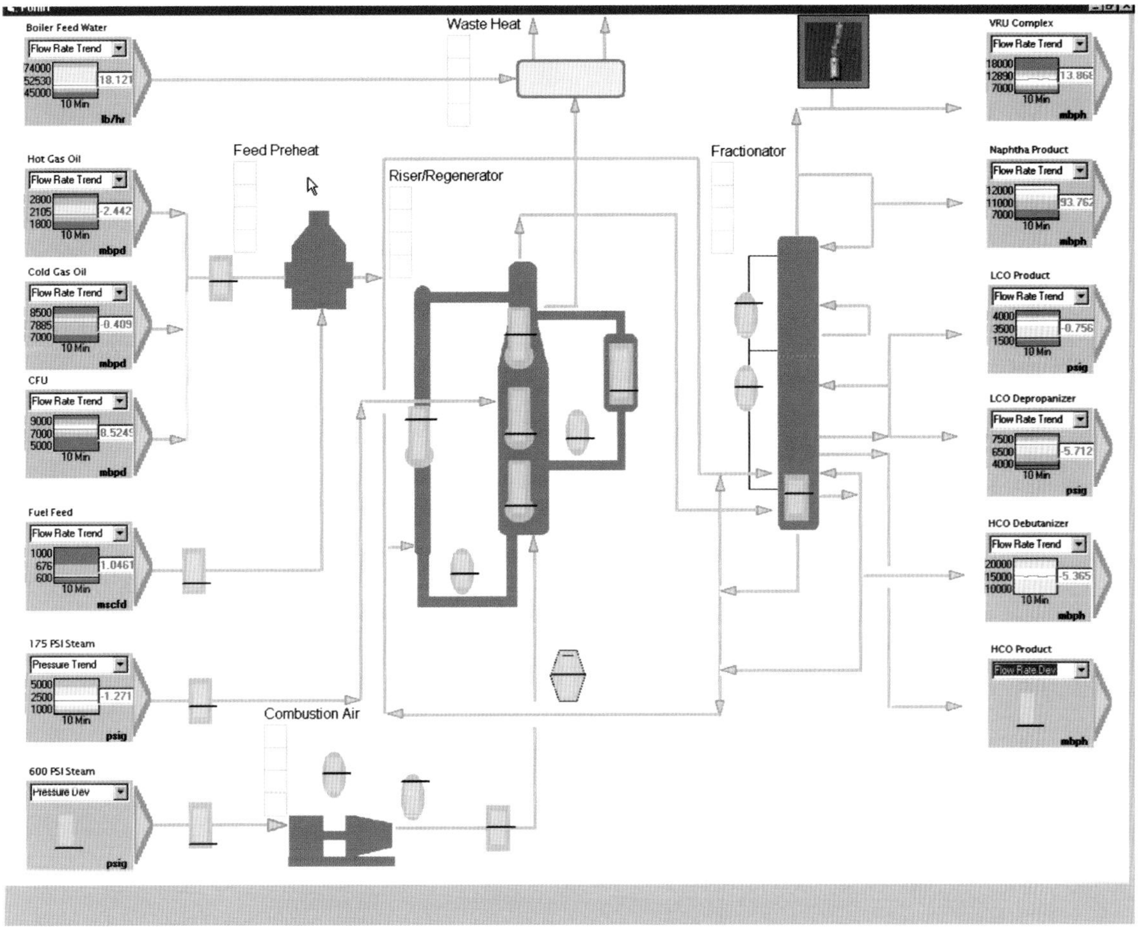

Figure 18.4.2. Overview display screen showing all of the important aspects of an operator's entire area of responsibility. Key variables are shown as icons. Alarms are summarized so their location and criticality are apparent. Trends are present for key inflowing variables coming from other plants. Similar trends are for the important product variables of this plant that flow to other plants or for shipment.

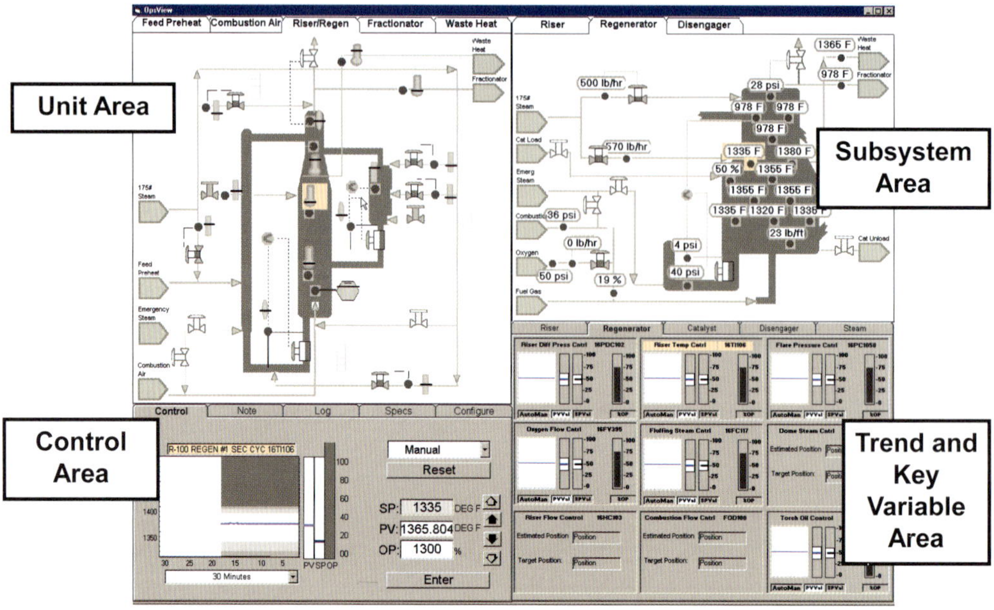

Figure 18.4.3. A situation-based secondary page that is pre-engineered to provide information to understand what the situation is and how to manage it. This display combines the window together with three additional windows to form a coherent view. It is constructed by the process of yoking (making a specific set of display screens or formats visible at the same time on one or more displays).

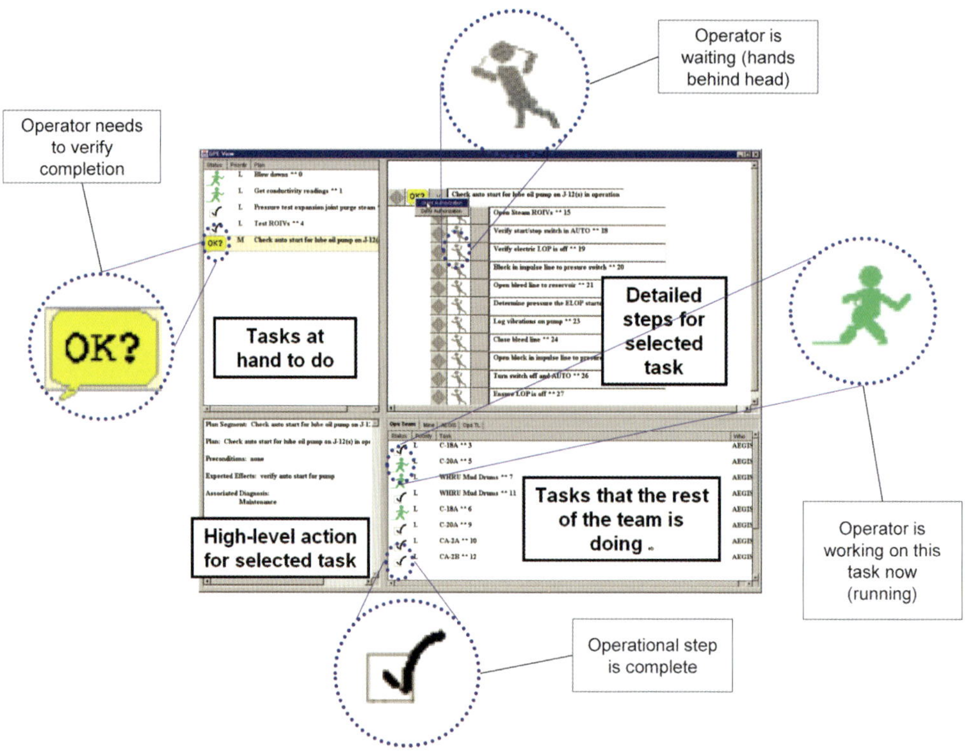

Figure 18.4.4. Procedure tracking is illustrated as a tertiary display screen. Each window depicts a different level of activity for the work being done.

Segmentation

Most tasks you work on have more information than can fit comfortably onto a single display screen. Segmentation divides or segments the total plant data and information into distinct areas of content on display screens. Every individual display screen must show enough essential information to understand what is on view. There can be no important information missing that you need in order to fully grasp the situation related to the content. Each segment is complete. References to closely related segments are clear, and navigation targets are present.

> Segmentation display screen design provides a focused view of a specific situation or operational task while you work on it. It ensures that all related items are at hand and organized for you in order to fully appreciate and manage a situation.

Style Guides

A *style guide* provides usability, performance, and uniformity. It covers all aspects of screen design. It is customized to meet your site-specific requirements.

> Style guides make sure that everything you see and need to work with on display screens is consistent and predictable from display screen to display screen throughout the entire HMI.

Your style guide contains all of the rules for constructing everything graphical that goes on the HMI for your enterprise. Except for the specific selected content for any given display screen, everything you see or do using the HMI will follow the decisions and requirements of your site style guide.

Graphics Library

The HMI *graphics library* is a complete collection of all pre-designed large and small components used to build display screens. It ensures that information and management control is conveyed uniformly in appropriate situational context. All needed process variables and elements are accounted for. All of it follows the style guide directives.

Dials and Gauges

Dials or *gauges* display consistent contextual information in a graphical or picture format. They provide important extensions of your ability to see, understand, and manage. In a dial, you usually see a single entity. In a gauge, you are seeing more than a single entity together.

> Dials and gauges are used to show you how a set of similar information relates to each other and a current situation. It is all in one compact place.

Dashboards

"A dashboard is a visual indication of the most important information, consolidated and arranged on a single display screen so the information can be monitored at a glance." [18.3] Dashboards use a compact, single information display screen format structure. They usually contain dials, gauges, and the like as the component building pieces.

> A dashboard is a compact coordinated view of directly related data and with its situational meaning and context. Everything you need to know about the overview condition of what it represents is there.

Figure 18.4.5 [18.4] shows an example of a dashboard. Like systems are grouped appropriately. Remember that dashboards are not intended to be closely studied. They are meant to be a type of overview snapshot. If they do not work that way for you, either they are not well-enough designed or you are missing how to use them. Work this out.

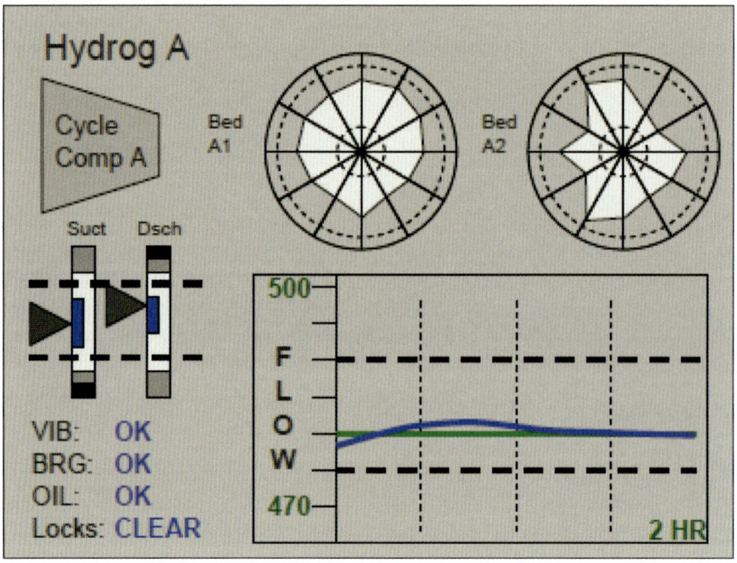

Figure 18.4.5. Clear and effective dashboard. Everything is related. Each dial, graph, glyph, and gauge is simple and understandable. Quickly, a viewer will gain all intended information.

Trends

A trend (or trend plot) is a visual representation of values for a variable of interest over time. Figure 18.4.6 illustrates the most simple of forms.

> Trends provide a compact view of the current value and history of one or more related variables. They make it clear how each is changing over time and how changes in one relate to the information in others.

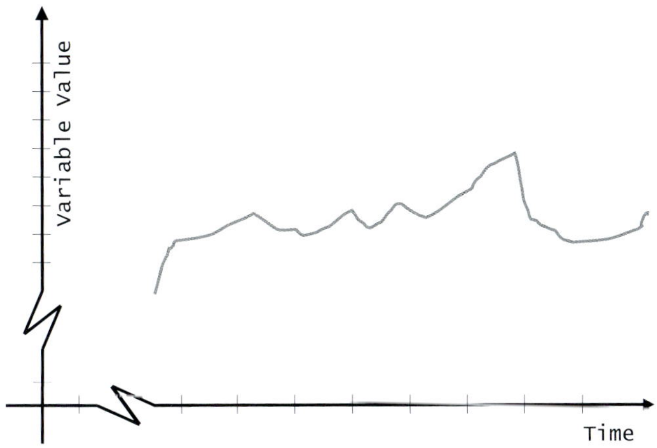

Figure 18.4.6. Single trend. This is useful when the time-changing values of the variable (the history) are important to see in addition to the current value.

Sparklines

A sparkline [18.5] is shown in Figure 18.4.7 without scale values. The purpose is to provide a miniature in-context view of data to support added understanding for an associated visual display. It is a mini trend that can be tucked into a graphic display screen at just the right place to be useful.

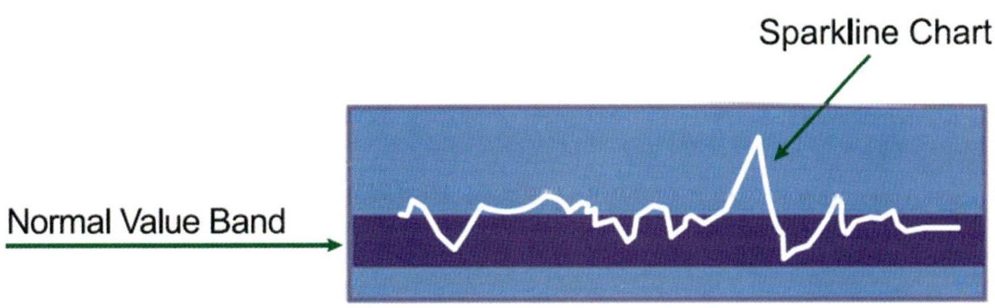

Figure 18.4.7. Sparkline chart showing key. This is a very effective way to show a real-time trend in a way that mimics an icon or gauge.

Fan Charts

A fan chart suggests likely future results projected from past history. Figure 18.4.8 illustrates one. [18.6] A fan chart can be an extremely useful preview of what future results might look like, given past knowledge. You can use this possible outcome range for hypothetical but useful information to reinforce operation or encourage your extra attention.

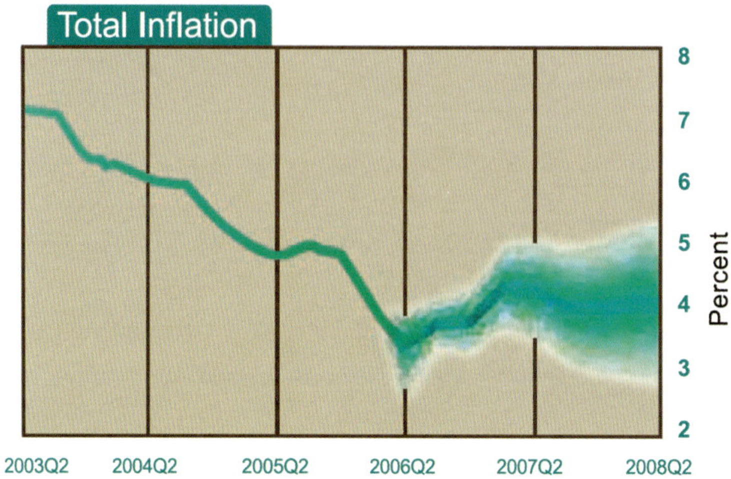

Figure 18.4.8. Illustrative fan chart (Bank of England "Inflation Report") that is extremely useful to compactly depict where future values are likely to be. The "fan" part begins at the beginning of the uncertainty period, Q2 2002 in this case (which may actually be the present, before, or a bit later).

Elements

Elements are the smallest building blocks of graphic displays. Examples of elements include glyphs, icons, dials, gauges, and charts.

Glyphs

Glyphs quickly identify content as important, meaning please pay close attention. You are probably familiar with their use in quick start guides and everyday traffic signs. Figure 18.4.9 is an illustration.

Figure 18.4.9. This glyph means, "Do not dispose of in ordinary trash." Easy to clearly see and understand exactly what is intended.

> A glyph is a crisp and completely clear visual notice of the context of a message or action. The same glyph always means the same thing. They are never to be glossed over!

Icons

Icons clearly communicate status. That is, they contain context. Take the thermometer-like icon shown in Figure 18.4.10. Clearly this information is about a temperature. In the normal state, it is depicted without adornment or much context. It immediately depicts normal. No more looking around is needed.

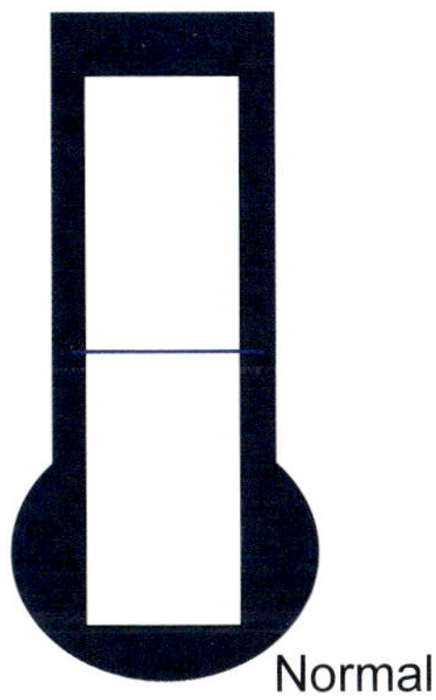

Figure 18.4.10. Example icon illustrating a normal temperature, in addition to the label.

> Icons are a clear representation of the real-time value and expected condition of a single variable or piece of equipment. It shows all that you need to see for that moment in time.

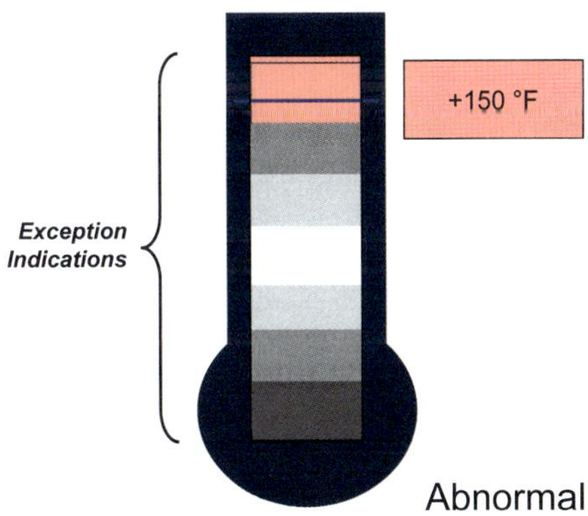

Figure 18.4.11. Example abnormal temperature icon. Note the change from the previous icon. Now there is a lot of context and relative importance limits. No problem seeing that this temperature needs understanding.

When that same temperature is abnormal, the icon form remains the same, but context is added (Figure 18.4.11). The additions provide information to relate the current value to important needs and limits. Again, you are quickly able to observe what is abnormal about this temperature.

Sound

Sounds are powerful stimuli. We are hardwired to instantly recognize familiar sounds, even those from a distant past. Caution is the guide. This is so easy to overuse. When overused, they tend to be ignored and treated as background, allowing an important use to be missed. Here are a few.

Cues

Audible cues are used as a pre-warning of an announcement. Because announcements startle, often both the beginning and the importance of the announcement can be missed. The cue is used to both eliminate the startle factor and prepare the listener for appropriate understanding. Just to be clear, cues are neither alarms nor messages. Typical ones are soft but noticeable. A muted chime is often used.

Announcements

An announcement is the control system or plant communicating a brief, clear message that you need to hear rather than see on a display screen. They should be preceded by appropriate cues.

Audible Messages

Most audible messages should be very infrequent and used only where information is likely to be not readily at hand during other operations and you must know about it now. Messages should be preceded by a respectful cue.

18.5 Navigation

Navigation is the process of locating something of interest on a display screen and causing it to actually appear in your view. Getting this designed correctly is one of the most important requirements for HMI design.

> The ability to use the HMI for effective navigation is an essential competency for operators.

The purpose of navigation is to find the right display screen to obtain needed information and execute operational commands. The ease with which you are able to use the tools provided by your HMI will significantly impact your speed and effectiveness. This requirement is poignantly (negatively) illustrated by the "Windows" navigation strategy that requires mouse clicks after mouse clicks to work one's way down a pathway. This is unaffectionately called "click forever" navigation.

You never actually stop navigating. Even with a perfectly designed HMI, there is always more to see or follow up on, based on a current situation. But you do not want to get lost. Checks and confirmations are done at every step of the way. These considerations should be part of basic operator training [**Q7**].

1. Confirm current location—is the navigation starting from the place intended?
2. Decide what is needed next (the purpose of the navigation).
   Before clicking, decide what it is that is being looked for.
3. Decide where is it—knowing what is being looked for, decide where to find it.
   At this point, either the specific "location" is known, or there is a good idea of what might be near what is needed.

4. Decide how to get there—using the existing tools and infrastructure, what exactly is the specific way to navigate?

   Use a map; do a guided search; follow a planned idea.
5. Mark where you are and get to where you want to go.
6. Confirm where you went is where you need to have gone.

   Did the navigation step(s) actually end up at the intended destination?
7. Ascertain that what you needed is where you went to find it.
8. Get what you need.
9. Return to Step 2 and repeat as long as needed.

Proper navigation can make a difference between an awkward situation and a disaster. Not always, not even often, but when accidents and disasters are examined for cause, it is almost always affected by poor navigation. Using this tool competently is a part of being a professional.

Module 19: Using the Alarm System

Alarm management is the design, implementation, and effective use of the alarm system.

19.1 The Alarm System

This module provides guidance as you use your enterprise alarm system. It is a cornerstone for effective operation. Alarms enable you to be notified of abnormal situations in time to take appropriate action to make the difference. This module guides you through the recommended approach to handle each alarm activation. If anything here differs from your enterprise protocols, of course use yours instead of what is here. For areas where you are unsure or which you find unclear or sketchy, the material here should be helpful. Discuss with your supervisor and get approval.

> Alarms are the last chance to put you into the solution path for the specific problem before all of the rest of the safety management systems are challenged.

An enterprise is not safe without a properly designed and maintained alarm system. [19.1] *Alarm management* is the design and implementation of the alarm system. The overall alarm management plan is called the *alarm philosophy*. Alarms are designed to call you to action to protect an abnormal situation from escalating to the point of causing harm or damage. They are not designed to warn you that the process is becoming abnormal. You rely on other situation awareness tools for that. You will get an alarm only when you miss something important, it now threatens operation, and it is configured in the HMI as an alarm.

Designing, Building, and Operating Alarm Systems

Alarms are built (configured within the Basic Process Control System or BPCS) for every abnormal situation that the design and operations experts could recognize in advance as requiring operator intervention [**M14.1**]. The total number of configured alarms, how they are determined, what you need to do about them, and how the enterprise works them into their entire infrastructure is carefully laid out in your enterprise Alarm Management Philosophy. It is your complete design plan. Once it is designed, specialists will modify or configure the control system for it. The next few sections cover design aspects that involve how to use it.

This design also includes alarm performance measurements to track how well it works and how well you do using it. These are used to report to management and any governing authorities. They also form the backbone for continuous improvement. To gain a better understanding of how alarm systems are built and how they are intended to work, explore *Alarm*

Management for Process Control. [19.2] It will bolster your in-depth understanding and confidence about using yours.

Alarm Fundamentals

Your alarm design involved a straightforward engineering process. Everything known about engineering, safe operation, system design, and project execution was used for its design. To get it right, all alarm systems are built on four fundamentals. They are listed below:

1. Every alarm requires timely operator action, and that action must be a necessary one.
2. Every alarm activation must occur in time to permit the operator to successfully remedy the situation, if that remedy is at all a reasonable outcome (given the realities of the situation).
3. Adequate information must be provided to the operator for how to work the alarm.
4. Only configure alarms for important enough conditions/situations.

No alarm can be considered to be properly designed and implemented if it does not conform to all of the fundamentals. It is why alarm systems are dependable and reliable. If you experience problems using the alarm system, it is likely that one or more of the items needs attention.

19.2 Alarm Activation and Activity

Alarm Activation

Each alarm configured in your control system is managed as a separate distinct entity. When an alarm activates, it must provide enough time for you to be able to manage the abnormal situation it represents. The actual time needed is the time it will take you to complete SUDA [**M13**] and the time it will take the process to respond enough to avoid reaching the unwanted status or condition. See figure 19.2.1. The alarm activation point is our way of providing enough time for you to manage that specific problem.

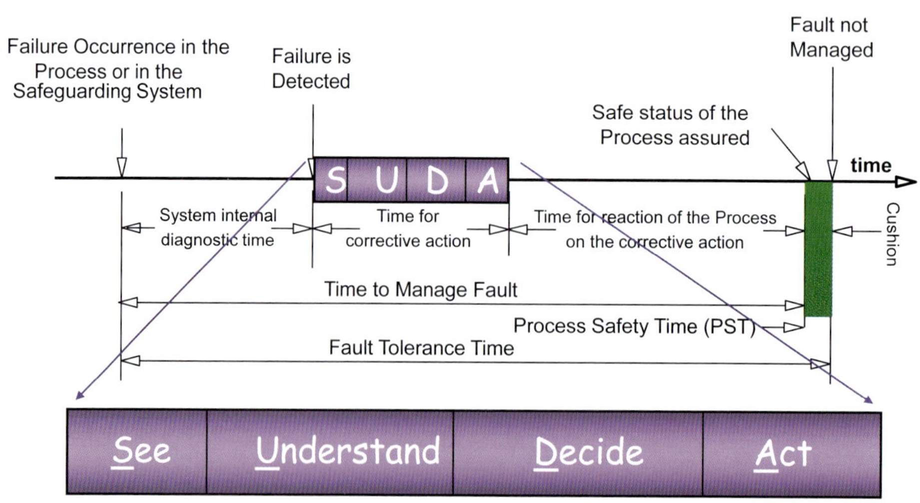

Figure 19.2.1. (Repeated from Module 13). Process safety time, fault tolerance time, and SUDA time (See, Understand, Decide, and Act), illustrating the basic concept that alarm management is time management. Alarms are designed to activate early enough to provide the operator enough response time to successfully manage that specific abnormal situation **[Q5a]**.

Every alarm must be configured (implemented in the BPCS) [**M20.3**] so that when it activates, you will have enough time to [**Q5c**]:

1. SEE: see it (on HMI graphics and alarm summary screen).
2. UNDERSTAND: be able to sufficiently understand what the alarm means.
3. DECIDE: decide what to do to respond to it.
4. ACT: implement your selected response action. And make sure it is working.

The actual plant equipment must also have sufficient time to respond so that the specific abnormal situation the alarm is designed to protect remains protected. The protected situation means keeping the variable from crossing the safe operating limit (SOL). If the situation cannot be managed, the variable will likely cross the safe design limit (SDL) and bad things will happen. If enough time cannot be assured by choice of the alarm activation point, the alarm is improperly conceived and implemented. **The enterprise must use another mechanism to manage this situation.** You, as operator, cannot!

Alarm Priority

Alarm priority captures the degree of risk that each specific alarmed abnormal situation, if unchecked, poses to the enterprise. Its level is determined directly from the requirements of your alarm philosophy. Alarms at the same priority level are assumed to represent an equivalent risk to the enterprise. This would mean that you would work alarms in priority order, from highest to lowest, until all available time was used. Alarms of the same priority are normally worked in chronological order of activation.

As a design goal based upon normal operational risk, expect the numbers of alarms configured in to fall around the following priority percentages:

- About 80% of the alarms at Low Priority (the lowest risk level)
- About 15% of the alarms at Medium Priority
- About 5% of the alarms at High Priority
- Less than 1% of the alarms at Critical Priority (the highest risk level)

The above distribution is only intended to reflect the priority ratios of configured alarms. It is neither intended nor useful to understand the numbers of alarms activating and their priority. Not in any way whatsoever! Please do not confuse *ready to activate when needed* with *actual activation.*

How to Use the Alarm System

There are five situations to understand: (1) normal, no alarms yet, (2) an alarm activates, (3) one active alarm and another alarm activates, (4) three or more active alarms, and (5) alarms are flooding in. Each situation is different, though the general approaches are consistent. Understanding what to do for each situation is important. Section 19.5 covers this. You are expected to work each alarm activation using the appropriate individual *alarm response sheet.* See Figure 19.3.1 for an example.

> There is no such thing as a *false alarm*. Every alarm activation must be **proven** to be of no importance by a preponderance of evidence [**M15**] gained by proving that **none** of the confirmatory actions are true [**M19.3**].

Tag ID:	**FI-2009**			Alarm Priority:	High
Point Descriptor:	PB Feed Water Flow			Alarm Status:	Enabled
Comments:				Operating Group:	12
Setpoint:	155	Units:	KLbHr	Process Area:	PB
High Alarm:	185		(0 to 275 range)	Trouble Point:	200
				Rate of Change:	40

Abnormal Situation:
1. High steam consumption
2. Unusual boiler operation

Causes:
1. Failure of LVC/LIC-2011
2. Plugged flow element
3. Incorrect steam flow measurement
4. Instrument failure

Confirmatory Actions:
1. Check steam drum level LIC-2011
2. Check strip charts, check "periscope"
3. Check BFW header pressure FI-2078/2077, DEA Pressure PI-2098/2099
4. Check BFW pumps
5. Check flow control valve FCV-2009

Consequences of Not Acting:
1. Boiler could overheat and seriously damage tubes
2. Insufficient steam flow, high press.

Automatic Actions:
None

Time to Respond and Process to React:
90 seconds

Manual Corrective Actions:
1. If steam drum level and steam pressure are high, **put feedwater valve in manual and begin to close it.**
2. If steam drum level is low, **manually increase opening of feedwater valve**

Advanced Alarm Considerations:
1. Disable alarm when the boiler is shut down
2. Enable alarm when boiler is started up

Safety Related or Testing Requirements:
HaZop 12-4453; Yearly testing required

Figure 19.3.1. Example alarm response sheet. This is a single page from a fully detailed document that you will use each time an alarm activates. It includes all the information and steps to manage the specific alarm. Every alarm in your operating area will have its own alarm response sheet **[Q8a]**.

19.3 The Alarm Response Sheet

A typical alarm response sheet is illustrated here. Your specific detailed form is determined by the alarm philosophy of your enterprise. As soon as an alarm activates, you will locate and open this sheet. Your management of the active alarm will be guided by the sheet, what you find out about the situation, and the rest of your situation management tools and resources. All information contained on the alarm response sheet was carefully determined by the rationalization team, as specified by your enterprise alarm philosophy.

> As soon as an alarm activates, you will locate and open this sheet. Your management of the active alarm will be guided by the sheet, what you find out about the situation, and the rest of your situation management tools and resources.

19.4 How to Use the Alarm Response Sheet

You are expected to follow a structured procedure for handling the situation, regardless of when an alarm activates or what else is going on at the time. Your enterprise may have a different plan. It should touch all of these bases, and perhaps more, and likely in the same order. This material expects that you have an alarm response sheet for the alarm, ready for your use each and every time an alarm activates.

> The alarm response sheet documents all operational knowledge about the proper handling of an alarm once it has activated. Use it to manage each and every alarm activation.

When you select the alarm to work on, you follow these steps to manage it. All of the steps refer to your use of the alarm response sheet information and the requirements of your alarm philosophy [**Q8b**].

Step 1: **Observe** the CURRENT VALUE of the process variable; <u>compare</u> it with the ALARM ACTIVATION POINT (alarm setpoint). Take careful note of the *Trouble Point* value. This is the value that the alarm is designed to protect (by your response actions). Next, **decide** how fast the abnormal situation is moving. Compare this value to the *Nominal Rate of Change* on the header of the sheet. If considerably faster than expected, make preparations to abandon working the alarm and instead *escalate* [**M34**] or invoke *permission-to-operate* [**M35**] protocol. You must have enough time to work the alarm response sheet, make all appropriate decisions, and effect all proper actions.

Step 2: **Verify** which of the ABNORMAL SITUATIONS is present (e.g., what caused the alarm to activate). The ABNORMAL SITUATION(S) must be readily and clearly visible. If none, be on notice that this problem may be much more than just the alarm activating. Move quickly to step 3.

Step 3: **Work** through each of the CONFIRMATORY ACTIONS to verify which, if any, lead you to believe the ABNORMAL SITUATION(S) you are verifying is true and caused the alarm to activate. Check all of them; do not stop early! One or more should be true.

Escalate or invoke *permission-to-operate* protocol if you cannot confirm anything and the situation remains abnormal.

Step 4: **Review** the CONSEQUENCES OF NOT ACTING so that you have a clear expectation of what might happen if the abnormal situation is not managed in time. This should give you a clear expectation of where the trouble will lead.

Step 5: **Check** to ensure that any AUTOMATIC ACTIONS have occurred and are working as expected. Note that usually there are none, so there is no problem if that is the case here.

Step 6: For each CONFIRMATORY ACTION that you verified as true, **select** the appropriate MANUAL CORRECTIVE ACTION, execute it, and verify it is working as needed. If not, revise and redo according to established practices and guidelines. If it still is not working, *escalate* or invoke *permission-to-operate* protocol.

Step 7: If the alarm does not return to normal (clear) within a reasonable period, you must **find** the specific reason(s) why and remedy. If it still is not returning to normal, *escalate* or invoke *permission-to-operate* protocol.

19.5 Specific Alarm Situations

Here are some of the situations you will experience.

Normal, No Active Alarms

This status is considered the normal situation for alarm operations. In a properly designed and run enterprise, with an effectively designed alarm system, it should be usual and expected to operate for extensive periods of time without any active alarms. This means that all previous alarm activations have returned to normal (no longer in alarm, and cleared). There is no need for you to do any routine monitoring of the alarm system. No alarms can become active without your knowing about them.

The only situations that require monitoring are those after the alarm returns to normal (clears). And this monitoring is informal. All you need do is to check now and then to make sure that the process variable that went into alarm and then returned to normal is moving back into the more-expected and normal operating range. Because there is not any alarm now, it only means that the variable value is inside of the alarm activation point value. If it does not seem to be behaving as you would expect since it is not returning to its desired normal value, you will need to find out why. The bottom line is that your keeping an eye on that cleared alarm is no longer an alarm response matter. You need to remind yourself to watch out for future trouble as part of your routine operations for a while.

Alarm Activates

Once an alarm activates, everything you are doing in the control room changes! Everything. Follow these steps (unless your enterprise has different ones):

1. Note everything you were doing up until the alarm activated. Any vital current actions must be rapidly parked, delegated to others, or abandoned for the moment. Go immediately to the alarm summary screen.
2. Identify the alarmed variable, its value, and its priority.
3. If the priority is the highest category, follow procedures to immediately notify others BEFORE beginning anything else.

4. Access the alarm response sheet (Figure 19.3.1) for the alarm. Begin the established process of working through the alarm response activities documented on the sheet and set by enterprise procedures. If you need assistance and know for sure that you have enough time, engage in collaboration or delegation [**M34**] as needed and allowed by your enterprise.
5. Ensure that your alarm response efforts are working and that the abnormal situation appears to be managed to the point that the specific trouble point will not be challenged. If not, follow your established procedures and protocols for *escalation* or *permission-to-operate.*

Alarm Appears Managed But Still Active—Review

When the alarm appears to be managed (the trouble point is not at risk) but not cleared, return to your other necessary duties including those just before the alarm activated. However, until the alarm clears, the current status of that process variable must remain visible on the HMI [**M18**] and under your constant monitoring.

Alarm Clears—Review

Return to your regular other duties before the alarm activated. However, keep a watchful eye on the variable in question just to be sure that it is moving properly into the expected normal operating region. If it is not, attend to that problem by working it into your current operating plans.

One Active Alarm and then Another Alarm Activates

You are in a situation with an active alarm that you are working on when another different alarm activates.

1. Pause everything you were doing with the previous alarm and go immediately to the alarm summary screen on your HMI.
2. Identify the newly alarmed variable, its value, and its alarm priority.
   a. If the new alarm is the same or lower priority than the one you are working, continue to work on the previous alarm. If you are concerned about the new one, delegate it to others or escalate it.
   b. If the new alarm is of a higher priority than the one you are working, current alarm response actions must be rapidly parked, delegated to others, or abandoned for the moment, so you can immediately begin work on the new alarm.
   c. If the priority is the highest category, follow procedures to immediately notify others BEFORE beginning anything else.
3. Access the alarm response sheet for that specific alarm. Begin the established process of working through the alarm response activities documented on the sheet and set by enterprise procedures. Only if you know for sure that you have enough time, engage in *collaboration* or *delegation*, as needed and allowed by your enterprise.
4. Ensure that your alarm response efforts are working and that the abnormal situation appears to be managed to the point that the specific trouble point will not be challenged. If not, follow your established procedures and protocols for escalation or permission to operate.
5. Return to the previous active alarm and follow steps as before. Make sure that you achieve proper orientation for it, since conditions will likely have changed

substantially. Do not assume that the issue will not have already gone beyond your ability to manage.

Three or More Active Alarms at Nearly the Same Time

This should be an unusual situation. When it happens, something is definitely abnormal. Even if the alarms appear to be related—that is, they might seem to be coming from the same part of your enterprise—it means that significant trouble is at hand. There are only two reasonable outcomes. Either you will get lucky and be able to work the alarms in priority order without anything major happening, or you are heading for significant consequences since you will probably run out of time to manage.

In any event, plan to access and follow established protocols to manage this situation. Immediately *escalate* the situation to the appropriate individuals. This does not mean for a moment that you are not competent and capable to manage. It recognizes that you need more time than you have. You can get more time by having other competent hands working with you at the same time. Any explicit alarm management activities done by you, or on behalf of you by qualified others, should closely follow the activities outlined in one or more situations above.

Alarm Flood

This is an extreme situation. This is not normal. Everything is changed. When this happens, something is disastrously abnormal. Dire trouble is at hand. There should be a clear and detailed procedure for you to follow when this happens. You will have trained for it. Follow it. Usually, immediate *escalation* is required. All available operational safeguards should be brought to bear including *getting help* [**M34**], *permission to operate,* [**M35**], *operator redeployment* [**M36**] and *operational safety nets* [**M37**].

19.6 Alarm Metrics

Alarm management practitioners use alarm operating measurements to better understand alarm performance. This should be approached with caution. The only measurements recommended relate to specific operational issues or alarm design issues. Specifically, measurements are (a) for proper alarm design and maintenance, and (b) to determine operator loading. There are two categories of these measurements: configuration and performance.

Alarm Configuration Metrics

The specific design of the alarm system is not important so long as it is specific to your alarm philosophy, and the working alarm system does not overload you. All alarms must meet all four critical design fundamentals [**M19.1**]. Alarms for security, extreme hazard detection (facility fire and gas), and control system integrity are not covered here. The general design targets are as follows:

- Expect about one alarm per control loop.

  A control loop is considered to contain a measurement, a desired setpoint, a feedback calculation mechanism, and a control output action. For most process industries, each loop usually has a control valve.

- Expect about one alarm for every two analog measurements (not a part of a control loop).
- Expect about one alarm for every five digital measurements (level switch, pressure switch, etc.).

At completion, the total number of configured alarms (of all types and priorities) should be between 900 and 1,200 for each one individual sitting in the operator chair. If any operating area has more than one operator sitting in the chair, the above numbers are increased proportionally. If the above numbers were exceeded, it suggests that the alarm load on you will likely be too much and/or the operational risks are not properly understood or applied. The alarm design needs a serious review.

Alarm Performance Metrics

If you are overloaded, make sure that your management knows about it and works to fix things. There are ways to check to see if the actual (as opposed to potential) alarm system performance might be overloading you. Regardless of these measurements, the only determination that has value is the one that properly identifies your true operator load. These are useful guidance.

1. Average alarm activation count of six per hour (at a mix of priorities; see priority mix examples below in Table 19.7.1)
2. Average number of standing alarms (have not returned to normal within a twelve-hour period); average number of stale alarms (have not returned to normal within two shifts)
3. Short-term average alarm activations of twelve per hour (for an hour or so)
4. Low-only alarm priority—ten per hour
5. Medium-only alarm priority—two per hour
6. High-only alarm priority—one alarm every hour and a half
7. Critical-only alarm priority—almost never (so they do not enter into loading estimates)

Alarm performance measurements are primarily used to ensure that the alarm system does not overload the operator. However, even a low number of alarm activations during the shift suggests that something is not going as it should. The enterprise may be at or near upset. There is something wrong with the instrumentation. There are important gaps in the alarm system design. You may be overtired or distracted—or worse, impaired.

19.7 Operator Alarm Loading

It is vital to keep in mind that you must always have enough time to handle alarms. Time is the most important issue in alarm loading. It assumes that your base load is between 25% and 35% of time. Base load includes all of your general and normal operator activities needed to manage a usual shift without any abnormal situation events. The rest of your available time can be used for alarms, training, or other discretionary tasks or activities. Table 19.7.1 illustrates a few example situations of alarm activations over 8 hours for a mix of three priorities. Other shift durations work proportionally. For each row, if you were to experience all of these alarm activations during your shift, you would have no discretionary time left over for anything else.

Case	Low Priority	Medium Priority	High Priority
1	38	5	0
2	40	8	0
3	33	2	1
4	16	7	2

Table 19.7.1. Maximum operator eight-hour alarm loading for a mix of priorities (assuming base load of 30%). Read each row across. An example taken from the third row would be that thirty-three low-priority alarms, plus two medium-priority alarms, and plus just one high-priority alarm activating during an eight-hour shift would fully occupy that operator in that shift. All of this assumes that the rates of activation are spaced out enough so that multiple alarms do not overlap. This example is showing only the lower three alarm priority levels for clarity.

> Once an alarm system is implemented and all configured alarms follow the four fundamental requirements [**M19.2**], if the operator alarm loading is too high, all remedies must be found outside of the alarm system. These changes may include adding more operator time, adding advanced controls and automation, improving existing controls, improving instrumentation, modifying process design or equipment, etc.

Alarm management is time management. It is that simple. It is that important. You can only work on alarms until you run out of time. If there are too many, some will not be addressed at all. Every alarm that was not addressed means that the abnormal situation that the alarm is designed to protect will not be protected. Your plant will take the hit for each. This reduces to a simple responsibility: system designers must have expended sufficient effort to ensure that you have the time (and resources) available to manage the alarms. Figure 19.7.1 illustrates a hypothetical hour of your time. Notice that this example uses a base load of about 30%. The situation shows an added alarm load that is just under the maximum available time. If this situation prevails throughout most shifts, the alarm load is manageable (but taxing). If the load gets worse, something must change.

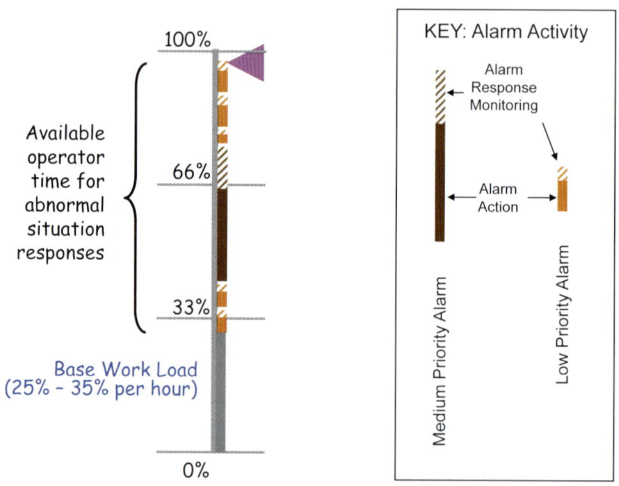

Figure 19.7.1. Illustration of operator alarm loading for a hypothetical one hour. Shown is the base load (in gray) that is taken up by planned and regular duties (reports, regular monitoring, shift changes, etc.). Added to that is the time needed to respond to five low-priority alarms and one medium-priority alarm. All of this activity leaves just a small amount of time left over (above the purple pointer at the top).

If the total number of alarm activations is not too high, but some alarms occur too often for you to have enough time to manage them, it might be possible to defer a part of your operator base load activities to another portion of your shift until the alarm load lessens. This assumes that your enterprise has a plan for this. On the other hand, if your actual base load was higher than the assumed normal, the alarm performance metrics will need proportional reduction. A simple example would be if the base load were to double. Now it is up from 30% to about 70%. All alarm loading metric targets must be reduced to half. You have much less un-allocated time.

19.8 Contribution of Alarms to Situation Management

Alarms provide the front line tools and technology for situation management. Proper alarm management is required for **all** abnormal situations that responsible and prudent enterprises are capable of identifying as part of good design and operation. Period. Extensive effort is required to ensure that the alarm system identifies all reasonable abnormal situations that operator action is required to manage. Each alarm will be accompanied by a fully identified, detailed procedure for management. Remember to use the *information fork* [**M11**] to make sure all information is always known information.

> Proper alarm management is required for all abnormal situations that responsible and prudent enterprises are capable of identifying as part of good design and operation.

All of the rest of the abnormal situations that arise must be found using your skill and intuition, coupled with the rest of the technology and tools in this handbook.

Module 20: Managing Automation

20.1 Benefits from Automation

Automation is designed into process controls. Appropriate automation can make a significant difference. It improves production, reduces the amount of manual control, and manages the load for close process monitoring by you. Enterprises use a number of partial or selective automation methods: basic process control, advanced regulatory control, advanced control, model reference adaptive control, and optimizing control.

Please remember that the material in this module (and all modules) is intended to inform and offer suggestions for how to understand and think about managing the complexities of operating your enterprise. Nothing here is intended to actually instruct you about what to do that is not approved and part of your enterprise operating policy. If you find gaps between what is here and what might be missing in your activities, ask your management to help you to figure it out.

20.2 Process Variability

The process you are managing must be working well. Nothing can be overly disturbed or questionable. This means that everything is where it needs to be and is not moving around; or if it is, it is moving very little and is completely explainable and under your control. The variability is low.

The general term for measurements moving around is called *variability*. The measurement may seem to be hovering about a particular value. But, upon closer inspection, the second-by-second or minute-by-minute readings may change too much to confirm that you really are closely managing it. The more the changes, the higher its variability. Low variability will allow you to operate things closer to a limit. For example, you are controlling the speed of a mixer in a reactor. The faster the mixer runs, the better the mix and the faster the reaction batch is complete. But if the mixer gets above a certain speed, it starts to entrain blanket gas into the mix and the reaction product becomes foamy. Foamy is not good. So if the mixer speed keeps moving around a lot, even under feedback control, you will need to run the average speed lower so that when any variation is going faster, it keeps below that foaming speed. On the other hand, if the mixer speed is tightly controlled, you can run the average speed higher.

20.3 Basic Process Control

The most reliable and best-implemented automation methods use the feedback control loops of the BPCS. No special code or programming is required. The design is completely visible to all having access to the control system. This means that operator graphics can include all information, including switches from automatic to manual, from on to off ratio, and from cascade to off cascade. All alarms needed are fully integrated by the BPCS and so require only conventional configuration and documentation.

Knowing How Controls Work

As operator, you will need sufficient knowledge as to how your controls work and what to do when they seem to be having problems. They may need to be turned off due to problems with a sensor or valve or when the enterprise is becoming upset when its operation goes outside the design limits of those controls. Training is needed to observe potential problems and how to manage them when they present. Proper documentation needs to be available to permit manual adjustment when practical. All of this requires periodic practice and monitoring.

> Any control loop in "manual" that cannot be kept steady enough by hand is a problem usually beyond your responsibility and ability to fix.

There is a very important point here. Whenever a process control loop is not keeping the measurement quite steady (e.g., it is moving around too much) and when you switch the mode to "manual," it still moves around, seemingly for no observable reason, you have a problem. Something very basic is going on. Get help soon. Watch the loop carefully until the problem is resolved.

Magic of a Control Loop

Let us start with the basics. Most of the control loops you manage are single-loop controllers. A single feedback control loop, properly applied, produces a remarkable result. By design, each delivers effective operations over a very wide range of operating conditions. You do not have to do much watching. It is a bit like magic. This magic is fundamental. And once you see it, it should forever change how you understand and operate them.

> A control loop moves all disturbances from where they are harmful to an acceptable place where they do no harm.

Feedback control loops work their magic by moving unwanted disturbances from where they cause harm or other unwanted effects to another place where they are not much in the way! The loop moves the trouble from where you do not want it to another place where it does not matter what it does. Okay, your imagination and interest are up now. But this probably still does not make enough sense for you to say, "Ah, I understand." So let us look at some examples.

Temperature Control Example

Consider a temperature control loop. You have a large drum partially filled with a liquid that you want to keep at a specified elevated temperature. Liquid is constantly flowing into and out of the drum. The drum is internally heated by steam that flows through a submerged coil. A control valve manages the steam flow. A controller sensing the temperature of the liquid in the drum manages the steam control valve. What can go wrong? The drum temperature can be disturbed by changes in inflowing liquid flow, temperature changes in inflowing liquid, changes in the pressure of the drum, changes in the weather outside the drum, or changes in the heating value of the steam itself. The place where you do not want the disturbances or variations is the temperature of the liquid within the drum. So to where does the temperature control loop move them? To the steam line!

The temperature control loop manipulates the steam flow in order to respond to temperature variations. Regardless of what might be the cause of the temperature variations (remember that list we just went over), the steam flow is manipulated automatically to provide additional heat if the liquid in our drum is too cool or less heat if the liquid is too warm. The act of control moves the temperature variations in the drum liquid into flow variations in the steam system. The steam system is designed to handle those variations. Unless the utility steam runs beyond its ability to provide enough steam, everything will be fine.

Flow Control Example

The steady management of materials flow is important to the uniformity of many manufacturing processes. The way this is usually done is to measure the flow and then use a flow-modulating device. This device (usually a valve for liquids) is used to increase or decrease the flow until it is in agreement with the desired flow value. On the surface, this seems like the practical way for this to work. Below the surface, where the actual activity takes place, we have an entirely different story.

All efforts to modulate a flow in one part of a series process will upset the flow both upstream (at another processing station where they have the same type of flow controller that you have) and downstream (again at another processing station where the steady management of flow is important and might be a different value than your upstream one). We have a paradox. Engineers have long recognized this. To deal with it properly, good equipment design includes places for the flowing material to accumulate between the flow control loops. They call those places surge. Surge usually takes the form of a separate surge drum, a tank, an accumulator, or simply a space at the lower end of processing or separation vessels for the liquid to ebb and flow.

In this manner, enterprise designers are able to move the flow-affecting disturbances, which get in the way of good production, into level variations, where it is presumed they will not bother the process very much. This is a good plan. Unfortunately, sometimes the designers of the level control systems are not privy to the master plan. Many control designers value level management in the same way the other control loops are valued—they want them to control perfectly. As a direct consequence, levels are controlled to be as close to a desired value as they can. But that desired value almost always has no process importance. In the end, you often see tight level controls interacting with other tight flow controls. These loops fight each other by trading disturbances back and forth. Each one works very hard on its own problem. None can work together as needed. Ask for assistance if you see this.

Important Message about Moving Disturbances

The important thing about all of this is to make sure that you can clearly see that each and every control loop is moving its disturbance to where it should be—that is, out of your way. Unfortunately, some process control schemes get it a bit wrong. They sometimes pass those disturbances around to any other seemingly convenient place. But that other place can be where they still eventually get in your way. Those moved disturbances are not far enough away from the rest of your enterprise.

There is a benefit for you to understand this concept, locating all the places where the disturbances are ineffectively or improperly moved, and asking engineering to try and move them to other, better places.

20.4 Advanced Regulatory Control

The steps up from basic regulatory control can take many directions. The most accepted and widely used is advanced regulatory control. It usually involves *cascade control* and *ratio control.*

Cascade Control

A *cascade control* loop is where one simple control loop (say, loop B) has its setpoint determined by another simple control loop (say, loop A). For example, let loop B control an inflow to a level-controlled area (managed by level control loop A). Loop B keeps the flow steady. As the level increases, the flow setpoint may need to decrease, but **only to keep the level from becoming too high or too low** in spite of the separate control on level (level A). If the level is quite high, the level A control tells the flow B controller to lower its setpoint slowly and just enough to bring it back in range. All the while, the flow B control loop keeps the flow steady (at the new, slightly lower value). And the other way around, as the level drops. Otherwise, each controller works independently. The advantage of having both variables under basic control is that both keep their measurements steady by the feedback mechanism independently until they need coordination. They coordinate through cascade control.

The cascade control scheme above is designed to do two things at the same time: keep the flow as steady as possible, while also making sure that the important level is not going out of its comfortable design range. All the while, they are reducing operational noise in both loops. Note that it is expected that the level be designed to move about within a range. After all, most levels are used to help smooth out flow variations. So when the level seems to be getting higher than the normal surge range, it gently informs the flow to back off a bit. And vice versa. All of this is to say that the controls are here to make sure that while the level and the flow interact, they coordinate with each other.

Ratio Control

Another common type of advanced regulatory control is *ratio control.* It is used where the desired values (controller setpoints) of two or more independent control loops need to be adjusted together. For example, if a mixing operation requires the manual adjustment of the primary mix component, all of the other mix components would then be automatically adjusted (their setpoints determined by a ratio from the primary one) so that the overall mix ratios remain proper. Certainly, each of the other mix components could have their setpoints manually adjusted at the same time as the primary one. But it takes added operator work; and any mistakes can be costly. However, if the ratios of each of the mixes are different, or change based upon the primary value, there is much room for operator error (wrong ratio, waiting too long, missing one, etc.). Here again, a little bit of control design provides a substantial amount of production accuracy and reliability. Unnecessary operator load is managed.

20.5 Advanced Control

Advanced control is a broad but important tool for reliable operation of complex and highly integrated operations. This type of control is quite different from the basic and advanced regulatory controls just discussed. Advanced control utilizes special engineering and process design information to coordinate various basic controls in a customized way. A few common examples would be the final blending of automotive gasoline into a delivery pipeline, controlling a chemical reaction that requires careful intervention during the various stages, or network management controls to accommodate node outages or wide changes in loading. These controls work in such a coordinated way that would not be straightforward for operators to observe or intervene when necessary.

As the operator, you are going to observe all monitoring points and advisory messages. But even well-designed advanced controls can go astray. Section 20.9 discusses a few ways you can keep on top of them. Get help when anything does not seem entirely proper.

20.6 Model Reference Adaptive Control

A specific arrangement of models used indirectly for adjustment of controls is called *model reference adaptive control* (MRAC).[20.1] MRAC uses the same model building technique described in Module 9: Models to track the process and adjust the internal parameters of the process control algorithms to better match with the process needs of the moment. In this case, the model is already built. Now, we are using it for process control purposes. The builder of the advanced control loop provides the calculation mechanism called "Adjustment Mechanism." The "adjustment mechanism" uses its knowledge of the model to compute the proper changes needed for the controller to work better by knowing the process needs at the moment. Other forms of this advanced control arrangement include multivariable process control (MPC) and inferential control.

This situation brings up an important issue for operators. This type of process management is quite different from the conventional control loops you are accustomed to managing. If this specialized controller stops keeping the enterprise in good control, outputs not in their proper place would be quite visible. Not so the causes for that trouble. Unless these advanced types of controls include specific diagnostic information for you about their inner workings and provide remediation suggestions, [20.2] you are left with two awkward alternatives: either (1) turn the whole control scheme off (sometimes bypassing, sometimes shutting it down) or (2) live with the problem until an expert comes and fixes it. Your enterprise will set out the proper ways for you to know what to do and when.

20.7 Procedural Automation

Most industrial enterprises have two general modes of operation: regulation and transition. In regulation mode, the enterprise is kept operating where it is desired. The control system and you work together to keep it there. Variations in controlled variables would be automatically accommodated for by the feedback and other controls. When the accommodation is too slow or inadequate, you make adjustments to help things along. Transition mode is where the current operation itself needs adjustment, say, to accommodate different raw materials, change over from one energy source to another, or to switch operating modes from peak to off peak. To make these sorts of changes, you must make many complicated adjustments. You utilize complex or involved procedures to make the necessary manual adjustments over a potentially long period of time. All of this opens up operations to a lot of extra attention and exposure to error. One approach is to automate the procedure. Figure 20.7.1 illustrates a process.

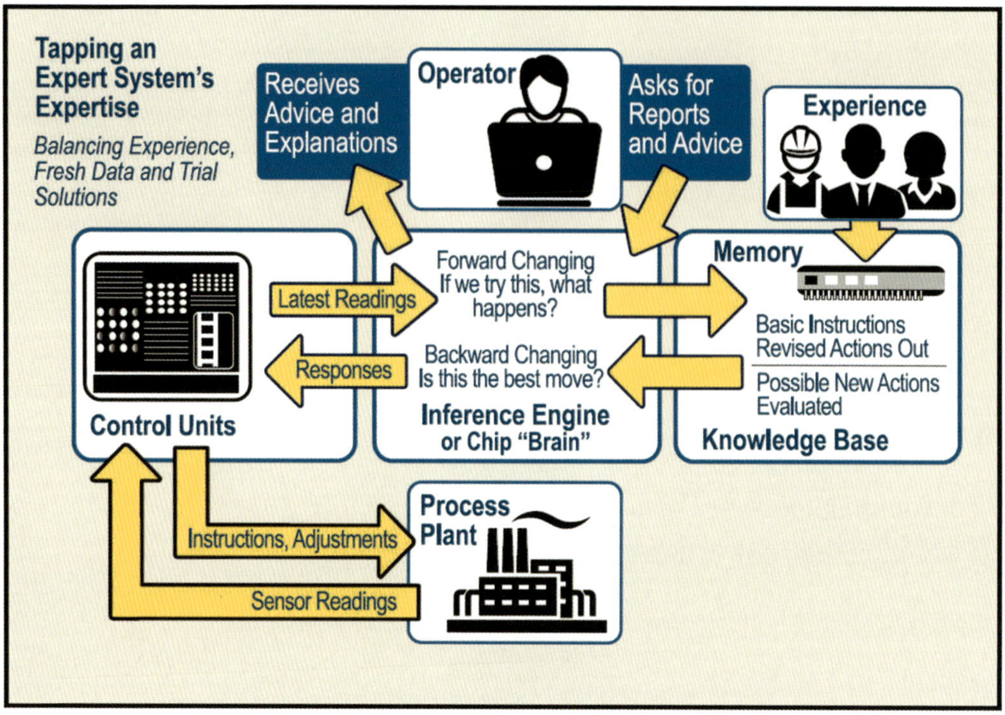

Figure 20.7.1. Concept of operator as coordinator—no direct hands on the controls. [20.3]

The procedures you are using should have been carefully designed by your enterprise engineers and experienced operations teams. They are spelled out in detail. Where sufficient instrumentation and remote control devices exist, those procedures can be automated. [20.4] An excellent example of how this works and its effectiveness can be found in an InTech article by Wilkins. [20.5] A portion of the article appears below.

> An example helps illustrate the concept [of procedural automation]: An oil refinery in Japan underwent crude oil feedstock switches two or three times a week. The efficiency of the operation depended on the experience and skill of the board operator running the distillation unit.
>
> With a skilled operator, the time to reach normal steady-state operations was typically five hours, but it could take more than eight hours with a junior operator.
>
> An automation supplier was called in; they interviewed the board operators from different shifts and were able to uncover and document best practices.
>
> It was also discovered operators typically had to make more than 100 adjustments to the process through the distributed control system during the switchover.
>
> Implementing procedural automation enabled the refiner to make significant improvements in the switchover time to a predictable four-and-a-half hours, regardless of which operator was on shift. The operators' workload was also reduced significantly, with over 100 control system adjustments reduced to 10, and more than 2,000 process alarms cut substantially by configuring the system to be operationally aware of process conditions. Finally, there was

increased process knowledge sharing, a significant reduction in operator errors, and reduced required operator training.

That was pretty neat! As is the case for other forms of automation for operators, the design will need to include any specialized operator refresher training and procedures for proper manual operation should that need arise.

20.8 Optimizing Control

Optimizing sounds like a grandiose reach for an operator working very hard just to keep things on target and out of trouble. You have a full-time job just doing your job. So why bring something extra? The answer is that enterprises usually need more to be successful at production. To get more, someone else will have designed all of the optimizing to deliver it. From your point of view, all that you need to do is understand how it generally works and what it is probably doing. Here, we suggest a few ways for you to do that. If the design is good, then all you need is to account for what it is doing as you work your shift. If it is not working or you are not sure, you will probably turn it off. Let someone know so it can be fixed.

Optimizing

In process control, the term *optimizing* means that a part of your enterprise is being managed, not only to be working properly, but also to meet an extra objective of improving something. It might be to reduce energy. It might be to speed up a critical reaction. It might be something else important to the enterprise. The only way something can be optimized is when it can be managed to work properly so that it produces the right product, but also there is enough flexibility for it to work in a number of different ways that do not affect the main deliverable product. So if one way is better than another—say, lower energy cost or higher quality—then the enterprise might provide the design and controls to do it.

How It Works

To get optimizing to work needs two specific things. A value function called the "goal" and the manipulated variable we call the "handle." The optimizer watches the goal and adjusts the handle to move it to the best value of the goal. Your job is to sit back and watch it work. It can get into trouble in lots of ways. The usual way is when one or more basic control loops are not working well enough (too much variability). Another way is if one or more of the measurement sensors go haywire or stop working altogether. Still a third way is when you need to change something basic in the enterprise operation and, during the change, the optimizing control scheme gets lost. For all of these situations, you will have been given specific roles to play and actions to take. Follow them. Whenever you do not feel comfortable, get help.

20.9 Protecting the Automated Enterprise

Over the past many decades, industrial enterprises have been operated under increasing levels of measurement instrumentation and controls automation. Part of this can be attributed to the increased availability of more reliable instruments to measure conventional process conditions like temperature, pressure, level, and flow. Their use has been enhanced by specialized sensors that are more accurate, more tolerant of wider process conditions, easier to install and maintain, and more affordable. Additionally, as separable components of an enterprise became more integrated due to production efficiencies, energy integration, and sophisticated production procedures, added instrumentation and more controls devices were required. It is not too much of a stretch to imagine the degree of controls to increase to the point of near full automation.

Even at present levels of automation, it is not unusual to operate for long periods of time without a need for significant operator intervention. Certainly, there are the usual snags and snafus due to uneven automation or faulty equipment or other operational annoyances. Most are quickly managed and result in more nuisance than operational threat. One of the unintended consequences of increased automation and the resulting operations variability reduction is that you have less to do and, by doing less, you are less able to test your skills. While you might be effectively trained to do a wide variety of tasks and operations, those particular skills are perishable without frequent exercise and validation. Consequently, enterprises now face increasingly complex operations, with operators less able to step in and take charge when they go amiss. If we take the leap to a nearly fully automated enterprise, the exposure to stale operators now becomes vastly more important.

There are three additional approaches to managing operator effectiveness in this situation: (1) use only automation that is very robust, (2) force periodic operation without the automation turned on, and (3) cease operations during upsets. All have their proponents.

Require Only Robust Automation

There is a share of enterprises, enterprise management, and enterprise operators who resist any significant degree of automation. They quickly suggest that it is too costly, needs too much maintenance, is too hard to understand, and is much more than they feel is actually needed. For them, hands-on operations are a personal enrichment. Unfortunately, this feeling is more an expression of wishful thinking than operational benefit. Operators in these situations find themselves burdened by the necessity to always be very close and remain overly diligent about watching for things amiss. They become highly susceptible to the monotony of repetition and boredom. In the end, there is really less operator satisfaction and even less operational benefit. The bottom line is that it is important to use automation as a tool—provide what is needed and no more or less, and provide only what will always work. Your enterprise should strive to implement automation that is robust to the point of the enterprise operating much better when it is on than when off.

Operate Periodically with Automation Off

Another way of training for operating without automation is to do just that: operate the enterprise with as much of the automation as can be properly supplemented by manual operations turned off. Not so fast. This is more than it appears. The enterprise needs to be in a production mode that is not overly sensitive to operational variability. There must be adequate procedures in place. There must be additional personnel present to monitor and assist with extra hands and additional coaching. All of this has the important benefit of developing operator skills as well as testing procedures, the alarm system, collaboration, and escalation. Carefully monitor the events to make the appropriate recommendations for improvement.

Before we leave this, it is important to identify a critical principle: if an enterprise cannot be operated without full automation in place, then it must not be. If the failure of any automation element cannot be satisfactorily managed by going to manual action, then either (a) the enterprise must be designed so that any such failure cannot occur, either by automated safeguards or replication, or automation to move to a safe state; or (b) the enterprise cannot be operated at all. There is no safe middle ground. You will find extensive coverage of this topic later on in *Permission to Operate* [**M35**]. The concepts and tools there underpin ALL safe operation.

Cease Operations During Significant Upsets

There is one other alternative. This is when the process becomes upset to the point that it should be evident to you that the automation is not properly functioning and the remedy is not procedurally evident and right at hand. At this point, you act to shut it down or move it to a safe operating situation. This decision timing is much earlier than what would be the case under the usual permission-to-operate process. Now, by normal operating protocols, the process will be directly moved to a safe state. The benefits are: minimal risk of abnormal operation escalating into anything significant, no need to have extra personnel present just to manage upsets, much less chance for equipment to be operationally stressed, and much less chance for off-specification product.

If you are thinking that this option is extreme, please stay with this a bit longer. Without proven and reliable procedures and the rest of it in place for you to safely and efficiently move the production process to a safe operating state (or shut down), the enterprise is really not safe. It is always better to avoid damage beforehand than repair it afterwards. You will need all sorts of spare parts you will likely not have. The expertise to repair may have to be brought in from outside or, at the very least, taken away from other scheduled and necessary work. And the outage will almost always come at a time when your enterprise would not have wanted it.

20.10 Using a Simulator

A very good way to maintain skill levels is to exercise them on a regular basis. This can be done quite effectively using high-quality simulators. You read about simulation in an earlier module [**M8**]. Here is an important use. Simulators permit skills testing through the point of complete failure without any actual risk except a little lost time and perhaps a degree of bruised pride. They allow repetitions of scenarios to hone skills. They also facilitate the ability to qualify required skill levels and compare relative mastery of individuals. All in all, when simulators sufficiently match the real thing, they are extremely valuable. Nor do all simulators need to be a physical mimic; thought simulations, sometimes called tabletop simulations, have a proven value.

20.11 Summarizing

Once the proper level of automation is set, it is imperative to ensure that it is properly utilized and the needed skills maintained to keep the enterprise safe and effective. Studies have shown that 20 to 30% of all instrumentation and controls are not working as designed and intended at any one time. The lower the level of instrumentation and controls availability, the higher the minute-by-minute requirements are on your presence, your very close attention to operating conditions, and your making all of the ready hands-on adjustments needed to keep things going right. Being under this extra load detracts from your attention to other things. Operating under this extra daily load is wearing on your motivation and job satisfaction. The few who take a sort of primal pride in being essential is far outweighed by the many for whom each shift is a relentless grind, with constant exposure to unnecessary untoward events. The bottom line is that having a proper level of automation more than pays for itself in the long term.

Module 21: Managing the Shift

21.1 The Shift

As a seasoned operator, you really do not need us to tell you what to do to manage your shift. So we will not. What we will try to do is to lay out general concepts and ideas to provide helpful ways to make your work easier and clarify and supplement existing company rules and requirements. You should consider these new ideas only if there is room for them. All of your existing procedures, policies, practices, and requirements come first. If there is room, this module offers useful ways to make your job easier, more comfortable, more effective, and more satisfying. Review any that interest you with your supervision.

21.2 The Shift at a Glance

We start with an entire shift layout. Again, this is not to be used to replace your requirements. We are using it to help everyone start on the same page.

Operator Duties

You are tasked with ensuring that the operation under your care performs as required. Requirements are varied, of course. They generally consist of the following:

- Orient to the current status and objectives of the shift; maintain this orientation throughout the shift
- Effectively monitor the totality of the operation and make appropriate adjustments
- Modify production to follow scheduling and necessary production modifications
- Evaluate results of routine testing and sampling and adjust operation appropriately
- Arrange for non-routine testing and sampling of production (including product quality, environmental quality, and resource utilization)
- Arrange for, monitor, and confirm maintenance activities
- Originate, authorize, and accept receipt of materials used in the operation
- Manage delivery or shipment of production products (produced chemicals, electrical power, rail car movements, etc.)
- Engage in ancillary aspects of the operation as needed (alternate spared equipment to even out wear, issue and monitor permits, control and manage visitor visits, etc.)
- Participate in training exercises, other "tool sharpening" activities, and management interactions and feedback
- Practice operator alertness and other fatigue mitigation activities
- Prepare reports including log entries, handover, and problem identification and resolution activities
- Prepare operation to conclude or be transferred to relief operator

The Whole Shift

The way you schedule your duties and responsibilities for operating does not just happen. You work it all into your shift plan. However, there is a lot of latitude in when you do what. You can choose and develop a comfortable working process for it all. We begin by taking a look at general activities as depicted in Figure 21.2.1.

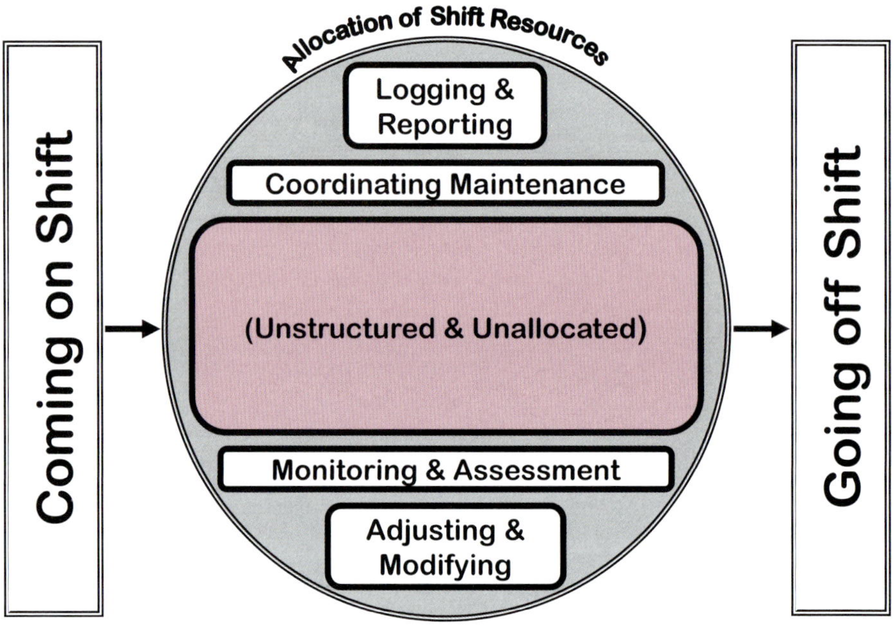

Figure 21.2.1. Traditional activities during a shift, illustrating the large unstructured block(s) of time where the operator has to invent working processes within them. This usually results in operators having their own methods or adopting unproven "common denominator" methods without sufficient enterprise involvement.

There are the specific parts of shift: orientation, reporting and logging, coordination with testing and maintenance, and monitoring and adjusting. Typically, these duties account for around 30% of *operator* time. The rest (light pink colored area) is loosely organized and generally not planned or structured. The operator just uses it to make sure things are okay. By being left so loose, operators will sometimes devote too much effort for a while and then leave it for probably longer than desirable—or the other way round.

A suggested enhanced framework for time allocation is shown in Figure 21.2.2. This arrangement provides a new operational practice opportunity. Using it enables a regular workflow that balances required structured activities with informal aspects in a way that ensures good monitoring, with time for the other support activities. The approach should enable you to feel confident that your efforts for keeping an eye on things are responsible, appropriate, and adequate. Here is where you might plan to actively look for *weak signals* [**M23-M26**]. Briefly, weak signals is a way for you to be able to observe small indications of things that may suggest problems and then find out what problem they might be. That process works to ensure that less slips through cracks. At the same time, it replaces the stress of your needing to constantly look for something that occurs very infrequently. This benefit should not be underestimated. Weak signals is covered extensively in **Part 5** (Modules 23 through 26). They are mentioned frequently in the remainder of this module. For this module, it will

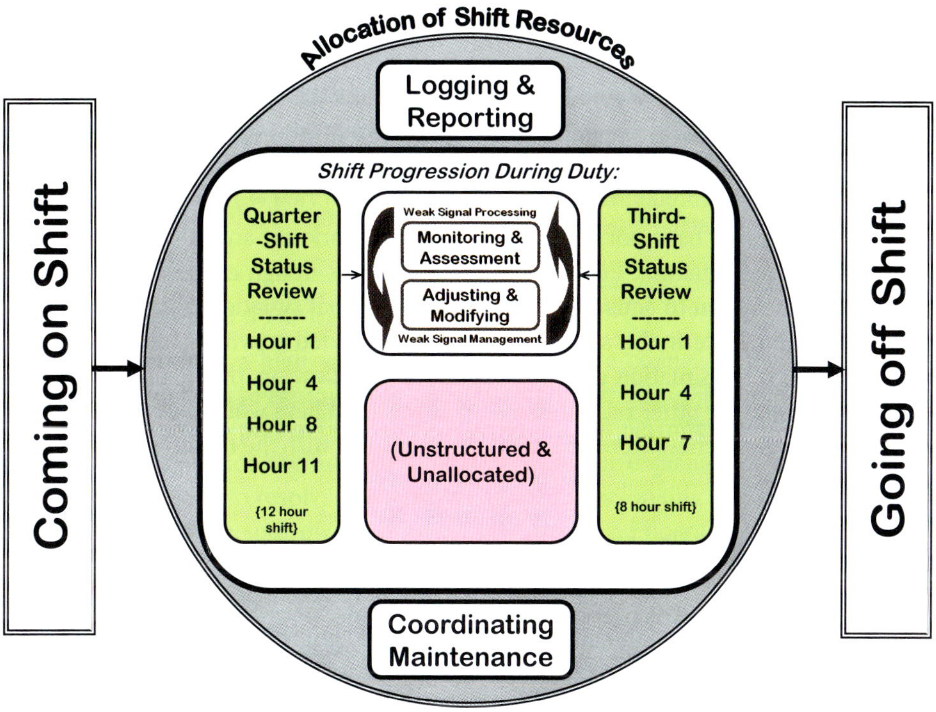

Figure 21.2.2. Suggested structure for the shift, illustrating the clearly structured block(s) of time to manage operator responsibilities. This is tailored to your approved methods for maintaining the proper balance between the various roles needed for good operation and ensuring you keep on top of both the short-term normal operating performance, as well as early developing problems that impact things later on.

suffice if you understand that, when you are using them, you are simply looking for small indications that could suggest that you need to look at something more. The looking itself, to be done properly, requires the use of specific methods and procedures, all explained in **Part 5**. If you are having difficulty following, you may want to skip ahead and read the introductory part of Module 23, particularly Section 23.3 *Weak Signals Work Process Review*. Then come back here.

Now your shift includes specialized and clearly defined timing for that "looking around" part. Always looking, looking always for everything amiss is not reliable or sustainable. Structured looking keeps weak signals under reasonable control. Now it is just a part, not a career.

General Shift Tasks and Duties

General shift activities include the conventional and necessary duties and responsibilities of the operator. But now, there is a regular and predictable review of the operational status using the technology of weak signals. At regular intervals, you perform what amounts to in-shift operational assessments. They include the following and more:

- Review all alarms that have not returned to normal
- Review all ongoing maintenance activities in progress
- Review all key production indicators

All permits to work must be cross-referenced to all the area(s) they refer to in the enterprise so that all verifications for safety and safe operation are taken into consideration and properly managed by all players. Understand exactly the full aspects of the following:

- What is the job they will be doing?
- Why are they doing it?
- For how long are they permitted to do it?
- What, if anything notable, are they not to do?
- Where are they doing it (including how important any surrounding areas are, to make sure that there is no spillover or intrusion there)?
- Who is doing the work?
  - Do they appear to be the expected personnel?
  - Is there any visible or suspected impairment?
- How might the work schedule be modified to account for
  - Early termination or completion?
  - Suspension?
  - Delay?
- What in the way of reporting do you expect from the permittees during the work or at completion?

All permits to work must be cross-referenced to all the area(s) they refer to in the enterprise, so that all verifications for safety and safe operation are taken into consideration and properly managed by all players.

Visitors

Visitors always present a special case fraught with potential distraction. You will want to be ever vigilant. Visitors are important and useful. They are an opportunity to show the workings of your control room and maybe portions of your enterprise. Visitors may be from the general public, government representatives, or guests from other parts of your larger enterprise (other plant managers, shareholders, and such).

When on shift duty and in charge of an operational area, you are not a tour guide, public relations representative, or spokesperson. To protect your primary operational responsibility, please consider the following guidelines:

- No guest or visitor should be permitted without a trained host—not you. The duties of the host are to ensure and manage: all aspects of the visitor walking around and being conveyed around; personal and enterprise safety; orientation as to where he is and why he is there; the request and acceptance of entry and any interaction with you.
- All aspects of your operational responsibility come first. Your eyes and attention must not be diverted from both the current task at hand and your ever-present requirement to maintain situation awareness [**M27**].
  - Only interact when it is assured that you are able to interact for very brief periods of time.
  - Pause interactions to attend to duties; only return if operations allow.
  - Avoid all attempts to impress or deride.

- You must fully control only your own very limited part including beginning and ending your interactions and, in some circumstances, the visit itself. The host must manage all of the rest.
- If you must take the lead in interacting with visitors (due to their high importance or if your expertise is required), your operational responsibility must be passed on to a qualified stand-in during the entire activity.
  - *Be sure to conduct a mini handover before, and a handback after.*

21.5 Visual Cross Reference for Permits and Activities

The benefit of the permitting system is how it provides procedures and safeguards for the conduct of activities requiring more than ordinary care. The permits issued or managed by you need to be kept in mind everywhere your activities and responsibilities are affected. Lists work, but are limited by needing to be consulted when the need arises and by being separate for each activity class.

This handbook strongly recommends that each enterprise develop and roll out effective ways to cross reference all permits with each other and between each class, to ensure the needed safeguards are realized. The permit-to-work system already requires this. Not so for all the rest. Nor do they require cross referencing between classes (permit to work, confined space entry, taken for maintenance, LO/TO, deliveries, and more). Let us look at how this might be kept in your mind as you operate.

To keep track, we advise your enterprise to add to your operational display screens the specific locations for all of these activities. See Figure 21.5.1. Now you have a ready tool to be continually aware of what else is going on. You will check the specifics for each activity for more information, of course. The display information helps keep it in your mind. Most, if not all, DCSs and SCADAs

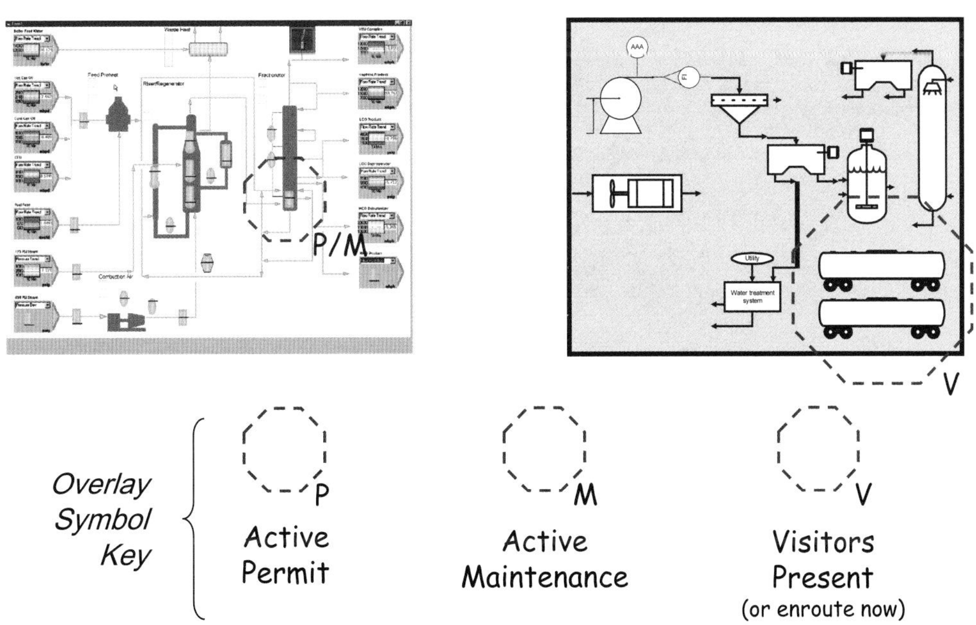

Figure 21.5.1. Example plant display screens, depicting overlays of permits and other activities locations that the operator needs to track and manage

are not currently configured for this level of graphic support. So you will want to suggest to your engineering and operations management teams to work out ways to get this job done.

Here is what the US Navy does to track aircraft on deck (Figure 21.5.2). As each plane is moved on the deck, the scale model is moved. As each plane's status changes, the tokens are changed. One could be graphically built for you to use at each operator station. For now, you will need to have and use some other good way of keeping track of all of those moving parts during your shift. Ask your supervisor for guidance.

Figure 21.5.2. Model mimic from US aircraft carrier flight operations tracking, showing all aircraft (to scale) on the flight deck, with special tokens (the colored pushpin heads) denoting special status conditions for each plane

21.6 Following the Shift Plan

If your enterprise has laid out a skeleton framework for you to use during activities that need to be done and has recommendations for timing, they should be followed. If not, or if what you are given is not enough, you will need to work one out on your own until one is in place. Good practice suggests that it be formed during or just after your coming-on-shift handover. Once you get it shaped up, please review it with your operating supervisor. This will provide the roadmap for your activities during your shift.

> Any omissions, changes, or additions to the approved shift operations plan need to be done following formal procedures in consultation with supervision.

Progress of the shift work plan should be checked at each shift review period (Figure 21.2.2). In the event that you are falling behind, pause and take a look at why. If there is flexibility to easily catch up, make proper adjustments to do so. Be sure that all of the shift activities are taken into account for this. If things seem tight, seek supervision. What you need to do during the shift needs doing. If that is in jeopardy, the enterprise needs to know.

Module 22: Shift Handovers

A shift handover is a communications event. It is one of the most, if not the most, important communication event that will occur during a shift. [22.1]

22.1 Overview of Shift Handover

Shift handover is the process for exchanging operators between shifts in a carefully controlled way. The shift is changed for a single reason: to provide a replacement operator to perform your operational duties where you will not continue in that role. For 24/7 operations, shift handovers will occur two or three times each day. It requires the leaving operator to fully assess what happened on shift, what is likely to greet the arriving operator, and to provide suggestions for handling those issues. The handover needs to be carefully designed and competently executed to ensure the arriving operator the opportunity to orient and come up to full speed. This needs to happen every time.

Reasons for shift handovers include all situations when the existing operator is required to be relieved due to:

- End of planned shift (sufficient hours of duty completed)
- Temporary shift change due to operational or personal need to leave the control board for a short time (where manning of the control board is required at all times)
- Inability to continue duty due to inability to remain vigilant, illness, injury, personal emergency, or unauthorized absence
- Inability to remain in designated physical area due to fire, physical security breach, personal needs, etc.

There are three types of handovers: board operators' handovers, field (or outside) operators' handovers, and operations supervisors' handovers. While the personnel differ [**M2**], the handover act and actions are remarkably similar. Thoughtful coordination will significantly improve their effectiveness. Each party to a handover knows what he knows and works to document and share it. A successful handover should result in all necessary information being shared. To help the sharing to be complete, always ask all participants "What do you know that I don't?" Fill in the missing. Ensure that all is fully understood in proper context.

Not all shift changes need to occur between individuals located at the same physical location. Operators and supervisors may begin duty at other locations prearranged and designed for them. One example is where day operators use one control room and the night operators use another. A second example is where remote central control rooms are located around the globe in a "follow-the-sun" configuration. Here, each control room can be manned during the normal day shift for their time zone. Another typical example is for emergency operation due to weather or other situation.

The actual contents for the reporting and reports prepared for these situations and the specific procedures to follow are left up to specific site style, requirements, practices, and standards.

22.2 Board Operator Handover

The board operator handover is the most basic. In addition to the details of the actual handover, there is the requirement for doing readiness assessments of fitness for duty and for travel home [**M5**]. There is also a strongly suggested requirement for a brief period of joint operation. This means that after the arriving operator takes active control, the leaving operator remains in the control room to assist for a short time. The reason for this is the mounting evidence that it takes a while for the arriving operator to fully incorporate everything to be managed; i.e., time for everything to be "put into mental place," so to speak. While the shift handover communicates what it needs to, the actual taking over is much more extensive. For this, both operators do a think through to carefully review all specific tasks planned for the next shift. Here is where they mentally review the activities and check their plans. For demanding operations or those in upset, more than ordinary diligence is needed. The assistance of a fully up-to-speed *leaving operator* will make the difference.

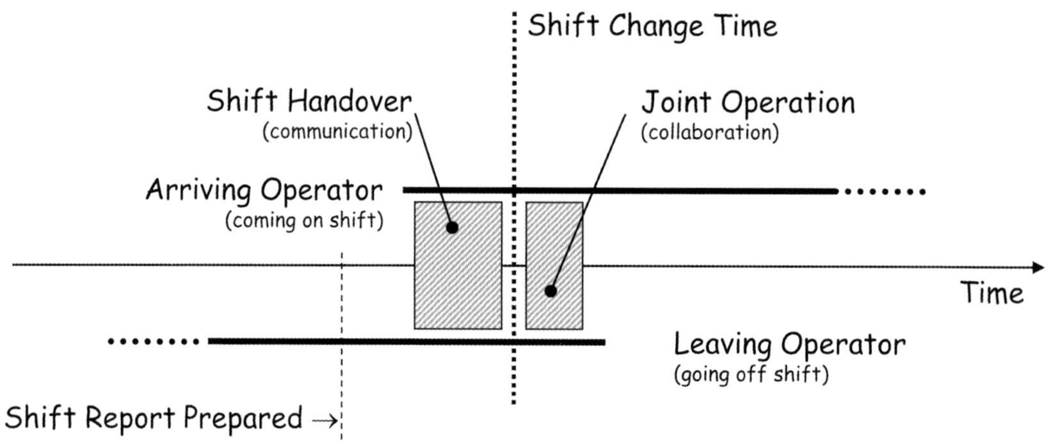

Figure 22.2.1. Shift handover timing activities, showing first communication to transfer knowledge followed by collaboration to transfer ownership

Figure 22.2.1 shows the time overlap periods for board operators, with responsibilities identified. Note that to ensure a smooth transfer of operating responsibility ownership, there is **communication** before the actual handover and **collaboration** after. During collaboration, the leaving operator briefly assumes the role of coach and second pair of eyes before going off shift. Timing and time periods are specific to your enterprise. General conventions might be fifteen minutes to a half hour overlap before shift handover and from five (a bare minimum) to fifteen or more minutes after. Of course, these times could be adjusted very quickly according to established procedure if there are operational situations or events to manage such as a startup, shutdown, or abnormal situation.

Overlap and Ownership Transfer

At shift change, ownership transfers from the leaving operator to the arriving operator. The shift change activity is one of the most valuable operating situations, BUT it is also one of the more vulnerable. At no other time in normal operations is it more important to fully understand what is happening. This understanding must be totally shared with another individual. This module lays out a process for the arriving operator to take proper ownership. This important step, when coupled with the extra step of actually taking over, cements the arriving operator firmly in role and function. Ownership is an important situation assessment and awareness tool. It also provides a natural segue into the topic on collaboration [**M34**], since it involves similar activities and goals.

> The leaving operator is responsible for transferring all shift handover information to the arriving operator. The arriving operator is responsible for taking full ownership and control of the operating chair.

The shift handover process begins at the point when the leaving operator has prepared all shift information needed to brief, and the arriving operator is present, fit for duty, and ready for the briefing. Follow the process depicted by Figure 22.2.2. Shift information includes all ongoing, just completed, and scheduled-to-begin-soon shift activities that are part of enterprise life. There are two classes of shift information: *objective* (direct observations and recorded data) and *implied* or *inferred* information (observations gleaned during the shift). This separation is important. Both are collated into a specified shift handover (or turnover) report and will be used directly in the shift change briefing.

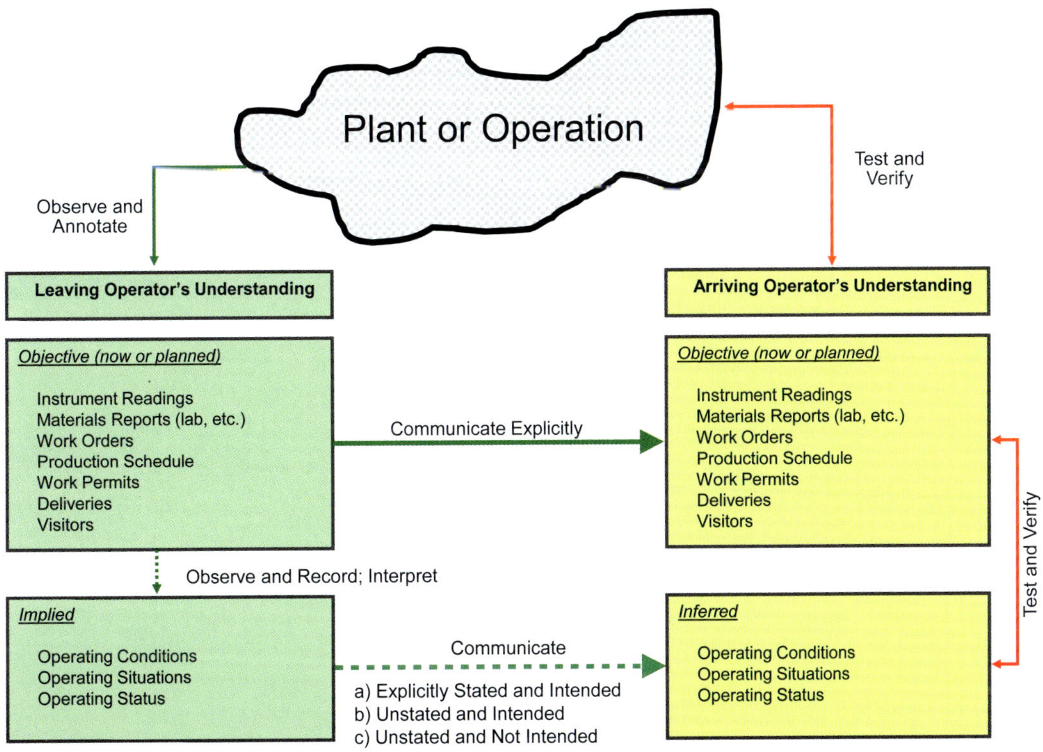

Figure 22.2.2. Information and decision flow for shift handover. The leaving operator gets everything ready and then shares with the arriving operator. The arriving operator must fully understand what is being communicated, as well as test and verify that it is true.

All shift information flows into the green boxes (□) depicted on the left in Figure 22.2.2. The objective information would come directly from readings and other material that required it to be recorded without modification or alteration. Differences from what actually exists in the plant or operation must be due to errors, such as: (a) instrument malfunction, (b) report-writing error [noting a lab value in a wrong location of the report], (c) transcribing errors, (d) intentional deception, and the like. Implied information is that which an individual conjectures as being relevant by personal supposition, conclusion, or conception. It may be anchored on or suggested from objective information. But these implications are not proof. This is an important and useful distinction.

Transferring and Receiving Information

In Figure 22.2.2, heavy green arrows indicate the transfer of shift information from the leaving operator to the arriving operator. The solid green arrow (→) labeled *Communicate Explicitly* indicates the transfer of objective shift information taken directly from the operation. The dashed green arrow (- ->) labeled *Communicate* is used to depict the transfer of *implied* to *inferred* material.

Moving to the yellow boxes (□) in Figure 22.2.2, observe the *Objective Information* box and both *Implied* and *Inferred* information boxes. Since the objective information flows between the two operators, it is presumed to have been accurately transferred. Transfer means read or seen on a screen or paper report and such. The word inferred is used to mean that regardless of what might have been intended by the leaving operator (who does the implying), the arriving operator (doing the inferring) takes it in through his own thoughts, expectations, and preliminary verification. Each needs to check the other for alignment and agreement. All shift information is now in the hands of the arriving operator (the yellow boxes in the figure). The stage is set for the next activities of verification and then ownership.

Verifying Information and Taking Ownership

At this step, the arriving operator has all of the shift information. Two important events need to happen: he needs to verify that what he has is the best evidence for the truth, and he needs to understand it and to take ownership. Verification is simple to say, but can be a bit involved to actually do. It is best done in two steps. Note that these steps are shown as solid red arrows (→) on the figure. The first step is to verify that the objective information is correct. Examine it for reasonableness. Spot check a representative sample. Make historical comparisons. In short, utilize tools in your toolbox of competencies. See the upper red arrow connecting *Plant or Operation* to *Arriving Operator's Understanding*. Once the objective data is verified, it's time to turn your attention to what it means. See the lower red arrow (labeled *test and verify*) between the *objective* and *inferred* yellow boxes at the right side of the figure.

Now the *arriving operator* is asking whether the objective data and the implied data are consistent. Basically, the question is "is the process actually operating as I think it is?" or, in other words, "Does my thinking about where the plant is operating match the objective data (that I have verified to be correct, so far)?" If so, then he reasonably knows where the plant is and what to watch out for and manage. He can take ownership of the operational chair (the shift). If the data does not match, then the arriving operator does not know to any reasonable level of certainty what is going on. This is the time for *escalation* [**M34**] or *permission-to-operate* [**M35**] protocols to begin.

During-the-Shift Handover Emulation

The production review and status monitoring you will do for a formal shift handover can also be useful for an operations check during the shift. Consider conducting an informal, in-shift handover as an additional situation awareness tool. It would be a streamlined version of a full shift handover, but done for your own situation awareness review. These are depicted as "shift status reviews" in Figure 21.2.2. It would touch the key points, but without the need for excessive formality. It would be used, as needed, during the hour-by-hour workings of your actual shift. Doing it this way has the benefit of taking a very good look at the details of the current status as well as an overview of operations. Each time you do one, you would feel confident that you understand what the process or operation is doing at the moment and whether anything needed special attention. Doing this periodically during the shift would also ease the time to get ready for the end-of-shift handover, since a lot of the information would have already been identified, noted, and processed. The purpose is to use a nice existing framework and repurpose it. Consider adding these in-shift evaluations as a part of your regular shift activities.

Actual Board Operator Handover

Figure 22.2.3 shows a shift handover timeline identifying the various activities and when they might best occur. Notice that this diagram is constructed so that time starts at the top and increases to the bottom.

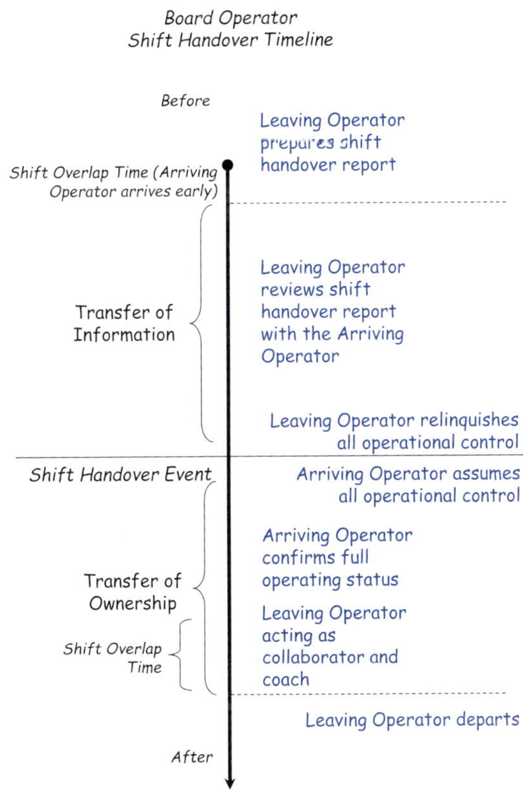

Figure 22.2.3. Detailed board operator shift handover timeline, identifying leaving operator preparing handover report, sharing the report with arriving operator, transfer of operational control, and joint operation and collaboration, leading to arriving operator assuming full ownership of shift.

Start at the top. The leaving operator finalizes the handover report before the arriving operator arrives. Information, notation, and observation activities actually occur throughout the shift. Operators will best remember the critical details if they are captured as near in time to the event as possible. These activities can be significantly aided by use of any of a number of commercially available shift handover reporting tools. The report and all other needed information are shared with the arriving operator in a give-and-take atmosphere. **Operational control passes to the arriving operator at shift handover event line shown in the figure.** This is followed by a period of operation with arriving operator as lead and departing operator as collaborator and coach. The arriving operator assumes full ownership of the shift at the end of this overlap period. The departing operator leaves unless the plant is upset and extra hands are needed.

22.3 Shift Supervisor Handover

Shift supervisors may also have formalized handovers. Basically, the purpose and general activities mirror what is done by board operators but at a different enterprise level. We will cover only the timing and any coordination aspects. The handover is shown in Figure 22.3.1.

This handover timeline shows the leaving supervisor's first responsibility is monitoring (and assisting where needed) the board operator shift handover. After that is complete, and with the benefit of both the control room shift handover report and the experience of the board operator

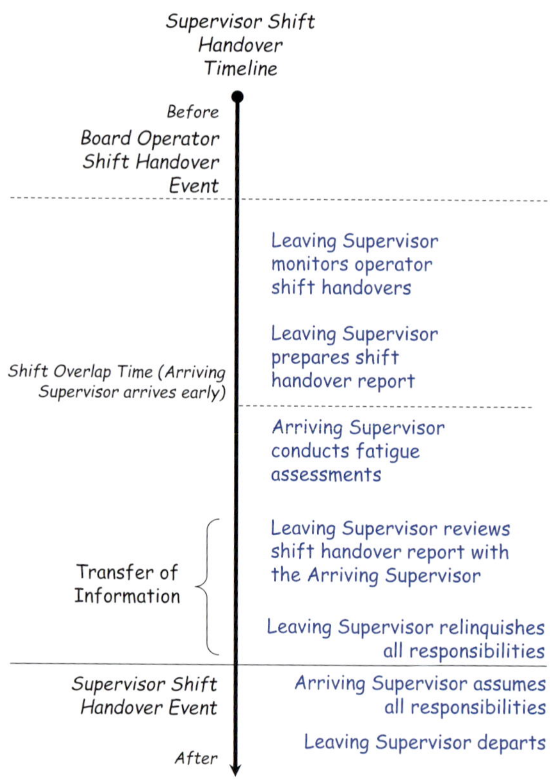

Figure 22.3.1. Detailed supervisor handover timeline, noting that the leaving supervisor first monitors the operator shift handovers, then prepares his own handover report. The arriving supervisor first monitors the fatigue levels of both arriving and leaving operators. Then both supervisors review the supervisor handover reports. Once that is done, the leaving supervisor departs, leaving the arriving supervisor in charge.

handover, he prepares his own handover report. This activity may continue a bit after the arriving supervisor comes on shift, since the arriving supervisor's first duty will be to assess the fatigue levels of the sitting (arriving) operator. Once this is done, the leaving supervisor reviews the handover report with the arriving supervisor. After the handover, the leaving supervisor departs unless the plant is upset and extra hands are needed.

22.4 Field Operator Handover

Field or outside operators also perform shift handovers. The purpose and general activities also mirror that process done by board operators. This handover is shown in Figure 22.4.1. The handover timeline identifies the leaving field operator preparing the handover report, sharing the report with the arriving field operator, and transferring operational field control to the arriving field operator. After the handover, the leaving field operator departs unless the plant is upset and extra hands are needed.

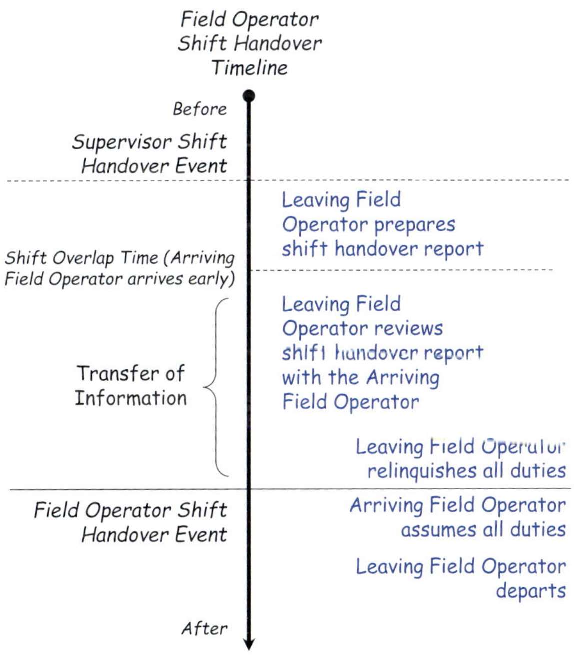

Figure 22.4.1. Detailed field operator shift handover timeline, identifying leaving field operator preparing handover report, sharing the report with arriving field operator, and transfer of operational control, resulting in arriving field operator assuming full ownership of shift.

22.5 Coordinating Shift Handovers

All handovers require appropriate time and attention to do them properly. Thus, handovers should not all be done at the same time, or even very close together. Otherwise, any issues or problems with one will spill over into the other. Also, if one handover needs extra attention, the other players are in place, fully functioning, and can assist.

Board Operator and Supervisor Handovers

The board operator and the shift supervisor shift handovers can occur in any one of four arrangements:

1. Board first
2. Shift supervisor first
3. Simultaneously
4. Randomly, without any need to coordinate their timing

Let us rule out both simultaneously and randomly. If conducted simultaneously, any problem in one would likely create more load for the other. While there would be more hands available, the advantage of more hands is outweighed by the added risk, complexity, and distraction. The choice of random creates its own set of awkwardness in addition to being difficult to create useful procedures in case of problems.

We are left with deciding either shift supervisor first or board first. This choice is not as clear as in the earlier two cases. Considering that the control room handover is likely to be more complex, it is suggested to do that one first. Let the new board operator get fully settled. Any earlier problems and issues would have already been familiar to the supervisor, so he would be up to speed and able to assist the board. Figure 22.5.1 depicts the suggested handover coordination timelines of board operators and supervisors. Note the coordination timing alignment of the two activities at the purple (barbell) line.

Shift Supervisor and Field Operator Handovers

The field operator handover will work better after both the board and the shift supervisor are done. At that time, the control room operations should be well settled in. Later, during the field operator handover, the shift supervisor can assist if any issues arise in the field.

Figure 22.5.2 depicts the timelines for both supervisor and field operator handovers coordinated at the orange (barbell) line. It is after this event that the field operator begins preparing his handover, using the information as needed from the operator shift report and handover results. This enables the arriving field operator to better understand what has happened in the control room (the primary activity) and put it in perspective with what is going on in the field.

Board Operator, Field Operator, and Supervisor Handovers

Figure 22.5.3 depicts the suggested coordination timelines of all three handovers. Here, the board operators and the supervisors coordinate at the purple (barbell) line exactly the same as before in Figure 22.5.1, with the follow-on of the field operators' handover at the orange barbell line of figure 22.5.3. The board operators' handover is complete before the supervisor handover begins, and the supervisor handover is complete before the field handover begins.

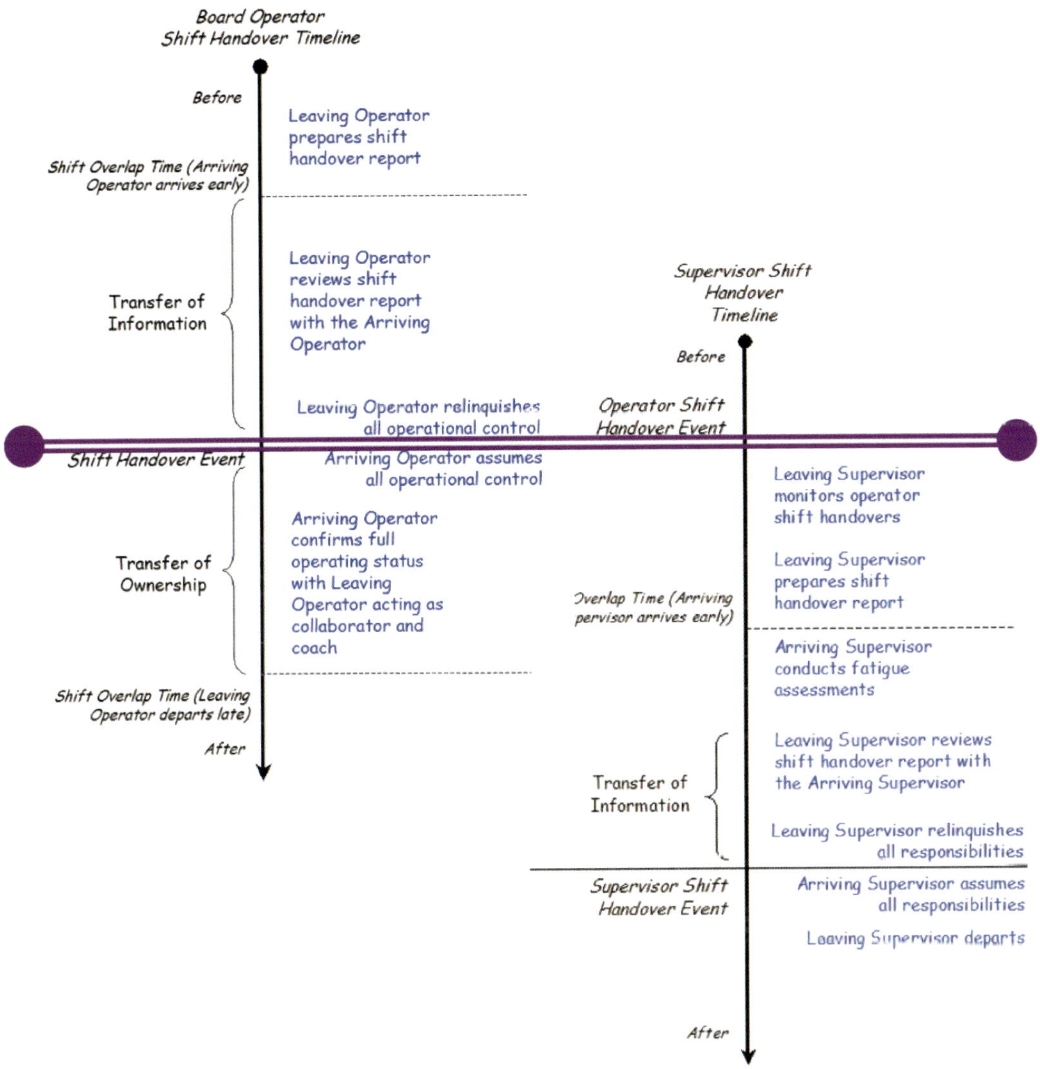

Figure 22.5.1. Combined timelines for both board operator and supervisor handovers. Purple (barbell) line depicts how the alignments of the individual handovers relate to each other. This arrangement allows the supervisors to monitor the joint operations phase of the control room handover, assess the fatigue levels of both operators, and assist the handover in case of problems.

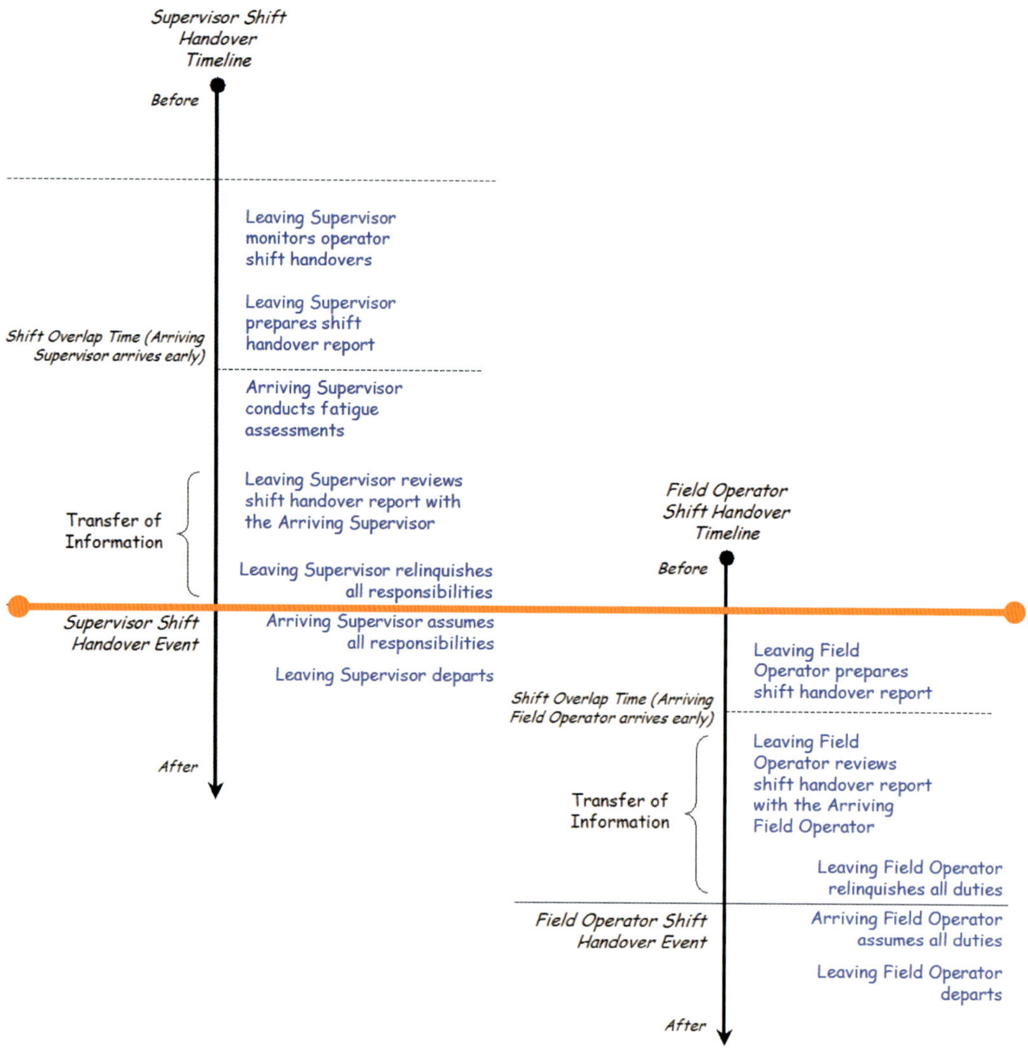

Figure 22.5.2. Combined timelines for both supervisor and field operator handovers. Orange (barbell) line depicts how the alignment of the handovers relates to each other. This arrangement allows the control room to become fully settled before the field does its handover.

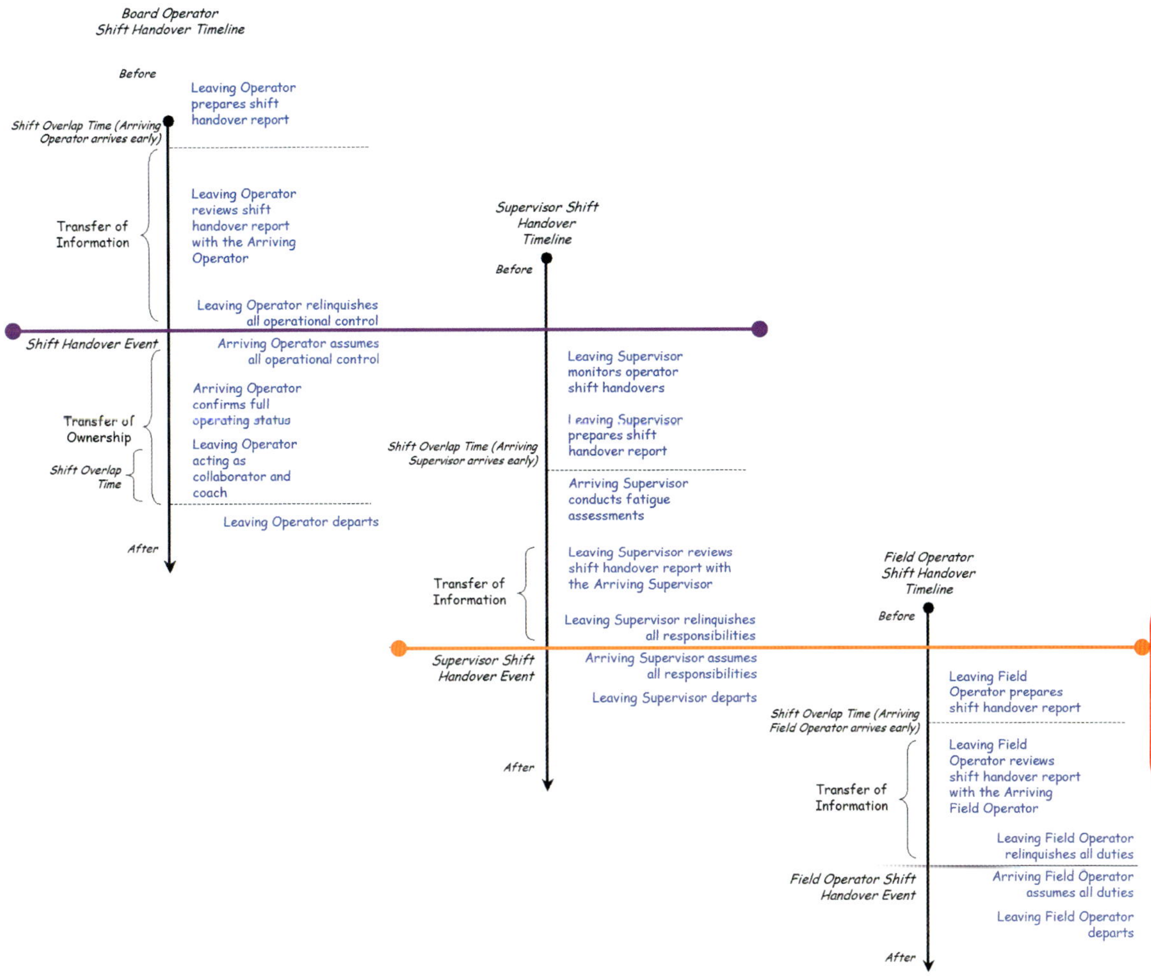

Figure 22.5.3. Combined timelines for board operator, supervisor, and field operator handovers. Purple (barbell) line depicts how the alignments of the control room and supervisor handovers relate to each other. This arrangement allows the supervisors to monitor the joint operations phase of the control room handover, assess the fatigue levels of both operators, and assist in the handover in case of problems. Orange (barbell) line depicts how the alignments of the supervisor handover and the field operator handovers relate to each other. This permits the supervisor to monitor and assist the field handover preparation and activity **[Q9]**.

Non-Contiguous Shifts Handover

A non-contiguous shift is when the next shift does not begin immediately at the end of a previous one. This type of shift is actually quite common for some segments of industry and manufacturing. It occurs when operations are not twenty-four hours a day or when operations are twenty-four hours a day but not seven days a week. These shifts certainly have need for continuity, responsibility, and effectiveness similar to contiguous shifts (24/7). There are two variations: (1) the same operator resumes duty for the next shift as was on duty for the previous shift; (2) another operator resumes duty for the next shift. The actual handover operation is the same for both.

> Shift reports require the same degree of completeness and shift handover events require the same degree of care, no matter the intervening time between shifts nor whether the same individual staffs both shifts.

Upon arrival for duty, the arriving operator (even if he was the same departing operator for the previous shift) completely reviews the last end-of-shift report prepared by the leaving operator. All end-of-shift reports must be sufficiently complete to be fully utilized by another operator different from the one preparing it, even if the same operator is expected to staff the next shift. The resumption of shift will require that present operational status matches the expected status. All anomalies and differences will be examined, explained, accounted for by remediation, if needed, and appropriately documented. The operational plan for the shift will be reviewed, if previously determined, or set if it was not already. If all is ready, operations can begin.

22.6 Maintenance Handover

Maintenance handovers involve coordination of maintenance activities with operators. They go considerably beyond a simple handoff and handback. They involve transferring physical and operational control of equipment from the board operator to maintenance personnel during operation and back afterwards. These changes require following a strict handover process to ensure that both ways are done properly. This handover, while appearing to be quite different from the ones from board operator to board operator, is really not. It may be more limited, but it is no less important. Incident investigations have shown, time after time, when this handover or handback was flawed, serious consequences resulted. All maintenance handovers require close coordination between board operator and maintenance.

Handover

The board operator is the responsible individual since he must properly remove equipment from service before placing it in the hands of maintenance. This responsibility includes directing everything necessary to effectively remove the equipment from functional service. Functional service means that no longer will the equipment be necessary or used for operation. Its absence must not create improper or risky operation. Therefore, all proper work by maintenance must not improperly affect current operation. All enterprise procedures for doing this will be referenced and followed. This is a checklist of functionality, not work procedure [**Q10a**].

1. Board operator identifies all equipment to be taken by maintenance.
2. Board operator reconfigures and adjusts current operations to remove the need for the equipment (and all ancillary parts like piping, instruments, etc.) to be taken from production.
3. Maintenance verifies, using independent means, that the equipment to be taken has been properly removed from production.
4. Maintenance takes (temporary) ownership of equipment.
5. Maintenance takes the proper extra steps (not related to operations) to render the equipment safe and ready for maintenance activities, e.g., fitting of blinds, electrical isolation, erection of physical protection barriers, etc.

Handback

When equipment is to be taken back from maintenance and returned to operational service, the board operator is responsible for the process. This includes directing everything necessary to effectively return the equipment to functional service. Functional service means that the equipment can be used for operation. Any limitations on full use are identified and accepted by the board operator. This is a checklist of process, not procedure [**Q10b**].

1. Maintenance identifies exactly what equipment is ready to be returned to service.
2. Maintenance affirms that the equipment is properly repaired and/or evaluated and ready to be placed back in service, including removal of blinds, electrical restoration, clearing of physical protection barriers, etc.
3. Board operator verifies, step by step, that the equipment is actually ready for service.
    a. This is an active process where the board operator and maintenance check and test and otherwise ensure that the equipment is ready to be back in service.
    b. Note that this involves limited or no verification that proper repairs have been made, just that from the perspective of what the board operator can see and test, the repairs seem correct and the equipment should operate properly.
4. Board operator takes back ownership of equipment.
5. Maintenance leaves the area.

PART 5
WEAK SIGNALS

An innovative new and powerful tool is introduced. The introductory module defines weak signals—small but important indications. It then describes how understanding and mastering them aids the operator to gain situation awareness.

The final three modules in this four-part series describe how to use weak signals, how to get more out of them, and how to troubleshoot problems that may arise with using them.

PART 5 — WEAK SIGNALS

Module 23: Introducing Weak Signals

You have seen the term weak signals mentioned in earlier modules. Here, you will find out what they are.

23.1 Weak Signals and Their Benefits

You notice things throughout your shift. Tank levels go up or down a bit. Sometimes they move more than you expect. Flows change enough so that you may adjust a valve or two. A temperature is a bit higher than it was an hour ago, even though you expected nothing to change. You see reports and documents. Once in a while, you notice a missing word or two, or a date that seems to have the digits reversed. There is a permit to review, and you seem to remember you did it last shift. You see lots of other things: small ones. Most often, these little things seem to be nothing—just normal things doing their usual moving around. Anyway, you are too busy, too tired, or too near shift handover to look more into them now. But even if you did look, you would simply be looking around again. Why waste that time? Anyway, when you did notice, it was natural to think about what it meant. Almost always, you thought you knew what it meant. Maybe you saw it before. Or it was new, but close to what you were thinking it was. If it was useful, you did something about it.

You move from noticing something to concluding what happened almost automatically. But automatic association can be sometimes right or sometimes completely misleading. To operate safely, you need to know which. So here is where you are going to make the mental adjustment. Push the *pause* button. Instead of *seeing and concluding*, now you have a new way to *see and confirm*. A way you can count on.

Let us give those little things you are noticing a name: *small indications*. A small indication is something you run across that may look like nothing much, but maybe you are not comfortable enough to just drop it and forget it. Something there suggests you might need to know what it means. *Weak signals* is a process you can use to examine a small indication and discover what it might be trying to tell you.

Before proceeding further, we clarify a slight issue with whether they are singular, as in "weak signals **is**," or plural, as in "weak signals **are**." The honest answer is we are going to sometimes think singular and other times plural. We refer to them as singular when we are discussing them as the tool. We use the plural when we are discussing the signals themselves if they are more than one.

> Weak signals is a clever but reliable tool for you to evaluate *small indications* of things that you are noticing or thinking about as you manage production.
> It is a way to figure out what these indications are very likely trying to tell you. It is a way to ask yourself: is something important going wrong, yet everything is still fine, but moving around just enough to sense?

Weak signals is a very, very new concept. It is new to the control room, and therefore it is new to you. When you understand and use it competently, it provides a sharp tool to identify situations that otherwise might have gone unnoticed and thus unappreciated. These are situations that may be going amiss with a real potential to cause trouble. Or, after checking into them, you could end up confirming nothing at all is going amiss. Weak signals is a tool in the box. It is just a tool.

The power of weak signals comes from its inherent ability to point to things that may be amiss to help figure out what the weak signals might be trying to telling you. It is done by always *forward extrapolating* to find the worst (biggest) things that weak signals could be telling you are happening. And then *backwards projecting* to look for any evidence that those things are or are not happening. When everything seems to be going right is exactly the time when finding what might not be right is important. There are no benefits from just having a preponderance of evidence pointing to everything being okay. What if no one sees the one thing that might be pointing to something that is not okay? Or, if it were noticed, it did not seem important at the time.

> Weak signals is not used to diagnose. It is not used to remedy. It is a true situation awareness tool.

As operator, you need to know that safe and effective operations are being achieved beyond a *reasonable doubt* [**M15**]. Small indications of doubt can be hard to see. But you need to see them. When you choose to find out what these indications might mean, they become a weak signal to be processed. When all have been processed and nothing untoward remains, they have served their purpose. Weak signals work extremely well for getting around preconceived notions arising from biases [**M16**]. It also helps you understand how and why the HMI screens were designed to help you find problems, both large and small. Weak signals is not used to diagnose. It is not used to remedy. It is a true situation awareness tool. It is used to aid you in finding operational issues early. Its single purpose is to identify and confirm potential operating problems.

Weak signals is very useful. But as powerful as weak signals is, please keep perspective. Seeing small indications and then choosing some to investigate as weak signals is a tool, not a responsibility. This means that a failure to notice any given small indication, or deciding not to process it as a weak signal after seeing one, does not in any reasonable way suggest a lack of due diligence on your part. If any small indication was known to be that important, the designers of the enterprise had a duty to have identified it and arranged for definitive and reliable ways to let you know.

There is nothing weak about using weak signals. It is a robust capability. The term weak implies that subtle changes or cues can provide relevant clues. These clues offer important avenues for assessment. But they first must be identified. A weak signal is:

> "A seemingly random or disconnected piece of information that at first appears to be background noise but can be recognized as part of a significant pattern by viewing it through a different frame or connecting it with other pieces of information." [23.1]

This tool is easily incorporated into your existing control room operations [**M21.2**]. It can fit seamlessly into shift changes and periodic walk-throughs during the shift. It can be used, as needed, when troublesome things seem to sneak up. Weak signals works to supplement alarms and expose much that might be subtle, inconspicuous, or otherwise elusive to you as you work to maintain adequate situation awareness. Weak signals does not take the place of everything else. Rather, it builds upon and depends on the rest.

23.2 Seeing Weak Signals

Weak signals are not alarms. They do not sound when needed, gain your attention, and provide most, if not all, of the necessary guidance to manage. They are subtle and vague. Every weak signal starts out first as a small indication. When you make the choice to find out what it might mean, it becomes a weak signal. For the most part, a weak signal comes from considering an off normal, but not yet abnormal, small indication. This is important. The word "weak" is used because it is not easy to see any underlying condition that may be present. You might happen upon them or discover them intuitively during dedicated activities designed for looking.

The more conscientious you may be, the more potential for all of this to cause anxiety, concern, and maybe overload. This is not to say that an underlying situation could not be dire and heading for disaster and havoc. It means that whatever that situation might be now, it is not currently easy to see. So, yes, a weak signal may lead to abnormal. But even in the extreme, what might be visible is only a hint of an indication. All the more useful to be able to find out early what is going on.

Please accept that looking for weak signals needs to be a part-time activity. It is not intended to occupy all of the time left over after doing the main activities of your job. If you happen across something that seems not to look right, do the weak signals activity. At planned times during the shift, actively look for them. Do a careful review as you prepare for shift handover. During the actual shift handover, do a re-review with the other operator and follow through them together. When you have done all this, please try your best to be comfortable with not spending any more time looking or worrying about them!

Weak Signals Announce

So how do weak signals actually fit into the picture? We have an intuitive understanding about what a weak signal actually is:

> "A 'weak signal' is something which, in and of itself, may be relatively insignificant and does not [necessarily] justify action.
> "Sometimes however; weak signals are early indications that something is indeed wrong." [23.2]

If there is a very significant problem in the making or danger lurking about, some small parts of it are going to "leak" out. The bigger the problem, the more likely there are to be leaks. Pictorially, it might look like a leaky bucket. Some possible leaks are shown in Figure 23.2.1.

Some of the small indications you might notice include an odd smell, an error in an instruction, or an uneasy feeling. You will see how to decide whether that small indication needs more attention or is just noise. When you follow up on each, you will be working it as a weak signal. This can be important. By the time it becomes clear to you that you are facing a serious problem that demands your best (at times, heroic) efforts to resolve, successful remediation may already be beyond reach. By the time the loud call of danger is heard, the problem may have reached beyond the tipping point. [23.3]

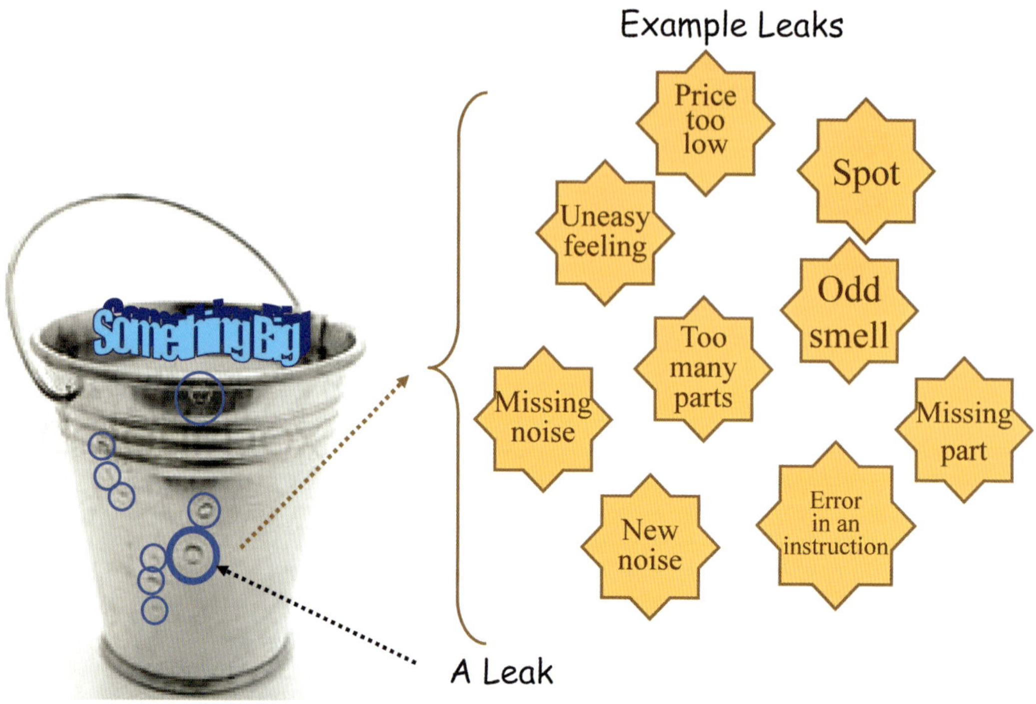

Figure 23.2.1. Big problems always have small bits and pieces (call them small indications) "leaking" out because the problem is so large. Working them as *weak signals* means finding the small indications and then using weak-signal analysis to figure out what they might likely mean.

> "[A tipping point is] the point at which a series of small changes or incidents becomes significant enough to cause a larger, more important change."[23.4]

The best way out is to know earlier. The only way to know earlier is to plan and implement effective ways to see them earlier.

23.3 Weak Signals Work Process Preview

We now take a quick look at the weak signals work process itself. It is going to be a quick overview. It is not intended to explain the whole thing; it is just to give you an idea so that getting started makes sense. Please do not worry. You will get the full details later of exactly how you work them. It will make complete sense because of the natural way it works. It is not complicated.

There are six basic steps, as illustrated by Figure 23.3.1.

Step 1: Notice something that catches your attention.

Call this a small indication.

Step 2: Decide that you are going to look into what it might mean.

Now it becomes a weak signal to investigate.

Step 3: Imagine that the weak signal is a "leak" that implies something very important is happening. It may be one or even more things. Imagine specifically what each of the **very important things** (problems) might be.

This is called a forward extrapolation.

Step 4: For each very important problem (in Step 3), decide what you need to look for—which, if you find it, means that the very important thing is actually happening (or not).

Since you know (in your mind) what is the worst that might be going on, you know what to look for to see if it really is happening. This is called a backwards projection. You are looking.

Step 5: For each important happening in Step 3, one at a time, look for evidence (identified in Step 4) that it does, in fact, exist (you look for any confirmation) or it does not exist (you find nothing suggestive of confirmation).

This is called finding evidence.

Step 6: Based upon what you find that is believable by a preponderance of the evidence, decide that (a) if you find confirmation, then the "really important thing" you were looking for is happening or about to happen, or (b) if you cannot find enough evidence, then either the really important thing is not happening (the small indication you saw was just everyday common noise) or nothing has gone enough wrong yet to really be of concern.

You have finished: you know what the small indication is trying to tell you. This is called reaching conclusions (a situation awareness judgment).

Here is what it looks like.

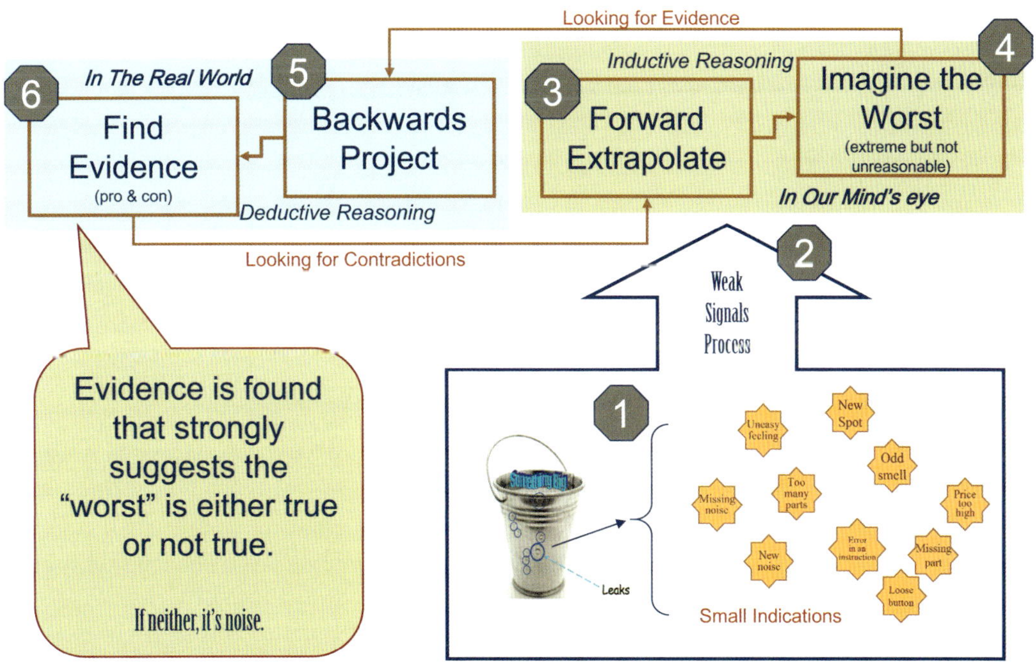

Figure 23.3.1. Overview of the six-step weak signals working process.

Remember, this is just an introduction. The next module [**M24**] goes into detail about exactly how this works and how you can use it. You really do want to wait to see the whole picture before jumping in.

23.4 Weak Signal Concepts

Small indications for weak signals are ever-present. They are visible to those who are sensitive to them. There are two important weak signal concepts: *importance* and *dominance.*

Importance: A weak signal can either announce a problem with maintaining the status quo or warn of a serious future event in the making. It is an operational necessity to identify which is which.

Dominance: "No weak signal ever rises to dominance by itself." [23.5] It is the co-occurrence of the weak signal and the potential exposure to risk that can lead to serious operational situations.

Said a different way, it is extremely rare for any weak signal to evolve so that it appears to become stronger and stronger. They stay weak; but there may be more and more of them as the underlying situation matures (gets worse).

What Weak Signals Look Like

Weak Signals always start first as small indications that appear in an operating observation that seems slight, small, or otherwise nearly imperceptible. Some are not considered useful enough to follow up. You cannot be expected to follow up on every hint or suggestion that something might be amiss. There needs to be a balance. You follow up on only the ones that you feel need it.

When different individuals work jointly, it becomes important for them to communicate and work together as a team to process what weak signals mean.

Expectations Interfere with Weak Signals

Weak signals and expectations do not mix well at all. You need to work each step in proper order and trust the working process. No shortcuts. Weak signals is a discovery process. Avoid looking directly for an explanation of what it means by matching a specific weak signal to a conclusion—this will almost always lead to a *confirmation bias* [**M16**] error.

Weak signals are not meant to be viewed or sorted in any probability order. "Most probable" or "most likely" are not what to look for or assess. What comes to mind is the working process used by the mythical character of Sherlock Holmes, created by Sir Arthur Conan Doyle. [23.6] It is the perfect thought model for what we seek: "When you have eliminated the impossible, whatever remains, however improbable, must be the truth."

Sherlock Holmes actively investigated anything that appeared to be just a bit unusual. Anytime there were none, he knew he was not looking hard enough. First, **observe**. Only **after** all of the clues are noted does the Holmes' deduction engine start working. We ask you to use the same plan. Detection clues are designed into the HMI. This is part of the power of situation management. The careful design of the HMI is explicitly done in a way that exposes as many of the irregularities (*small indications*) as good understanding of the enterprise will permit. Being able to understand keeps you better focused on finding out what might be amiss without getting in the way of your keeping an open mind to gather confirming or disconfirming evidence for it.

Weak Signal Analysis Models

To do weak signal analysis, you are going to use good mental models of your specific enterprise [**M9**]. Those mental models come from your experience and training. All of weak signal analysis depends on your having a really good understanding of how your enterprise works and how it can get into trouble. This is fundamental.

> All of weak signal analysis depends on your having a very good understanding of how your enterprise works and how it can get into trouble.

When you see anything of concern, use your deep knowledge, your understanding, to "suppose" what the real enterprise might be doing at its limits. You then use that understanding to suggest where to look for confirming (or disconfirming) evidence. You continue to use that understanding to evaluate what you find when you carefully look. Weak signals reinforce your need for good enterprise understanding. It relies upon it.

23.5 Weak Signal Characteristics

Weak signals will:

- be based upon something real, something easily related to something real, or something felt that could not be readily dismissed.
- depict a reasonable difference between seemingly normal and seemingly off normal.
- look normal when things really are normal.
- indicate likely off normal when there is a reasonable likelihood that things can be really off normal.
- accumulate, in a reasonably repeatable way, as additional off-normal situations add more clues for the new situation that is related to the same or closely-related, earlier off-normal one.
- have the ability to be forward extrapolated (to be fully explained later on).
- be backwards projectable to useful opportunities for investigation (also to be fully explained later on).

23.6 Weak Signal Types

Weak signals come in three types: *immature*, *parametric*, and *transformative*.

Immature Weak Signals

Immature weak signals are what natural variations are made of. Noise. They could be normal variations in energy or raw materials. They could be control loops that are working, but could be improved with a bit of analysis and effort. They could be your tinkering a bit too much. So far, it is all a matter of degree. Just as in real life, noise can be just noise. Or it can be something more. If it rises out of the noise box, then something is really happening. Deciding when the noise is more than just noise is divided into two parts:

- The first part is making immature weak signals visible. This is aided by context *icons* and *dashboards* [**M18**] and, to some extent, the *alarm system* [**M19**].
- The second part is the need to see whether they actually are immature weak signals or something else. If an immature weak signal matures, it should mature into a parametric weak signal.

A weak signal does not become more apparent with time. *Maturing* means that if you find confirming evidence, it strongly suggests that an abnormal situation is actually developing. If it does not lead to confirming evidence, it stays *immature*. It is rare for something very significant to suddenly rise out of noise. It is really noise. If you remain concerned, put it on a "post-it" note to remind you to keep an eye on it for a while.

Parametric Weak Signals

If it is not readily clear that your immature weak signal is noise, you should assume it to be a *parametric weak signal*. These are close to, or have crossed the boundary of, customary and usual. Their importance is that they are likely to be moving towards trouble. These signals represent process conditions that are not normal and likely to change in ways that have a proportional impact. Usually, they are individual variables, but not always. Small, slow changes should have correspondingly small and slow changes in operational vulnerability. Faster changing or larger changes of weak signals (meaning more of them; easier to find evidence) suggest faster or larger changes in operational vulnerability. Their detection requires good understanding of process operation and the careful setting of the mental boundary between noise and what is potentially important.

If confirmed, those situations should represent manageable problems. Depending upon the enterprise operating policies and the available operator resources, early action would be suggested. That is good operation protocol. This also underpins the operation of the alarm system. The alarm system exists to ensure that no weak signal capable of maturing to cause problems is allowed to cause problems, if these problems can be avoided or managed by earlier operator action. The other alternative is to wait until the situation actually becomes abnormal. An eventual alarm is our best hoped-for result.

Transformative Weak Signals

As in the case of parametric signals, transformative weak signals are the weak signals that are close to or have already crossed the boundary of customary and usual. Of special concern is that they are likely to be **moving very fast** towards trouble. Their early detection requires good understanding of process operation and the clear understanding of the boundary between noise and potentially important. These signals represent process conditions that are not normal but likely to change in ways that have a **dramatic** and **unpredictable** impact. Even small, slow changes are expected to cause significant changes in operational vulnerability. Faster changes or larger changes announce that the enterprise is imminently exposed to potentially serious operational vulnerability. Those situations could represent manageable problems. However, since these weak signals represent actual, very fast-moving abnormal situations, figuring out and implementing proper corrective actions would need more time than is available. The best case would be that an alarm was configured for it. Once in alarm, the alarm response procedures of your enterprise become active [**M19**].

23.7 Weak Signals Indications Come in Categories

This entire module is about small indications of trouble. So let us firm up exactly what small means. Here are labels for four different categories of indications. The first category is *not visible*. Next is *barely visible*. This is followed by *visible but not apparent*. Finally, there is the *evident and distinct*. The fourth category is not weak signals, nor is it small indications. They are included in the categories to illustrate that, while they differ, all are useful and can be processed within a consistent frame. See Figure 23.7.1.

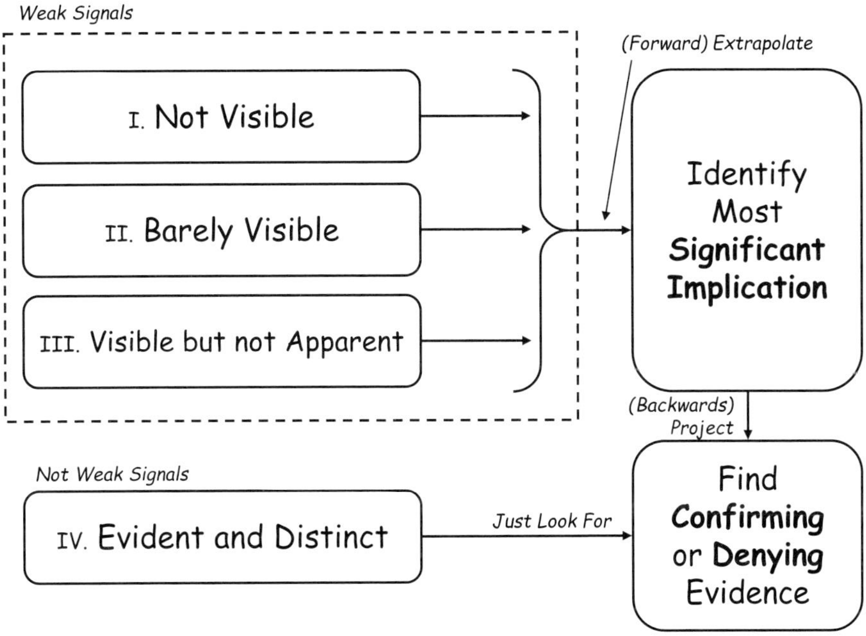

Figure 23.7.1. There are four categories of signals observable by operators. Here is how they are used to find abnormal situations needing attention. *Not Visible*, *Barely Visible*, and *Visible but Not Apparent*, are all handled as weak signals. The fourth category *Evident and Distinct* is not. The weak ones are extrapolated and then projected to find evidence. The last is only projected to look for evidence.

Not Visible Indicators (Category I)

Not visible indicators, category **I**, are just that. No one can actually see them. They are things like hunches, feelings, and indistinct or vague concerns. They emanate from the back of our mind. We do not actually see anything. When you try to focus on where they might be coming from, you may see reminders or thoughts. Or you are reminded by nothing at all. There was nothing to notice. You just sensed it. Something did not feel right. Or it felt too right. Everything was vague. It lingered.

Barely Visible Indicators (Category II)

Barely visible indicators, category **II**, means there is something you actually observe that is out of the ordinary. You thought you heard something. You thought you saw something. Or you actually did hear or see something, but it was vague or indistinct. Nothing is clear, but something seems there. A valve is a bit more out of its expected position than you thought it would be. Something just does not look right. Something you see is off. You notice. You have a place to put a finger, but you are not sure it means anything.

Visible but Not Apparent Indicators (Category III)

At this level, visible but not apparent indicators, category **III**, you know specific things. Maybe a lab report is missing. You see a valve closed that should have been open. You read a message and understand what it is saying, but not yet what it means. You see a reading that is definitely out of normal even though you may have doubts about its accuracy. You know that something is (or a few things are) not where it should be. What you do not know is why. And you do not know what it means.

Evident and Distinct Indicators (Category IV)

These category **IV** indicators are evident and distinct. You can see enough to be confident that something is amiss. A spare pump will not start; maintenance left the scene early; a downstream source reduced its take by 20% without notice. You know you have a problem, a situation, or an opportunity. You are not totally sure, but know this one is not going away. You need to identify what is really happening.

23.8 Getting Ready for Weak Signals

The next module, *Using Weak Signals* [**M24**] covers how you will actually use weak signals for situation awareness, finding out if they mean anything or not. It will explain the work process in detail. *Getting More From Weak Signals* [**M25**] provides ways to find them from alarms and the HMI. There are other places, but these two are the primary ones. *Troubleshooting Weak Signals* [**M26**] ties things together by covering the many ways to enhance your working process to troubleshoot your work through some of the rough spots.

Module 24: Using Weak Signals

24.1 Weak Signal Management

You now have a good starting idea of what weak signals look like from the previous module. It is time to learn how to use it. It does not take much imagination to see that lots of weak signals could be very easy to find anywhere you look. While they do need to be easy to find, you do not want to find more than are there. This is a skill, not an obsession or mission. This is why your HMI tools and views are carefully designed and implemented with useful clues. It is a major reason for style guides and consistent design.

Every chosen small indication requires processing. The single objective is to confirm there is sufficient evidence to reasonably know what the indication means. Weak signals involves forward extrapolation to potential significant implication and then backward projection to find confirming or disconfirming (denying) evidence. When you find evidence, you know either something is happening or, alternatively, not happening. When you do not find confirming evidence, you know nothing related to the small indication suggests a problem. At least for now.

The Work Process

The weak signals management work process is actually quite direct. It starts with the need for an effective way to see small indications. Once you select a small indication, you will follow this work process to examine it as a weak signal. It is a six-step procedure, illustrated by Figure 24.1.1.

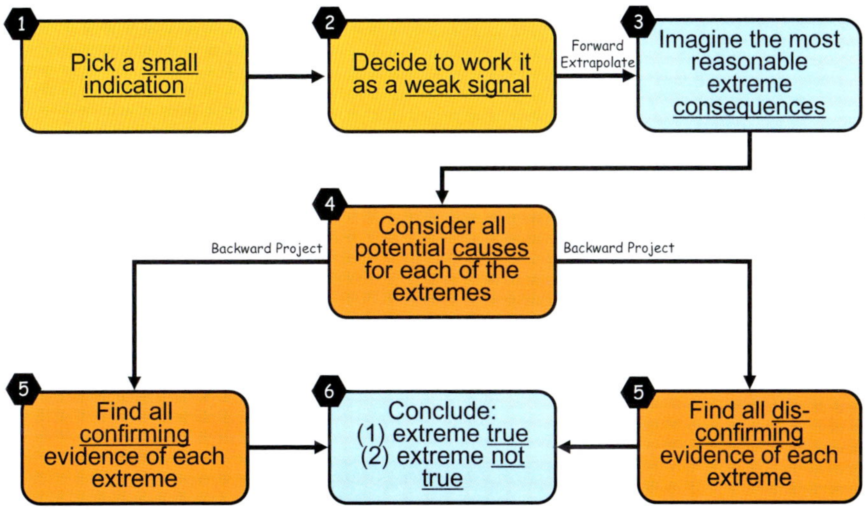

Figure 24.1.1. The *weak signal* six-step working process to prove or disprove whether a *small indication* means that there is something going on that needs attention. If neither, it is noise **[Q11a]**.

From the list of your small indications that you select to treat as a weak signal, do the following [**Q11b**]:

Selection

Step 1: Pick a small indication that you notice and decide might be useful to you.

You have seen something, thought something, or otherwise become concerned about something small (for now) and are curious about it.

Step 2: Decide that the selected small indication might be an indicator of something you want to know more about.

Take that small indication and consider it to be a "clue" of something important that is affecting your operational area. You now make it a weak signal, by definition.

Forward Extrapolation (to Find Significant Problem)

Step 3: Using the process model you have in your mind, imagine what is the reasonable but most significant concern that might be happening if this specific supposed weak signal is actually a true early warning.

Remember that you are looking for the significant thing (quite bad; really unwanted) that might be happening if it is not fixed or worked on.

If nothing can be imagined (identified), then you are probably forgetting something important about how the enterprise you are managing works and how it gets into trouble. Rethink things. Get help.

Backwards Projection (to Find Evidence)

Step 4: For each worst thing that might happen, make a list of all potential causes (possible things that go wrong) for that imagined bad thing to actually be happening.

In Steps 1 and 2, you have noticed something unusual. In Step 3, you were able to imagine something that your enterprise was heading for and is important to know. Now, you imagine all of the ways that that "something important" can go wrong and give itself away to show the bad thing actually happening. This list includes your best ideas for what could have started to go the wrong way earlier. And if they did, the situation would eventually materialize as the significant problem you envisioned. Make that list here.

Step 5: For each potential cause on your list from Step 4, look for everything that would confirm (or disconfirm) each supposed thing is actually happening now.

*Use the list from Step 4 above to tell you exactly **what** to look for. Now figure out **where** to look for specific evidence to prove or disprove whether or not each possibility exists.*

Confirm Problem

Step 6: For each "confirmed" weak signal in Step 5, you have identified a confirmed problem you will need to resolve. You have reached a conclusion.

Confirmed means that the weak signal was noticed, forward extrapolated to a worst bad thing, backwards projected to look for evidence, and actual evidence was found. So the original forward extrapolated problem is probably real. It will need attention and/ or action. The weak signals part is complete. If there is no confirmation (or there is dis-confirmation, which means that the bad thing really is not happening), then the weak signal that started this whole search is classified as immature (for the moment). Let it go for now. The activity to follow for a confirmed weak signal is covered in Situation Management [**M29**].

This ends the weak signals work process overview and activity. Remember that it is only a situation awareness tool. It will not diagnose a problem. It will not advise you about what to do to resolve any problem.

Why it Works

There are two very powerful reasons why this actually works. First, it is a straightforward way (via forward extrapolation) to identify early potential problems for investigation. It does so by not asking you to directly figure out what it means. Rather, you move your focus all of the way to possible end situations. For most of us, this move-to-the-end will reach over our preconceived notions, guesses, and suppositions. Second, it prepares the way (via backwards projection) to know what to look for and where to do your looking, to investigate the possible existence of the specific potential problem. Now you know where and what to look for. *Backwards projection* is the way to find underlying evidence for the problem. This is a focused way of what to look for and where to do the looking. If you find enough evidence, it will confirm that something really is going wrong—or confirm that it is not.

Before, without a small indication suggesting something might be amiss, you were likely unaware that something might be going wrong. Problems may have been there, but you did not know to look. Now you know. "Looking" this way, you will have specific things to look for and find, if there. You will be using weak signals as the "announcers" of a problem that just might be happening. You are looking for all indications that might be suggested by the specific current content of HMI icons, dials, dashboards, or by other observation.

Degrees of Evidence

Let us visit the reality of what you might find when you (backwards project to) look for evidence. In other words, how convincing is any evidence you find? When you looked (*backwards projected*), you might have found a clear indication that the supposed worst situation was actually true. The evidence you found should be enough for you and others to conclude that the supposed worst situation is exactly what is going on right now. No doubt. No ambiguity. In fact, had anyone seen or observed this from the outset, it would have been clear what is happening. It worked out this way because the small indication that started this off really was an abnormality that "leaked" [**M23**] out of the actual worst situation. You noticed it because you had a very good understanding of your enterprise and were curious.

> Evidence cannot be weak. Confirmation (or disconfirmation) must always be convincing.

To recap, you do a backwards projection to look for confirming (or disconfirming) evidence of an abnormal situation existing and needing to be found. What you find can be strong (clear and evident) or barely sufficient (just convincing enough). On the other hand, what you find may only be enough to cause you to feel unconvinced, but not so much that you can leave it alone entirely. Or what you find is pretty much useless (inconclusive), so that you either feel that you might have done something not quite right—or there is nothing there and you are imagining the whole thing. You must find only believable evidence. Care needs to be exercised so that only the weak signals are the weak part. Evidence cannot be weak. Confirmation (or disconfirmation) must always be convincing.

No Evidence Found

If no evidence is found (inconclusive, or actually nothing), it is likely that no real problem exists. An example of evidence might be a specific flow rate that is trending higher than normal, but is not yet abnormal. Or it is abnormal, but you did not notice it until weak signals suggested a place to look. Another example is a temperature profile you would not even have considered looking at before. Now that you are looking, you see a more pronounced bend where only a slight one should be. This can mean that a separation column might be malfunctioning. Since you are looking for evidence that it was not going well, the bend might be enough evidence. These are reasonably definite things. You get the idea. On the other hand, if you cannot find evidence to either confirm or disconfirm, it most often means that that small indication that you examined as a weak signal represented nothing more than ordinary noise. Or it might be something, but it is way too early for it to be really "leaking" out. Either way, for now you have found nothing. Move on.

Only Weak Evidence Found

What if only weak evidence is found? What must you think when all of the evidence from doing a backwards projection seems itself to be weak? Yes, you did find things that did not seem to be where you expected them to be. And they were enough out of the usual to raise suspicions, but not strong enough to confirm the problem you were looking to confirm (or disconfirm). What now? Do you have anything of value? The answer is "maybe." What you actually have is not evidence. What it is, instead, is a new set of weak signals to work with. Work them just as before. We call this a two-cycle weak signal process [**M24.4**]. You will treat them as new weak signals. If nothing clear comes from doing that, stop working this as a weak signal. Something might be going on, but you will need another way to find it. You still need to know. Collaboration may not be enough. Escalate the situation [**M34**].

Categories of Evidence

Looking for and finding (or looking for and not being able to find any/enough) evidence is a critical step. But it must be good evidence. Even if you are great at imagining what a weak signal might mean when you forward extrapolate, if you are hijacked when you look for evidence, this whole thing is not going to work. Evidence hijacking will occur always when evidence you find is not *actual*. Each step of weak signal processing is independently important. Watch out if you are careful with forward extrapolation and equally careful with backward projection, and then let your guard down and get sloppy looking for evidence.

Actual Evidence

Actual evidence is what you can directly observe from BPCS data:

- on screens, logs or reports.
- from laboratory reports.
- from your outside operator reporting evidence that he sees.

Fabricated Evidence

Evidence is *fabricated* when it is correct and actual, but for someplace else. It does not apply to you. Somehow it made its way to you and you found it. The test for *fabrication* is to carefully check everything closely around it to make sure that it all makes sense and should be the way you see it. If in doubt, drop it.

Forged or Counterfeit Evidence

Forged evidence is the stuff that you can observe but is not true. It could be the result of deliberate errors in reports or BPCS data or an outside operator being too busy or stressed so it is made up. The test for *forgery* is what you do for all information: ask yourself it makes sense and fits into other things to which it should be closely related. For example, if a temperature reading does not look right, and it is not going to harm a touching hand, ask someone to touch it and verify.

Never Assume the Problem

The first step in weak signal processing is to ask, "What is the worst reasonable thing that this weak signal, when considered by itself, might suggest could be happening?" That question is very different from "What do you think might be wrong?" It is too early for the "what do you think is wrong" question. Even thinking about it that way invites preconceived notions. Asking it is too susceptible to pitfalls that mislead. This type of shortcutting is not a good idea. Here are a few thoughts as to why:

> "Humans tend to judge too quickly when presented with ambiguous data; we have to work extra hard to consider less familiar scenarios.
>
> "Whenever multiple pieces of evidence point in opposite directions, or when crucial information is missing, our minds naturally shape the facts to fit our preconceptions." [24.1]

A misled operator leads to missing or mishandled important problems. The weak signal work process is designed to overcome these very natural biases and pitfalls. Work it as designed.

24.2 Finding Weak Signals

Small indications can be derived, built, or exposed from a variety of places. They can be anywhere and everywhere. They can appear even if nothing much is amiss. Real enterprises can vary their performance as raw materials change, weather changes, equipment ages, valves stick a little, pressures vary, and all manner of other things move a bit this way or that. Mostly, these are normal variation. Yes, you use control systems to try and keep things steady. You try to operate with as much predictability and sameness to keep the variations small. However, variations, and their cousins, disturbances, seem to be everywhere. Some will be obvious. The less obvious, the ones you actually do look for, can take a lot of time and effort to find. After finding them, each will need to be looked into to discover what they might mean. It all sounds promising. Seeing everything might at first appear to be a gift, but on practical reflection, is really a never-ending task. Clear understanding is needed.

Consider two ways to do this. First, you will want to be receptive to those small indications that come to be noticed during the normal course of doing what it is that you actually do: checking on production, operating values, reviewing laboratory reports, preparing or validating work orders and permits, and all the rest. Second, we suggest that specific times during the shift be reserved for making weak signal rounds of the plant or enterprise. *Managing the Shift* [**M21.2**]

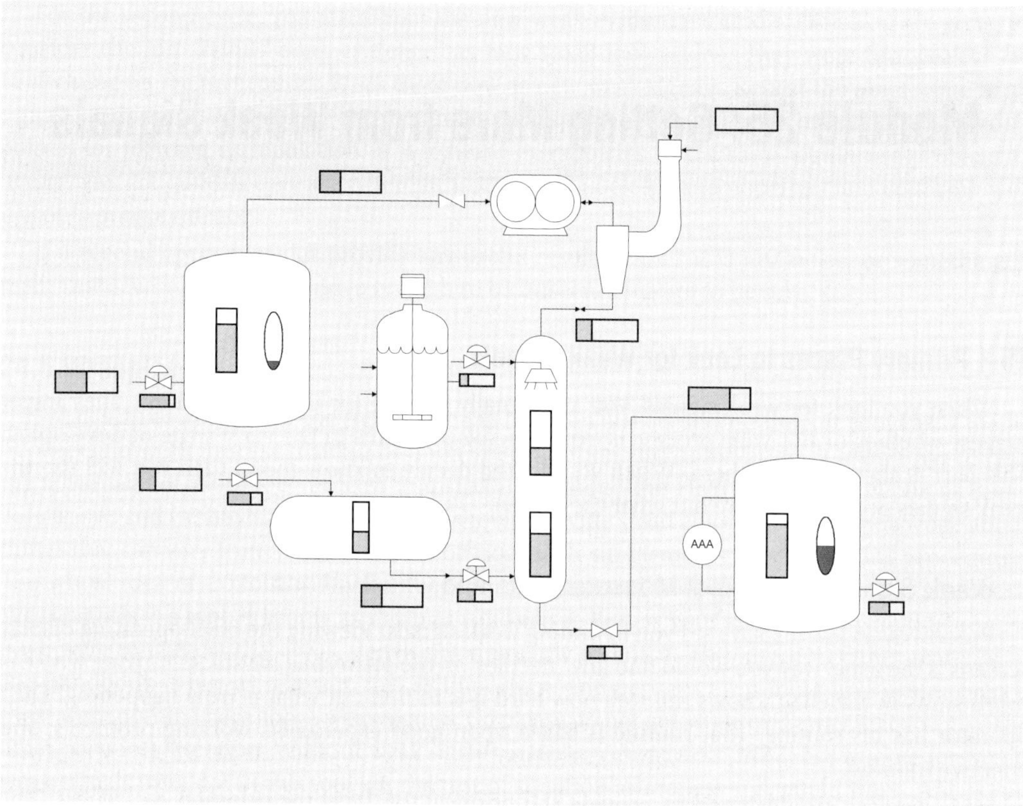

Figure 25.1.1. Current operation display with icons for key operating variables shown. The key variables are depicted by the graphic rectangles and ovals with embedded value bars. This encourages the viewer to readily see what appears to be off normal or abnormal in context and in relationship to other key parameters.

The Role of Icons, Dials, Gauges, and Dashboards

Icons, dials, gauges, and dashboards (IDGDBs) are the visible front lines for weak signals. These terms were explained earlier in *Using the HMI* [**M18**]. Their design would include as many potential small indications and other awareness information as can be identified by reasonable means. You will use IDGDBs to help find potential problems. Remember, all non-normal situations that are not exposed sufficiently will be very difficult for any operator to find. Please realize that the examples shown in this handbook are provided to illustrate the tools and concepts. They are not necessarily to be considered as the only way or even a best way.

25.2 More Places to Look for Weak Signals

Now we build on what you have learned about weak signals in order to add more depth and insight. There is also a helpful module on troubleshooting [**M26**]. When all weak signals have been processed and nothing untoward still remains, they have served their purpose. Here are a few more useful places you can look for small indications.

Weak Signals and Checklists

Checklists (*Operator Tool Set* [**M8**]) can be extremely useful to reduce the chances that nothing important for a task is left out, incomplete, or done in the wrong order. Their worth has been proven in everything from aircraft cockpits to hospital surgical operating rooms. Even with checklists, there is ample room for small indications to arise. Look for conditions that might be off normal. Look into them for possible problems, because even activities as formalized and straightforward as a structured checklist can be suggestive. Normally, each item on the checklist appears as a binary ("yes" or "no") choice. Real life is not always that sharp and clear. As an illustration, consider the following line item on a checklist:

☐ **Turn on the auxiliary pump and make sure the discharge pressure is over 42 PSIG**

The line is checked after the pump is turned on, and the discharge pressure is observed to exceed 42 on the gauge. But suppose that the pump was observed to be slower starting this time than normal? Or that the pressure was above 42, but just barely; or this time, it was moving around from 48 to 70, where such movement is not usual. This is where a small indication can alert you.

Intuition and "Raw" Information

Intuition is an interesting thing. We all have it. Not all of the time, and likely almost never when we really need it in our personal lives. Nevertheless, we do have it. Most of the time, it is useful. But in the control room, for purposes of situation awareness, intuition must usually be put aside. The reason is quite simple and fundamental. You are, by the very nature of being in control, proffered lots of information—information from sensors, analyzers, and the like. You want it that way. You want to be able to see everything that is going on. You are expected to be able to find what is right and what is not right within your enterprise. You want to find what is actually there, but nothing more. You want to look, but not before it is appropriate, and find the meaning of what is of concern.

There is one exception to putting intuition aside. If you are having trouble dismissing something you sense or feel about what is going on, and you really cannot let it go, consider treating it as a weak signal. When you do and if you find something, then your intuition was a useful guide. If you find nothing, let it go for now. Letting it go means that you will not use that nagging feeling to get in your way of thinking about the rest of your work. Come back to it later if it still bothers you.

You will also need to use the alarm system as intended. You will need to use the weak signals tools as they are designed. These two capabilities, the alarm system and the weak signals tools, used together will form the backbone of detecting abnormal operating situations in time to manage them.

Persistent Weak Signals

What if a weak signal or a few, either related or not, seem to hang around for a while? You see that each is definitely weak—a suspicion, not definite—yet noticeably is there, and it is there for a period of time that has been longer than just some transient thing. The problem is that when each was processed earlier using the weak signal tool, nothing came of it. Each was judged immature and would have been relegated to either noise or put on an informal watch list. What now? How should we use this situation to advantage? A persistent weak signal that is also too weak to confirm (or cannot be disconfirmed) might suggest one of three things. The persistent weak signal is:

1. Just part of the regular variations real plants can exhibit during normal operation
   While the magnitude or noticeability can be at the edges of normal, nothing is clearly visible or can be found upon looking.
2. At the very early beginning of a situation or problem
   Too early to know whether or not it might fix itself or need later attention
3. A situation or problem that is progressing and likely to be important, but not necessarily fast enough to be clear now
   While it is too early to know for sure, this case needs clarification to make sure it is recognized and work is started to ensure proper understanding and/or management.

The weak signals work process considers persistent weak signals to belong to the third alternative above, until ruled out. It is important to make sure that any persistent weak signal is not suggesting anything important. From the point in time when a weak signal is classified as likely persistent, you must go through the activity to make sure that trouble is not actually lurking and, hence, must be dealt with. So how do you do it? It is not enough to simply take another look to see if forward extrapolation and backwards projection will hopefully turn up anything different from the first time. To be sure, you can do it again. If nothing shows up, do not force it. Now is a good time for collaboration and/or escalation [**M34**]. It needs to be done without your taking the lead role in the thinking. Let the partner collaborator or escalatee lead the activity. Keep as much independence as you can here.

If the weak signal still persists, or goes away and then returns, without anyone being able to provide a sufficient explanation, it is time to get to the bottom of it. Involve engineering. Call in experts. Leave the explaining to them for now. Keep an eye on it while they are working.

25.3 Weak Signals as Flags

Each weak signal also can be thought of as a flag (like those that athletic sports referees use to identify a player error in American football). During your operation, if the number of active weak signals starts to increase, you should assume that something unusual might very likely be happening. We generally use three as that number. This situation is particularly important when you feel that nothing important is wrong. The more flags, the more likely something truly is going amiss. Watching for and understanding accumulating flags is a useful operational practice. We need to find out whether these signals are growing in a related way or just numerically. This can be especially important when responding to an alarm. It might suggest that a wrong or ineffective approach is being used. It might suggest that another important problem needs attention.

> If small indications (potential weak signals) begin or increase during alarm response activities, you must pause and ensure that they do not relate to what you are doing or suggest problems with your managing of the alarm situation.

When the number of visible *weak signals* grows, each individual weak signal takes on an entirely new importance. So important, that when the number of *flags* grows to three or more, we expect you to pause your work and directly begin (in the order listed) the following: [**Q18**]

1. Inform your supervisor at once.
2. Do the *flags* appear to be related?
   a. If *no,* continue working each.
   b. If *yes,* the relationship should directly suggest a new problem. Add it to your list. *Escalate* if needed.
3. Reconfirm that each current *flagged* problem you are working on is an actual problem.
   a. If *yes,* and they are getting resolved, continue work.
   b. If *yes,* and they are not getting resolved, either the solution is inadequate or unknown problems exist that need to be found.
      i. If no more problems are found, then the working solution must be modified or changed to fit the needed situation (this usually requires *escalation* or invoking *permission to operate* [**M35**].
      ii. If more problems are found, identify them and work them.
   c. If *no,* stop work on the wrong problem and look for more than one concurrent new problem that has not been worked on yet; find each and work each. *Escalate* if needed.
4. Is the number of *flags* growing?
   a. If *no,* then continue work to resolve each.
   b. If *yes,* then other problems are becoming important; *escalate,* find them, and work on them.
      i. Invoke *permission to operate* if you are running short of time.
      ii. The problem is getting worse much faster than expected or more problems are arising, AND you have time.
         1. Reassess to make sure the problem being worked is still the real problem.
         2. Reassess to make sure the solution being worked is the correct solution for the supposed problem.
5. Is the current number of flags diminishing?
   a. If *yes,* continue working.
   b. If *no,* then the wrong problem is being worked or the wrong remediation efforts are being used. Reassess the plan and revise as needed; otherwise *escalate.*

There are a lot of steps in the process above. They are important. Working it this way ensures a careful analysis. A good way is to have a work-process sheet listing the steps at the ready, and pull it out and work down the list [**Q18**]. It may help to have someone nearby read the list and check it off as you go along.

25.4 Weak Signals from Alarm Activations (Retrospectively)

The pattern of alarm activations can be considered a form of flag. A strategic application of weak signals would be to use alarm activation patterns as an early warning tool, possibly early enough to avoid more alarms activating. These weak signals can be uncovered by retrospectively examining alarm activations after a proper implementation and deployment of the alarm system has been completed. This examination would be done later by engineers or supervisors, not by you during your shift. Any alarm that activates signals a real problem in need of resolution. This means that behind every alarm activation there was trouble brewing earlier. There was a specific, defined problem with production, with raw materials, with energy management, or something else going wrong. Looking for and finding them, and providing useful early operator information will help with situation awareness.

Alarm activations should regularly be examined after the fact on a scheduled basis. This may be an engineering activity or an operations one. Examine those alarm activations and identify the root cause for each problem. For each root cause, determine if it makes sense to identify that cause early. If it makes sense, it will usually show up when the process is off normal but before abnormal. Weak signals would be the way to do this. Look for them as part of your usual activities of situation awareness. Roll appropriate lessons learned back into your suggestions for improving the enterprise operational protocols. No longer will nuisance alarms distract. No longer will alarms that are appropriate in one operational situation be permitted to activate in a different operational situation where they are not required.

25.5 Weak Signal Overload

Everything unusual can be useful. But not everything useful is useful all the time. Small indications can seem to be everywhere. The more attuned you are to this idea, the more often they can be found. The more success stories your fellow operations team tells about finding and using them, the more everyone is encouraged to look. And, in part, this is what keeps a winning team winning. But this can be carried too far. Keep everything in moderation. Weak signal management is only one tool—make sure it is not overused. You should find a comfortable working approach. We cover two ways to do it.

Weak Signals at Regular Planned Intervals

The first recommendation is that looking for weak signals should not be job one. As we mentioned before, most of your chair time should be taken up by your usual shift duties. In between, at regular intervals, make the rounds of plant operations to ensure that everything is going according to plan. See *Managing the Shift* [**M21**]. It is during these regular intervals that you perform the "small indications into weak signals" process. Look for and work those you find. Do as much as needed to identify them and work on them, but only at clearly identified parts of the shift. This looking and understanding is structured and limited. There should not be any weak signal overload problem here. If there is, this means that too much is going wrong all at the same time. This means significant portions of the ordinary infrastructure are out of line; e.g., maintenance is not what it needs to be, procedures are broken, you are inadequately trained, process controls are poorly functioning, etc. This is a management problem, not an operational one. Escalate.

Weak Signals as They Are Observed

The more likely way weak signals can overload is for you to be overly sensitive to them or over-aggressively looking for them. Ensure that you maintain a perspective during ordinary monitoring. Recognize that an essential part of your operator training must prepare you to fully understand the differences between truly normal variability and off normal operation. Expect that things can be a bit out of normal without announcing any problem in the making. Maintain a tolerance for that. In weak signals terminology, they would all be defined as less than immature weak signals—thus not meriting attention.

25.6 Weak Signals That (Seem to) Lead Nowhere

Weak signals can be too easy to find. A well-designed HMI [**M18**] purposefully contains a wealth of information to guide you to find them. Gaps in HMI design or procedures or operator experience can easily provide misleading ones or get in the way of deciding what they mean. In the previous module *Using Weak Signals* [**M24**] you learned how to see if they lead to meaningful information by using the process of forward extrapolation (supposing the "worst") and then backwards projection (to find evidence for the worst being likely). Periodically, some weak signals seem never to lead anywhere. What does this mean? Are you doomed to just keep cycling them about or should you consider something else? The short answer is that something can be really afoot in your enterprise. Weak signals are whispering a message, but they are not going to lead anywhere by themselves. There are several suggestions for what to do next. It is very important they be pursued.

Getting Help

Escalate or Collaborate

The two-cycle approach discussed in **Module 24** also suggested that you consider escalation or collaboration. Basically, consult with others to see what you might find together. And that should be an option here.

Review Your Model

Weak signals that appear to lead nowhere might just be caused by:

- weaknesses in your model.
- weaknesses in your understanding of the model.
- weaknesses in the design of the HMI, in that not enough necessary information for understanding is visible, or it is visible but too obscure to be useful enough.
- weaknesses in your ability to work the weak signals tool carefully enough.

Review your mental model of the enterprise—your understanding of how everything in the enterprise works. Because you, as the operator, understand what your enterprise is all about, you are able to determine what seems to be okay and what seems to be not. The "not okay" things that do not cause alarms to go off are most likely weak signals. Use it to assist you to evaluate what seems okay and what does not. The better your training and background is, the better your model is, and the more effectively you are able to use it. Proper use will make the difference.

Operations management should have an ongoing program to identify those weak signals that do not seem to go away. Each should be examined for cause to find the underlying issues. This activity is an engineering and supervisory one; please do not just let them leave it to you to figure out.

Looking Too Hard

Weak signals can be a strong tool. As you are learning about it and working it into your tool kit, it is easy to confuse useful with use it often. This is one of the important reasons why processing weak signals should be done at planned times during a shift [**M21.2**]. Pay attention to be sure it is not overused.

Weak Signals Among Strong Signals

Strong signals are not subtle. They are evident, obvious, and clear in their presence. They may not mean what they appear to, but they themselves cannot be missed. Most times, you see them and work out what they mean and deal with them. But there can be a sinister side. Whenever weak signals exist scattered among strong signals, the weak ones can get overlooked or ignored, while the strong ones catch your attention. This is part of our human nature, but that does not make good practice. When weak signals are among strong ones, it just might be the exact time they may be the most important.

Before acting on strong signals, unless there is a clear and present need, it should be part of your protocol to always look for weak ones. Those weak signals may suggest something different may be actually happening. Process the weak signals to see where they lead. If they reinforce the strong ones, okay. If they are secondary to the strong ones, you can bookmark them for later attention. But if they contradict or obscure the strong ones, be careful. This just might be the cautionary indicator that needs very careful attention.

Summarizing: the search for weak signals (no matter how brief) must be an integral part of recognizing important operational situations. Finding out what they mean needs to be part of managing every important situation. It is already built into alarm management, where you look for confirmatory actions [**M19**]. Build it into your situation management.

Weak Signals Might Not Persist Very Long

Weak signals sometimes have a limited life span. The weak signal may fade because the subtle problem has resolved itself. Or the problem has changed a bit so it now may appear somewhere else as a different small indication. Any useful evidence to examine (via a backwards projection) could be gone. On the other hand, any persisting underlying abnormal situation may progress to an early manageable problem or to an alarm. They need to be timely evaluated to rule them in or out as an indicator of a likely abnormal operating condition. Any operating protocols that did not include timely evaluation would not be effective enough.

25.7 Weak Signals as a Training Aid

An important place for training activities is during weak signals collaboration. This is a natural fit. And there are many variants. The training might be done with a supervisor or trainer looking over your shoulder and both of you actively looking for **and processing** weak signals. The training might be just hypothetical. Or it may be at the control console of a plant. You may be using a simulator. Coaching includes locating weak signals, forward extrapolating them, and backwards projecting them. This is the perfect time to identify weaknesses in basic process knowledge. It is very useful as well to expose weaknesses in how to conduct weak signal management. The participants discuss and agree on the weak signals, the forward extrapolation, and the backwards projection outcomes.

We are suggesting a particular step-by-step consensus type of agreement. That is, at each step, study, examine, and agree on what weak signals are present. Examine and agree on how each would forward extrapolate and to what. Backwards project and look for and agree upon all evidence found or not found. Of course, if a problem were confirmed, then cooperatively handling it could be part of the active training, but we are getting beyond the scope of weak signals and its training.

25.8 Alarms Come First

Alarms always take precedence over weak signals. Always! Weak signals belong to situation awareness [**M27**]. When things settle down and you have time and opportunity to get back into looking for things that might be going wrong, you are getting back into the situation awareness business.

There is a single exception to the alarm precedence. Recall the earlier discussion about weak signal flags in Section 25.3. When a critical number of related flags accumulate (say three to four or more), even during the alarm response process, it would be advisable to do a quick aside and have a look at the likely meaning of these flags. Accumulating flags usually indicate that something might really be going amiss. That something might be a problem with the current management plan for the alarm. Or it could be different or unrelated. A quick reconsideration might be a good plan before continuing.

25.9 Wrapping It Up

We close this third module on weak signals by reminding you that weak signals is a tool. It is only as sharp as you keep it. It is only as effective as your ability to use it. It is one of many tools available for you to operate successfully. When you have some time, you might want to look at what soft weak signals look like (Appendix 3).

The next module provides helpful tips on troubleshooting the weak signals work process.

Module 26: Troubleshooting Weak Signals

26.1 Troubleshooting Note

Weak signals is a powerful situation awareness tool. But no tool or guidance is perfect. Situations are different. So here we review what to do if you experience difficulty with processing weak signals. As you read, you will also encounter occasional recommendations to seek additional help through collaboration [**M34**], escalation [**M34**], and invoking permission to operate [**M35**]. Consider these assistance recommendations as important and supportive to your role as operator.

26.2 Special Cases

More than One Weak Signal Extrapolates to the Same Possible Worst Problem

If more than one weak signal extrapolates to the same or a very, very similar *possible worst problem*, you have a special case. For this situation, there is very likely enough evidence to begin to presume that the possible worst problem conclusion reached from these weak signals might well be actually happening. Do the backwards projection now to find evidence confirming or disconfirming each potential problem. For each confirmed one, start working on it. If evidence is not so obvious, go back to carefully working the weak signals process again.

Before we leave this case, consider that you may have a more complicated abnormal situation where you may only have found a part of it. There can be more going on. Do this by looking for any more small indications in the close proximity of the weak signals you are working. If there are none, the *confirmed worst problem* is pretty likely the only one for now. If there are more, get assistance. Something unusual is likely going on.

A Weak Signal Extrapolates to a Seriously Possible Worst Problem

There is another important nuance to this otherwise straightforward, step-by-step procedure. In the extraordinary case that any weak signal clearly extrapolates to a really, really bad possible worst problem, your immediate attention to verify and manage that outcome should take priority over all other weak signals of the moment. Separate out this weak signal and proceed directly to Step 4, Step 5, and Step 6 in Module **24.1.** Be even more disciplined as you do this to ensure that you do not permit the perceived seriousness to shortcut your careful proof. You are likely working on a transformative weak signal [**M23.6**]. This is a way to find them. Get to work. Stay objective.

Backwards Projection Leads Only to the Original Weak Signal and It Alone

In this case, a backwards projection leads only to the original weak signal. You are not finding any other evidence of something amiss. This weak signal is most likely immature or noise. Unless the forward extrapolated problem is a serious one, keep it on your list to look for the next time you do your rounds. If it is serious, seek assistance. See also the topic persistent weak signals [**M25.2**].

Backwards Projection Does Not Include the Same Weak Signal That Was Used for Forward Extrapolation

In this case, backwards projection does not include the same weak signal that was used for forward extrapolation. This is cause for concern. It is always possible that the forward extrapolation may have led you to an improbable result. Or you made an error. Carefully assess how the forward extrapolation was done to ensure that what you extrapolated to is a reasonable worst problem for that specific weak signal. If you reach the same forward extrapolated result(s), you must also be able to find other potential source clues for the problem when you backwards project. If other source clues are there and you find them, things are still working. Go with what you find. If they are not there, either you have made an error, the process understanding is too weak, or the work process is flawed. Seek collaboration or escalate.

A Single Weak Signal Continues to (Backwards) Project to a Single Insignificant Implication

When you backwards project to look for confirmation, if you can find only inconclusive evidence, do a second cycle extrapolation and then projection. If you are really suspicious, do a third cycle if you still find inconclusive evidence. We consider this because either the earlier extrapolations might have been imperfect or the projections might have been imperfect. If you keep finding nothing convincing, you have probably found noise. Move on. If it worries you, just keep an eye out the next time you look for weak signals.

Several Weak Signals Extrapolate to Several Significant Implications

This is the situation where several weak signals are found in a related process area and extrapolate to several different significant implications. But when you backwards project each one to look for confirmation, no useful confirming evidence can be found. You have weak signals that appear to lead nowhere. These cannot be relegated to noise. Escalate this situation or treat it like *weak signal flags* [**M25.3**].

Different Weak Signals Over Time Extrapolate to the Same Significant Implication

This is the situation where different weak signals over time extrapolate to the same significant implication. When you backwards project each one looking for confirmation, each time you look, no useful confirming evidence can be found. You have a persistent significant implication that cannot be confirmed. Something is working hard to let you know that it might need attention. Escalate this situation or treat it like weak signal flags.

A Weak Signal Directly and Unambiguously Extrapolates to a Specific Serious Problem

Consider a situation that forward extrapolated from a weak signal to a potentially significant problem. You examined that problem and found it to be completely and quite apparently true, without any weak signal management. Pause. Remember that the benefit of weak signal management is to avoid bias and keep from jumping to conclusions that are not warranted. However, in this specific situation, once that specific potential problem was proposed, it was clear and apparent that, without any intervention or luck, the related worst outcome would happen. No need to backwards project. It is clear from just looking. Direct management of this situation is called for.

Here is how it works:

1. If a weak signal points directly and clearly to a unique problem, and
2. If there are no other reasonable ways needed to confirm that that problem is actually developing, and
3. If your managing the problem now will not unduly put other operations at risk, and
4. If you know how to manage that problem, know that you have enough time to manage, and
5. If that problem can be managed directly—do it.

This often is the only way to manage natural disasters in the making. Remember the transformative category of weak signals? In these cases, the points of intervention are so few and the power of progression is so enormous that reacting at a weak point of Nature's plan may be all we have. But make sure that you are not unduly working under any biases!

26.3 Number of Weak Signals

Let us examine issues with too few or too many weak signals during your shift.

Finding Too Few Weak Signals

If you are finding too few small indicators, it suggests several situations: the plant you are managing is operating extremely well, or you are not skillful enough to find them, or you are not looking for them, or the HMI design is too weak to suggest them sufficiently to you. Except for the very first, you have a real problem. Resolving it has more to do with infrastructure and training than weak signals. Discuss this with your supervisor.

Finding Too Many Weak Signals

When you are finding too many weak signals, it can mean that there really are too many or you are looking too hard. Remember, looking for weak signals is not job one. Certainly, if one is visible, take a look. But rather than hunting around for them in every nook and cranny, keep it under control. Remember you have a schedule for looking. Try to follow it. On the other hand, if they really are all around, it is not too much of a stretch to suggest that the enterprise may have significant design issues, operating procedures are not up to the job, your training is not good enough, there is poor maintenance, or some other significant problem exists. Escalate this up the enterprise soon.

Finding the Same Small Indication(s) Again and Again

When a small indication catches your interest, you process it as a weak signal. It would be unusual for the same small indication to appear very often. But when it did, you worked it as a weak signal. Each always ends up being immature. What changes if they continue to appear, and if they continue to resolve as immature? What does it mean? This is not at all as unlikely as it first appears to be. Most everything going on in a plant moves. Signals are not rock steady. Pumps make different noises when a pressure or viscosity changes a bit. Temperatures, though they normally move ever so slowly, do move. What does it mean when you see some of the same small indications often? Maybe they are coming in a pattern, e.g., every time a reactor is charged; often in the late afternoon; a couple of times and then nothing for a while, and then a couple of times again. Maybe they come seemingly at random.

Since they do not seem to be disappearing, you only have three alternatives: (1) start to ignore them for a while; (2) keep working them as they come just to be sure that this last one is going to really be important; or (3) find some other way out of this cycle.

Ignore them

Not going to work. You will miss the very situation you are waiting for! Take this off the table.

Keep Working Them

You are not going to miss "the one." But really, can you keep your heart in this and stay objective and inquisitive enough so that it works all of the time? Likely not. And you might even start to ignore one now and again, especially when you are busy or think you know better. This is not going to work either. Resolve them (see next option).

Resolve Them

That gets us down to only one real alternative. Some underlying equipment or procedural issue is not "nailed down." The downside is that you are not going to be the one to resolve this. The upside is that you are going to be on the team that will. It now needs to be an engineering solution. Something about the plant is telling you all is not fully okay. Some control loop needs attention. Some setpoint or operational sequence is not up to task. A measurement transmitter needs adjustment or to be substituted for a better type. Piping or other equipment might work better if rearranged or modified in some other way. You get the idea. Note it to the necessary people. Watch it until it gets done.

26.4 Problems with Forward Extrapolation

Forward extrapolations are designed to overcome the normal tendency of being biased by preconceived considerations. But there can be problems getting it to work well. Let us focus on how the process of starting with a weak signal and supposing its most significant abnormal situation might get into difficulty.

Cannot Forward Extrapolate a Weak Signal

This is a situation where the operator is having difficulty forward extrapolating a weak signal. It means that his process understanding is too limited or he has very weak skills. This is basic and needs immediate attention.

Finding Too Many Problems When Doing Forward Extrapolation

This situation is where the operator forward extrapolates a weak signal, only to find that he has a large number of potential problems. It likely suggests that his process understanding is confused or he has limited diagnostic skills. This is basic and needs immediate attention.

There is also another concern. If, in actuality, the plant is in the very early stage of serious trouble, it might be that only one small indication would be visible. It might be too early for others. When imagining the worst from that weak signal suggests lots of significant problems, take one and backwards project to find confirming evidence. If you find it, go with it directly. It is likely that you have found a transformational weak signal. If you do not find anything, take a second or a third look. If you still find nothing conclusive, then limited process knowledge is the most likely problem. Figure out what is missing from the weak signals tool kit. Find a remedy. Get help.

Many Weak Signals Forward Extrapolate to the Same Problem

Here, many weak signals forward extrapolate to the same problem. This strongly suggests that the potential problem really exists. You found it. Go to work.

26.5 Problems with Backwards Projection

Backwards projection is the weak signals procedural step of taking that supposed worst situation, assuming it is true, and then looking for all of the places that should be abnormal and discovering whether they are or not.

Cannot Backwards Project from a Potential Problem

When an operator is having difficulty backwards projecting from a potential problem to look for evidence, it suggests that his process understanding may be too limited or he has very weak skills. This is basic and needs immediate attention.

Many Potential Problems Backwards Project to the Same Evidence

When an operator finds that backwards projecting a number of potential problems ends up with the same confirming evidence, it suggests that that potential problem really does exist. Get to work.

Confusing Evidence Obtained from Backwards Projections

When any operator finds that backwards projecting a potential problem yields confusing evidence, this suggests that the potential problem being projected from might not be the correct one. Try a second time. If things are still confusing, this suggests that something is going on, but cannot be properly identified. Something is amiss, so it is best to escalate right away.

No Evidence Found from Backwards Projections

If you find that backwards projecting a specific potential problem yields no evidence, it suggests the following:

1. It is likely the potential problem may not actually exist.
   It is either noise, so forget about it for now, or keep an eye out and look again later.
2. Your mental model for the process is too weak to be of proper use.
   Seek out expert assistance with understanding the process.
3. You are not working the procedure correctly.
   Seek out expert assistance with working weak signals.

26.6 Problems Working Weak Signals

Here we go into what might be troublesome with the actual nuts and bolts of your working the six-step weak signals tool.

The Weak Signals Procedure Does Not Seem to Work at All

If the weak signals processing procedure does not seem to work at all for you, something basic is going on. Stop. Do not force it. Keep your focus and temporarily work out other ways for finding problems.

Bring this situation up to your management so that they can work this out with you. The weak signals process can be effective and powerful. Let them figure out how to get it working for you.

The Weak Signals Procedure Does Not Work All the Time

If the weak signals processing procedure does not seem to work well all of the time, not to worry. It is just a tool. If the weak signals are so weak that sometimes you find nothing and other times it helps, then things are just fine. However, if sometimes it works okay and other times you get lost, or worse, get a wrong result, something basic is probably going on. Stop. Do not force it. Keep your focus and work out other temporary ways for finding problems.

Bring this situation up to your management so that they can work this out. The weak signals process can be effective and powerful. Let them figure out how to get it working for you.

The Weak Signals Procedure Is Too Hard to Use

If you are finding the weak signals processing procedure too hard to use, we both have a problem. Our part is that we may not have done a good enough job at explaining how all of this should be really easy, intuitive, and empowering. We may have led you astray by all of the definitions and extrapolations and projections stuff. So, we ask you to pause, take a deep breath, and give this a brief fresh start.

There are only six moving parts: (1) see clues that something does not look quite right; (2) decide which of those somethings that do not look quite right you want to know more about; (3) imagine the worst that the not-quite-right things could be; (4) assuming that the worst is really there, choose what to look for and where to check; (5) for each, do the looking to find evidence that confirms it or disconfirms it; and, finally, (6) reach a decision that the worst is true, untrue, or indeterminate.

When this is still difficult or is not working enough for you to be comfortable, ask for assistance. Coaching and practice will help.

26.7 Wrapping It Up

This module brings to a close the handbook part covering weak signals. Always keep in mind that this is only one of the many ways you can use to gain and maintain situation awareness. Seek competence and comfort with all of this.

PART 6
AWARENESS

Part 6 introduces the concept of situation awareness: the ability to fully appreciate what the operation is up to at any given moment and where it is headed.

It covers the psychology of situation awareness and how it contributes to a proper situation assessment.

Also included is a discussion of additional information that comes to the operator in the form of alerts, messages, and notifications to assist in achieving situation awareness.

PART 6 — AWARENESS

Module 27: Situation Awareness

Never let your guard down!

27.1 Situation Awareness Is

Situation awareness is all about your knowing where your operation is right now and where it is headed. Both are important in order to know whether or not what is happening right now is a problem. It is a process rather than an outcome. You engage in situation awareness activities, but are you fully situationally aware? If you assume that you know completely what is going on, it is more difficult to keep an open mind. It is hard to see differences. You downplay other information that might not support what you think is happening. Consider the following general definition of situation awareness:

> Situation awareness involves being aware of what is happening in the vicinity, in order to understand how information, events, and one's own actions will impact goals and objectives, both immediately and in the near future. One with an adept sense of situation awareness generally has a high degree of knowledge with respect to inputs and outputs of a system, i.e., an innate "feel" for situations, people, and events that play out due to variables the subject can control. Lacking or inadequate situation awareness has been identified as one of the primary factors in accidents attributed to human error. Thus, situation awareness is especially important in work domains where the information flow can be quite high and poor decisions may lead to serious consequences (e.g., piloting an airplane, functioning as a soldier, or treating critically ill or injured patients).
>
> Having complete, accurate and up-to-the-minute SA is essential where technological and situational complexity on the human decision-maker are a concern. Situation awareness has been recognized as a critical, yet often elusive, foundation for successful decision-making across a broad range of complex and dynamic systems, including aviation, air traffic control, ship navigation, health care, emergency response and military command and control operations, and offshore oil and nuclear power plant management. [27.1]

Operators do not own outcomes! They own only the responsibility to be there, be competent, and act prudently.

Duty of Situation Awareness

It is the enterprise's responsibility to ensure that adequate information is available and understandable. No operator should be judged to have lost situation awareness or failed to attain situation awareness if, at the time of his looking, the information could not have been reasonably seen, or was not accurate and understandable, and seeing the information and understanding it was essential. Therefore, the enterprise and its management must ensure that proper knowledge and the tools to understand all are incorporated into the culture and present in control room.

27.2 Psychology of Situation Awareness

We are not guided by what is true, but by what we believe to be true. What we believe to be true can often be mistaken for the truth. In our personal lives, being closer to the truth may make us happy or sad, proud or ashamed, discouraged or on top of the world. It all depends upon what we think that truth is telling us. In the control room, being closer to the truth will mean the difference between a chance for success vs. heading full speed ahead into a problem. You need truth.

Ownership

As operator, you do not own outcomes! You do own the need to expend your best efforts. You must only own the responsibility for action within your sphere of training, influence, expertise, and experience. You own that work process. Owning the responsibility happens at shift. Owning the competence happens by doing the training, arriving for shift prepared, and being on the job the whole shift. By not owning the outcome, you will not be expected or allowed to even consider that the end justifies the means.

> "people do not come to work to do a bad job. So when there are bad outcomes, you [as the investigator] must look beyond those people, at the conditions in which they worked at the time. You and [the enterprise] your organization may well have helped create those conditions" [27.2]

Therefore, it is the enterprise that owns the outcome. This is why it must follow design standards and best practices to provide safety, appropriate personnel selection, training, equipment, protocols and procedures, and all of the rest.

"Relative" Prime Responsibility

Whenever an enterprise or an operator gets ownership confused, bad things happen. This topic about relative prime responsibility is meant to reduce the chances that you get confused about ownership. *"Relative" prime responsibility* is the effect upon the operator of management that places higher value upon a forced enterprise goal instead of the proper operational one. When safety is job one, profit, throughput, and schedule must always take the back seat until safe operations are assured.

Accepting Reality

There are always problems when you accept reality selectively. Thinking "I can do it" is not always a confirmed reality. Some are illustrated here:

- Misunderstanding of facts (*confirmation bias* [**M16**])
- Failure to always follow procedural principles [**M35**]
- Failure by management to do a proper and complete plant design, build it that way, and properly maintain it [**M2**]

Leadership and Cooperation

Your style of approaching things, and the ways you think and behave, are important parts of your individuality. You cannot easily reshape personality. Rather, you can facilitate effective behaviors in the control room. The message is that well-intended individuals acting in good faith will work better when they use good tools properly. Proper tools better shape how you can relate to others. That is important because it will lead to good operation. Let us look at two aspects: leadership and cooperation.

Leadership

Leadership is the way groups are guided by one or more individuals. An effective leader is able to move the group in a collective, organized way. A good leader makes choices for the betterment of the group. A leader might lead by using a thoughtful consensus or lead by taking full and authoritative control. If that authoritative control is taken in an overly bold manner, unless the leader is right the first time, significant problems arise. [27.3] It is not the bold part that is the issue. It is whether they are right or wrong. When they are wrong, everyone goes the wrong way. Bold leaders leave little room for question or dissent. That does not work for nations. And it does not work in control rooms. So you will want to replace "bold" with "cautiously decisive." That includes collaboration and cooperation with others who have experience and provide expertise. The ways that collaboration and cooperation happen is through the proper use of *situation management* [**M29**].

Cooperation

One of the core factors for success by individuals is that we can and do cooperate. Even the most rugged individualists would not fail to recognize that not a one of us could do it all alone. Rhetoric aside, here is the fundamental truth: whenever an individual hits a really hard spot, if the right tool, skill, or knowledge is absent, that individual is out of luck. On the other hand, if the individual is part of a cooperative group, if anyone in the group has the right tool, skill, or knowledge, the situation can be managed. This is the entire driving force behind *crew resource management* and *escalation* [**M34**], and *shift handovers* [**M22**]. It is the important part of the why of this handbook.

The Triple Package

We now visit some learned human traits that add to your ability to successfully operate. When an individual possesses these three traits, referred to as *the triple package*, they are more successful. [27.4] These traits are:

1. a strong sense of (honest) pride and the feeling of belonging to something exceptional (but not exclusive).
2. just the right amount of insecurity to encourage that success must be sought (not taken for granted).
3. impulse control.

1. Pride and Exceptionalism

When you identify with a maintained and responsibly operated enterprise, this is pride. When you are treated as a professional, this is respect. When you are given the tools and processes to be successful, this is responsibility.

2. Security

When you know why you are here, you are better able to understand the risks and responsibilities of your position. When you are trained, evaluated, and retrained, you know your strengths and weaknesses.

3. Impulse Control

When management stands behind your actions, you can take the time to do things the way they need to be done.

27.3 Situations to Be Aware Of

This section is about the importance of being fully aware of the operational status of the enterprise entrusted to your care. You need to know how it is operating. Is it doing what you expect? If it is not going well enough, how much astray is it? Where is abnormal operation happening? To be clear, knowing where things are okay is important. However, it must be just as clear that situations can change, and change in ways that might be subtle, inconsistent, and otherwise hard to see.

Problematic Situations

Trouble does not usually come from nowhere. For all but the most unexpected, there are always clues, innuendos, and other subtle suggestions and information out there. Your task is to find them and make sense out of what is found. Some situations are more likely to provide problems than others. When they are present, you should be forewarned to actively look for trouble brewing. The following example situations might appear on your radar screen and point you to what should be looked at:

• Equipment	Broken equipment or equipment whose operational range is restricted by not being in full operational condition
• Maintenance	Maintenance operations are delayed, ongoing, just completed, or soon to be started
• Procedures	Not working exactly as expected or not in the same way they did when they were last used
	Performing any multi-step operation without a specific procedure or annotated checklist to follow
• Plant Documentation	Missing or of concern because of error, omissions, or aspects that cannot be verified either correct or incorrect
• Data	Missing or having suspect values or behavior (laboratory, HMI screen information, on forms or permits, etc.)
• Deliveries	Any deliveries, routine and non-routine, that have started or are soon to start, or just ended
• Personnel	Presence of unexpected or unauthorized personnel in plant or operation area
	Substitute operational, maintenance, or support personnel presently working
	Any expected, yet not present, individuals or teams
	Any individual whose workplace location is at your site, but whose immediate present whereabouts are unclear or unknown to you
	Any personnel injury

The above list is a partial one. Please view it as a starting point for your list. It should help you understand the need to exercise more than ordinary care when these situations exist. You will add to this list from your own operational experiences.

Operational Situations

Trouble can also come when your operation is near the boundaries of normal or usual. There is a temptation to assume that things are okay. Often, that is true, but not often enough. Be aware of the "not often enough" part. Focus upon being attentive to:

- more than just minor changes in operation (10% or more rate of production change, product changes, on-line equipment changes, etc.).
- any alarm activation.
- any challenge to an interlock or permissive that did not result in a direct shutdown.
- any spill or loss of integrity, even if it is water or some other seemingly harmless material (unless a part of and fully expected during routine operations).

Please use the above list as a starting point for your own list to help understand the need to exercise more than ordinary care whenever these situations occur.

Things That Are Not Visible

Any time something is not going the way it should, you have a problem. Your constant job is to look out for precisely those situations. It is what you do. This is why you pick out the small indications that suggest things might not be the way they should be. It is the reason that you use weak signals to find out what they might mean. But what if the thing you need to see is something that is not there? You cannot always see everything going amiss. The brief coverage next could just be an important way you can.

Setting the Stage

Sidney Dekker in *The Field Guide to Understanding Human Error* [27.2] introduces this situation:

Things that do not happen are not meaningful or informative in the monitoring of dynamic processes. Something that is not happening is not a good trigger for human (or even machine) intervention . . .

> In a number of automation-related accidents, the machine kept on doing something for longer that it was supposed to (or at least longer than the human operators had intended it to). This may be a climb in an airplane that gradually loses energy and stalls at a high altitude, the continuation of a heading in a cruise ship that takes it miles off course and into an obstacle, or the supply of 100 percent oxygen to a patient when it was no longer necessary. In all of these cases, the question is how people are to notice a non-event. How can they see that something is not happening? So instead of people not noticing changes, those accidents are about people not noticing an absence of change.

Some Notices Will be Present

For sure, you will never be able to see things that truly do not exist. The rest exist, but are seemingly hidden. What could be the things that are not evident? How can they be important? Well, here is how: what if the thing that is not there is *something that should be there, but is not*? To be absolutely clear, this situation is about your needing to know things that are not happening but should be. Say a pump on automatic operation is running too long. Or a highway lane that is

Never walk past a mistake. Notice it. Process it to put its significance and impact in context. Figure it out. Do not overreact. Do not underlearn. If it is straightforward and clearly part of your personal kit, work the lesson and insight back into your thinking and ways of operating. If it touches any part of the site's infrastructure, including procedures, operating materials, and the rest, then it is time to work with your supervisors.

29.2 Design for Good Situation Management

Good situation management comes from doing the hard work to get it. It involves understanding the success factors, recognizing what can get in the way, and exploiting the ways that help. Figure 29.2.1 illustrates the way it works. Each check mark bullet (√) is a requirement for a part of the safety net. Your enterprise should take specific steps to make sure they are properly in place. Each x-mark bullet (x) recognizes a problem area that your enterprise should take specific steps to prevent or eliminate.

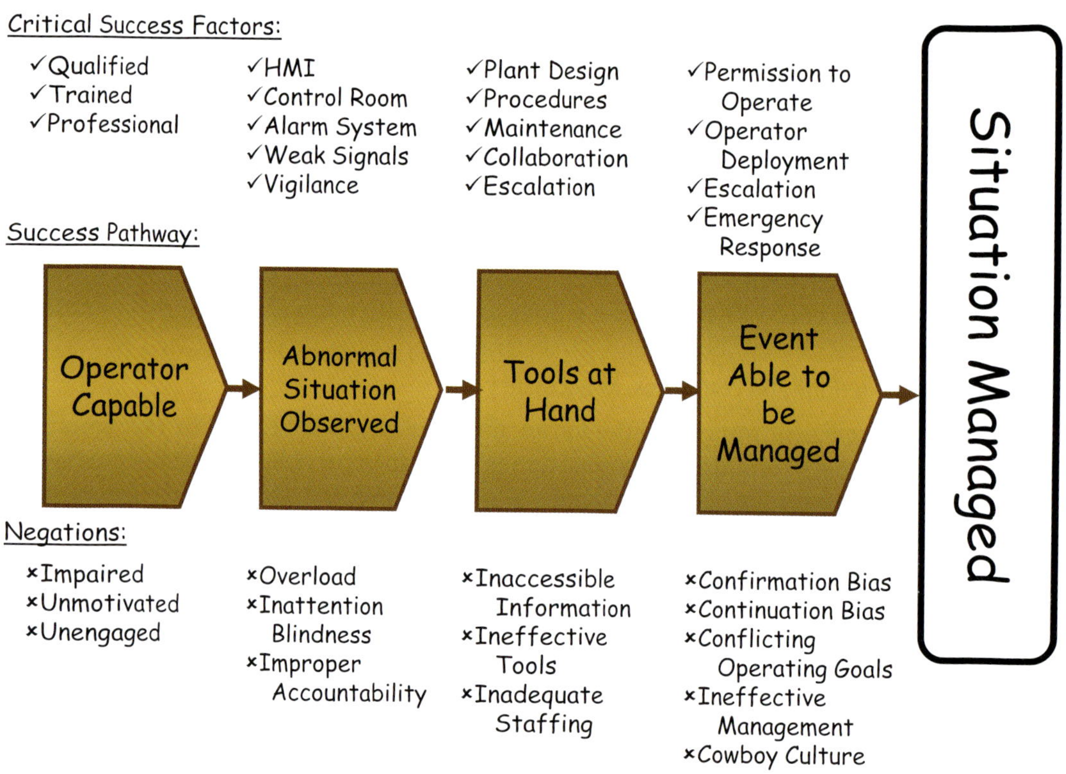

Figure 29.2.1. Situation management success pathway, illustrating both enablers and detractors for it all working together. We need to enhance the success factors and work hard to eliminate the things getting in the way (negations).

29.3 The Situation Management Process

Plant safety and responsible operations are not going to get better just because everyone thinks they should. The activity needs to be done carefully and deliberately. You use four activities: *observe, confirm, remediate,* and *confirm again.* These basics are as simple as they are vital. Each activity represents a necessary key component to do the job with as much effectiveness as nature, responsible plant design, and human management will allow. [**Q3b**] All are necessary (see Figure 29.3.1; introduced earlier [**M4**]). Any lapses will be costly.

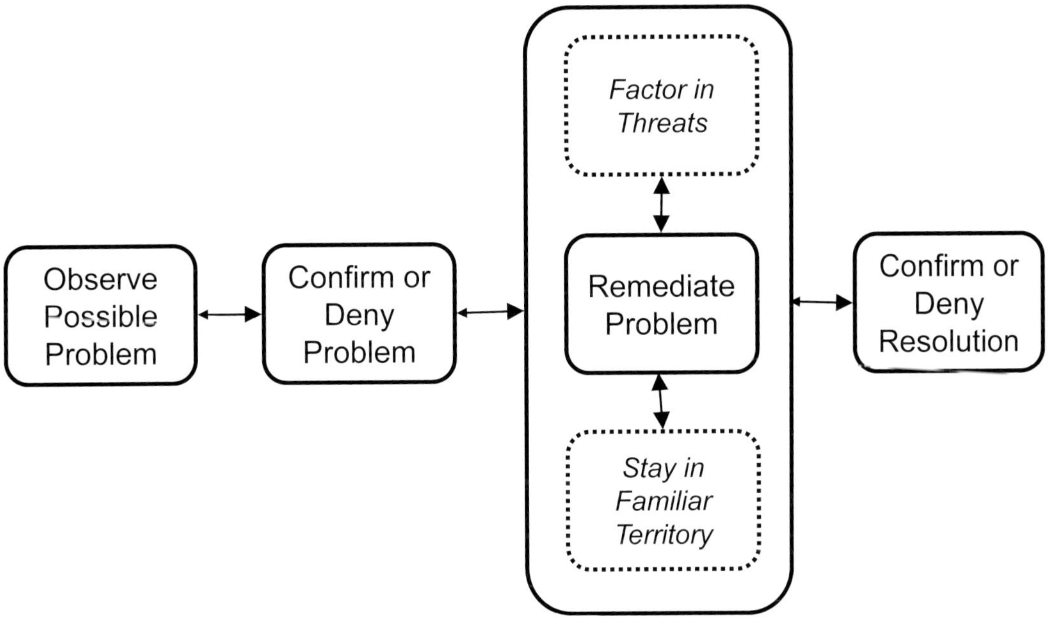

Figure 29.3.1. Basic four-activity situation management process, depicting the progression from "observe" to "confirm" to "remediate," with some possibility to move back a step when evidence is not there or there is not enough to accept. The figure illustrates the situation management process applied to a specific problem. Escalation and collaboration may take place at any step.

1. **Observe**: *The operator will explicitly observe to identify the current operational situation.* This means that job one is to always know what is going on in the entire enterprise under your responsibility. Not only must you know where things might be going amiss, you must also know, to a reasonably high degree of certainty, that the rest is operating properly. The absence of easily observable problems cannot be presumed to signify that everything else is okay. Take in the whole picture. You do not decide anything or fix anything at this step! You are just looking for problems.
2. **Confirm or Deny**: *All information about the current operating situation must be confirmed.* As you look explicitly or during your rounds to see how things are going, each of your conclusions about proper operation or off-normal operation or abnormal operation must be explicitly confirmed by examining proper independent aspects. Independent aspects include laboratory reports, production values, mass and energy balances, and the like. As you make your rounds (coming on shift, periodically during the shift, and preparing and doing the handover), all observing must go hand in hand with confirming. If you cannot confirm, you must be able to justify that the problem is not present (disconfirm, deny). Seek help if you cannot do either.
3. **Remediate**: *The operator is expected to remediate or resolve all identified problems and abnormal situations.* This is not a shotgun approach. Only the problems and situations that you have explicitly identified (and confirmed, of course) should be resolved. Your doing something just to feel needed is out. Each intervention has a price. Do not pay for anything unnecessarily. Just to be sure we have it right, all exploratory intervention done while explicitly investigating proper problem resolution is proper. If a sticking valve were suspect, attempting to move the valve would be proper.

There are two critical behaviors (below) to always keep in mind during remediation:

Factor in Threats (identify & accommodate risk): *Expose and identify all risk to the enterprise coming from outside the immediate problem area.* These are the things that a proper resolution of an abnormal operation or situation requires or depends upon, but is outside of the control of the specific problem resolution to manage directly. You must factor all of them into your intended solution, and as you implement it.

Stay Familiar: *Never let the process or plant take the operator to a place where he has never been before.* This means exactly what you think it does. You should not ever be operating your enterprise in a situation where you are not fully familiar, where you do not know where it is operating, and do not know what to do to keep it safe and in good operation. Upsets count. You are expected to do what needs to be done to respond. But if your training, experience, guidance, and understanding are not completely familiar with where things actually are at this moment, you are not permitted to figure it out anew. Operators are not inventors. You are not an innovator. Yes, you are talented and, of course, experienced. You are not tasked or permitted to "figure it out on the fly" during the actual situation.

4. **Confirm Again**: *Now that the remediation action(s) have been initiated or concluded, the operator is expected to carefully monitor the results to ensure adequate remediation or resolution of the problem or abnormal situation.* This is not a time to make an adjustment without further consideration. Lots of things might go wrong. To assume that your remediation actions were in the right direction and should have done the job is simply not enough. It might actually require a little more of one thing or a little less of another. On the other hand, if the needed remediation action were different from the one used, it needs proper assessment and modification. There is a third aspect. It is when the situation changes during your remediation process. What was the correct thing to do earlier now needs modification. All of this means that you are expected to remain vigilant and ensure that the right things happen in response to all of your actions and activities.

29.4 Following Rules

Having lots of rules can be useful. On the other hand, they might seem to get in your way. They can be too detailed—or vague to the point of desperation. They can be difficult to remember and hard to find. But they really are necessary, useful, and important. We visit them here with you to help to better understand their purposes and to assist your efficient use.

Rules must be complete, accurate, and available when needed. You will use them when they are required, and you must be adequately trained to use them. However, and this is a big **however**, no rule can be followed blindly. Every rule requires thoughtful compliance. [29.1]

> "Personnel are expected to follow rules and procedures. However **personnel are also expected to think** [*emphasis added*] about what will happen if the established rules and procedures are applied to the current situation. If they believe the risks of implementing [following] the rules and procedures are unacceptable they are expected to stop and seek advice from other knowledgeable people."

This thoughtful compliance is extended to include most of the situation management tools and processes in this handbook. All are meant to either augment your ability to do work or safeguard against losses. There is an important additional tool that belongs in the kit. Each enterprise needs to write the rules for following their established rules. This is the guidance of looking for, understanding, and managing all conflicts between following basic rules and anything that does not appear to be doable or safe. The best way to do this is by setting out a general protocol for following rules. They involve the following:

- How to know that a rule is available?
- When is the rule to be followed?
- How to manage parts of the rule that do not seem to apply or do not seem to be doable?
- What to do when actions appear to be needed but are not in the rules?
- How to know that following the rule should be terminated and how to do it?

The actual working procedures for meeting the needs of this activity are left to the enterprise. If your enterprise does not have a formal process, consider using this list to guide you.

29.5 Operating Situations

You certainly know more about your enterprise than we authors ever could. This handbook will help use that knowledge. Not all operating situations and modalities are even close to being the same. To begin, we will identify situations and add structure to them. This structure will be used later to define situation boundaries to assist in the implementation of the operator's permission to operate. Use what works. Ask your supervision and management for anything you might want to use. Ask them for anything more that is not here and that you think you might need.

Normal Situations

This part is your "bread and butter." All of your training and experience gets working here. Normal situations are typical of everyday operations. Things happen unexpectedly. You need to find out what happened and why. Then figure out and execute a resolution. Things break. Products might need to be changed or modified. These normal situations are characterized by their somewhat limited nature, their familiarity, and their ready management. This is not to say that they are trivial. Their impacts vary from the barely noticed to a nasty shutdown that might have been avoided had maintenance been better. So while we are not too happy when they come, we accommodate them without undue add-on risk.

Major Situations

Major situations come very infrequently, and when they do, they risk extensive plant damage, personnel injury, and environmental degradation. They really are serious operational problems. They are very different from normal operations and require concentrated activities. The practices for them need to be carefully understood and followed. Yet, properly managed, their impact and cost can be mitigated. Managing them will be covered later on in *Managing the Control Room* [**M31**].

Operating in Uncertainty

You are quite proficient in the day-to-day management of your production units. Automated controls and other equipment help maintain production. When variations appear and you respond appropriately, normal operation is quickly restored. Effective process management is maintained by your experience, procedures, problem-solving skills, and collaboration. Yet there are a couple of basic ways this comfortable situation might become a problem.

- Problem incidents begin with a whimper or perhaps a small bang somewhere; then, for a while, progress slowly, then take a turn and get complicated, often without your knowing; and suddenly get out of control.
- Problem incidents begin with a big bang and quickly become out of control (known by you or not).

Explosive Events

Some damaging events seem to happen almost without notice and move extraordinarily quickly into a catastrophic state. In this situation, we must rely on either your quickly identifying the situation and initiating an emergency shutdown or the automated activation of emergency shutdown protection equipment. In both cases, it is imperative to ensure that the shutdown procedures and equipment are adequate and effective.

Surprisingly enough, however, such an event almost never arises without some sort of prior symptom. If the right sort of notice were taken, the event could likely be effectively managed. Therefore, industrial best practices for reducing the frequency and consequences of these events depend upon a few key parts:

- Having comprehensive and effective equipment health monitoring programs
- Recognizing the role that operating stresses place on both personnel and equipment
- Putting into place technology to detect both subtle as well as incipient faults in the process and related equipment
- Mastering the process of *weak signals* [**Part 5**]

Unique Events

Most of the seriously damaging incidents never happened before in your enterprise. Nor were they really envisioned. For such unique events, experience and specific procedures are of little use. Situation-based training will not be of much use either. Industrial best practices now rely on three key aspects to respond effectively to unprecedented events:

- Competency-based operator training, where abnormal situation recognition and management are one of the core competencies
- Robust abnormal operation detection and identification tools
- Effective abnormal situation-management procedures and technology

29.6 Single Point of Failure

A single point of failure (SPOF) is a special case. We introduced this earlier [**M14.2, M27.4**]. They are useful here. Before attempting to dig in and manage anything, do a quick check for it. If you are experiencing a SPOF, unless your actions are proper, done in time, and work as planned, something bad will happen. That bad thing will not necessarily be protected by anything else but your good action. Every time you come anywhere close to a SPOF, take the following important steps in the exact order listed:

1. Escalate the situation before anything else [**M34**].
2. Determine if there is enough time to figure this out before doing anything more.
   a. If not enough time, invoke *permission to operate* [**M35**].
   b. If you know you have enough time, proceed to the next step.

3. Follow approved site practice for managing SPOF. If none exists, use permission to operate [**M35**].
   a. Estimate the amount of time you have to work the problem.
   b. Completely understand exactly what the SPOF is.
   c. Completely understand the SPOF's impact on the safety and integrity of your enterprise.
   d. Identify exactly what you are going to use to protect from the SPOF if your actions are not effective. Get that ready to use.
   e. Work to remedy the situation.
      i. Take the proper actions.
      ii. Ensure that they are working.
      iii. If they are not or they fail, invoke *permission to operate*. That backup you checked on in Step 3(d) might be ready to work, but it is just that, a backup. It is only there in case you are not able to wind things down fast enough or properly enough!
4. Ensure that all of your actions are successful.

Each SPOF represents a design or operational vulnerability. None should be there. When you observe one, take the extra steps of documenting and reporting it up.

29.7 Understanding Catastrophic Situations

Situation management is not crisis management. It is not disaster management. For either of these, as soon as you see it, initiate action(s) to move everything to a safe state and ensure that these actions have been properly and fully taken. Even if it is likely that a safe state of the plant is not attainable, your operational role now ceases. You may have a support role, but others will direct you. From this point forward, the problem is shifted to crisis management responsibility or disaster management responsibility. Your plant will have a complete plan and process for it.

Module 30: Operational Boundaries

30.1 Boundaries and Responsibilities

To ensure effective *managing the control room* [**M31**], it is useful to understand that specific operational boundaries and responsibilities exist and how they are adversely challenged. This module contains important concepts and considerations not usually explicitly recognized. Operators and management should review existing procedures and processes to ensure that it is always clear who is in charge, what the limits of their activities must be, what and how any coordination will occur, and who will do what.

Your overall enterprise may have several operational areas, each with its own dedicated operator. It is taken for granted that you, as the operator, are charged with ensuring effectiveness in your operational area of responsibility. An operational area is the physical extent of equipment and activities for which a single operator is responsible. You. Figure 30.1.1 schematically depicts this. There is the interior portion (green) noted as the Operational Area. Only a single one, yours, is depicted for clarity. The operational area is contained within an enterprise or plant area (blue). The enterprise area is contained within the universe (light yellow), which is shorthand for everything else outside of the enterprise (community, etc.).

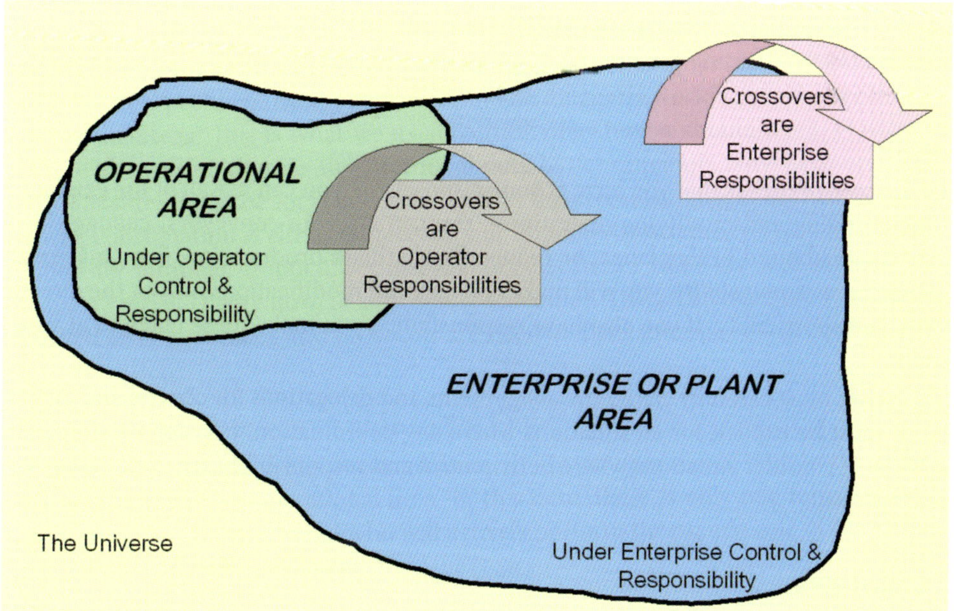

Figure 30.1.1. Boundaries and responsibilities that identify who is responsible for trying to prevent operational excursions from a particular region of operation and who is in charge once an excursion has occurred

not have available information at hand to fully understand the risks and properly respond. This situation is explicitly recognized by the mandated requirements for municipalities and governments to have disaster management plans ready. Thankfully, most municipalities will never have to execute any such plan. In any event, the enterprise will have and must follow the protocols and properly coordinate with the external management process.

30.3 Control Room Coordination with the Field

Proper coordination requires that both the field operator and the control room operator are always managing exactly as the other expects. All locations, all tasks, and all observations need to be locationally confirmed by both and in agreement. Effective coordination ensures that both are working on the same thing, at the same time, using the same expectations, and for the same purpose.

Field Maintenance

Coordination of maintenance activities goes considerably beyond a simple handoff. It involves transferring physical and limited operational control of equipment from the control room operator to maintenance personnel and back. These changes require following a strict coordination process to ensure that both ways are done properly. No maintenance work should be started unless the control room has authorized it, made all of the operational adjustments and accommodations for it, and passed control to maintenance for the work. No operational use can be made of the equipment passed back from maintenance until maintenance has certified that it has been properly managed and is ready to be returned to service—and the operator has independently verified its operational readiness. See *Shift Handovers* [**M22.6**] for specific additional guidance.

30.4 Maintaining Responsibility

It is important to maintain the chain of responsibility in the control room. This reduces delays while deciding who is responsible for taking action. When a problem arises in a group situation that all observe, each individual may believe that someone else is going to take responsibility for it. Even where there should be a clear chain of responsibility, if individuals of senior organizational authority are present with those of junior or equal responsibility, there is a tendency for the junior to wait and defer to a senior individual. In this situation, where time to act is important, action is subject to excessive delay. Time is a pervasive foe. This will need your careful management.

This handbook has many of the tools for recognizing and managing this problem. They form the backbone of the requirements for a qualified operator, that is, the formal operator qualifications that are a prerequisite for sitting in the operator chair [**M4**]. Maintaining responsibility is one of the key reasons for the formality of *delegation* and *escalation* [**M34**].

> The sitting operator has the permanent responsibility during the shift to make sure that someone is always in the role of responsibility; that someone knows that he is; and that someone is performing the duties.
>
> So long as the sitting operator has not been formally relieved, this responsibility exists, unaltered.

Where responsibility has formally shifted to another, your action would be to ensure that the new individual in charge understands and accepts operational and control responsibility and maintains it (aka *crew resource management* [**M34**]). When senior (even if very senior) individuals are present but you are in charge, courtesy would suggest that you announce that the situation has been noticed and you will manage it. All current established protocols for managing major situations may be used, as needed, including *escalation* [**M34**], *permission to operate* [**M35**], *sterile control room, restricted control room* [**M31**], and others.

To be clear, at all times when you are present and are the operator on shift (not relieved), you are responsible for making sure that someone is always in charge of all of your responsibilities, if that individual is not you. Even if you have escalated operation to another qualified person and that person has accepted operating responsibility, as the original operator, you must retain the explicit responsibility for ensuring someone is always in charge. Without individual enterprise rules to the contrary, if you initiated escalation and that escalation was accepted, you still retain the responsibility to ensure the escalated individual remains in charge until control is returned back or assigned to another—and, if necessary, to order it back if the situation requires it. Even if someone else initiated the escalation, you must be responsible for ensuring someone is always in charge.

The previous point is extremely important. Because of your training, experience, and regular schedule of work in the control room, you are fully familiar with the entire range of the job. On the other hand, the person escalated to possibly is not. Certainly, that person has special skill and should be fully qualified to operate. However, maintaining operational control for long stretches of time (a shift or more) would not be so familiar. If the situation were to stabilize or even be resolved, there is a natural tendency to (mentally) shift away from the "in charge" part and step back. This presents a serious continuity problem. It might appear clear to one party who should be in charge, yet it might not be the same for all. This ambiguity cannot exist. Therefore, someone must be tasked with ensuring that this does not happen. Someone must always be in charge. This is called the "in charge rule." [**M34.2**]. As the sitting operator, you must make sure that someone actually is.

Module 31: Managing the Control Room

Your handbook now moves into the specific ways that the interpersonal relationships in the control environment are adjusted for special circumstances. In the previous modules, we covered the basics, most of the tools, and a significant number of methods that you use to observe, understand, conclude, and plan your actions. Now we focus on the control room itself to see how planning and coordination aids its occupants to work more effectively as a team.

31.1 Control Room Situations

Operations in the control room are very different during normal operation vs. abnormal operation. Activities can vary, from boringly routine all the way to difficult and even life-affecting. All need understanding and proper protocols. Bottom line: no activity should be permitted to take place in the control room for which the control room has not been explicitly designed and personnel properly tasked [**M17**].

Normal Operation

Normal operating condition is when there are no current, recently passed, or anticipated upcoming threats. It is no reason for relaxing *due care*. This is not that everything is perfect. All duties and responsibilities are still to be properly conducted. Nor does it imply that usual or expected variations in weather, personnel, and equipment are absent. Normal operation focuses on what is happening now. It does, however, include one-off or unusual tasks or operations.

In a well-managed control room, a wide variety of activity is usual. During normal operations, control room activity can range from slow-moving for hours at a time to very hectic brief periods. As operator, you may be watching the board, talking on the phone, consulting other operators, and more. When you take brief breaks, you may sometimes ask another operator to keep an eye on your board. You take meals, so food may be delivered or prepared. If more than one operator is sited there, each of you will be managing your own assigned area, or you may be jointly operating the same area. You each have your own schedule to follow and your own ways of doing things.

Supervisors, visitors, and other authorized enterprise personnel (arriving/departing relief operators, field operators, maintenance personnel, and others) may be present. There may be light background music playing, production supplies being refreshed or removed, and all manner of other direct support activities going on. All of this happens with the understanding that distractions are managed so as not to be in your way. It is also important to note that no individual should be permitted to be in the control room not explicitly allowed by operations protocol. Those times are situationally dependent. The enforcement of these practices should be the responsible duty of all those present, including supervision (not necessarily present at the time).

Abnormal Operation

An *abnormal operating* condition in the control room exists when operation is no longer normal or when still normal, but significant threats are anticipated. During these times, it can be extremely important to manage behavior. When any part of the control room is engaged in significant non-routine operations of any kind, it can affect the entire "background." Managing will ensure that those charged with managing the operation can devote their proper attention to it. These are situations where even normal distractions can interfere. All unnecessary distractions need to be eliminated. All needed resources must be at hand or immediately available.

31.2 Control Room Conditions

There are certain times when it is important to adjust the entire atmosphere in the control room to better enable work. The two protocols designed specifically to manage unnecessary distraction in the control room are ***restricted*** and ***sterile***.

Restricted Control Room

The *restricted control room* is a first-stage condition designed to position the control room and its necessary occupants in a focused status with all unnecessary distraction at a minimum [**Q12a**]. It is a specialized operational mode. It shifts the internal functioning of the control room into a supportive environment to better respond to any significant operational situations present or coming. This mode is designed to support operational activities: preparatory, diagnostic, and reactive. During these situations, the enterprise functions according to pre-established plans and protocols, with the control room environment specifically positioned to support that. Examples include major weather event coming or arrived, unusual stresses in operation that might be caused by significant curtailment or increase in product demands, state of emergency due to security issues, and the like.

Restricted control room conditions do not specify duties or procedures for managing the enterprise. Those are always provided for by enterprise-specific instructions based upon requirements and operating practices. Here, we adjust the background so that all of this may play out in a way that provides enhanced support.

Initiators of Restricted Control Room Condition

Here is a partial list of situations that could be used to suggest initiation of a restricted control room condition. When any of these situations exist, a transition to a restricted control room status is recommended. The specific initiators would be determined by your enterprise protocols.

- Severe weather arrival sometime within the current shift, already arrived, or arriving imminently
- Significant change in operational methods, conditions, or goals, requiring more than "due care"

  (examples include startup or shutdown, initiation of a significant audit requiring specialized cooperation of control room personnel and the possibility of altered operational activities)
- Medical emergency in the control room
- Medical emergency in the controlled part of the enterprise (outside the control room), for which control room operators must accommodate operation and/or protect personnel and/or facilitate rescue or recovery

- Loss, or a failure to arrive for duty, of the **minimum reasonable number** of operational personnel needed in the control room
- Declaration by any authorized individual of the need to be in a restricted control room condition

Restricted Control Room Conditions

A restricted control room condition generally requires the following:

- All visitors out (to a safe and secure location while observing accepted security protocols).
- All scheduled personnel must be in attendance: if assigned supervision personnel are not available, predetermined backup individuals must step in.
- All essential personnel at their designed station(s).
- All operations are conducted with heightened attention.
- All non-vital activity in the field, including maintenance, construction, survey, etc., are expedited for early completion, paused, or delayed (always in a safe manner).
- Where access to special or detailed documentation information is needed that is not normally readily available, a specific dedicated individual will be responsible for opening access and its use.

Termination of Restricted Control Room Condition

Ending the restricted control room condition is via the authorized individual declaring that the condition has ended or by consensus of the attending operators (where supervision is not in attendance; and initiating conditions are over). All suspended activities are again permitted, but the necessary protocols for them will be followed.

Sterile Control Room

The *sterile control room* is implemented when operating conditions deteriorate strongly or if significant threat is anticipated. This mode stops all optional outside operations, summons additional resources to the control room, and restructures the activities of those in the control room to better focus everyone on the current situation. These are situations where even a restricted control room is insufficient to position personnel and restrict unnecessary activity. This situation benefits by changing the operational mode to a *sterile control room* [**Q12b**]. A sterile control room might be necessary when the operating situation deteriorates beyond the benefits of the restricted control room. Like a restricted control room, the sterile control room configuration does not specify duties or procedures for managing the plant or enterprise. Those are always governed by the protocols, procedures, and knowledge and skill of the operations team. Being able to do their work in a special control room configuration amplifies their ability to concentrate and bring to bear the needed resources. The state of being in sterile condition is more restrictive than restricted control room condition. It is designed to move the control room into an operational state that simultaneously eliminates all non-essential distractions and ensures that sufficient essential participants and resources are present.

Initiators of Sterile Control Room Condition

This is a partial list of sterile control room initiator situations. When any one of them exists, a transition to a sterile control room state is recommended. The specific initiator situations would be set by the enterprise.

A sterile control room would generally be set if any one of the following are present:

- Current operation (planned or unplanned) has a likely potential to risk life or cause significant disability, serious economic loss, significant environmental exposure, or other significant risk to the enterprise or surrounding community.
- Activation of any single highest priority alarm.
- Activation of multiple second to highest priority alarms.
- Partial but significant loss of operational control (power outage, substantial communications loss, severe weather, etc.).
- Security (cyber or physical) attack or breach.
- Declaration by any authorized individual of the need to be a *sterile control room* (either from a *restricted control room* or directly to *sterile*).

Sterile Control Room Conditions

The following requirements of a sterile control room are in addition to restricted control room requirements. When the control room is not already in a restricted control room condition, first activate the following restricted control room requirements (repeated below), seamlessly followed by the sterile control room conditions.

First Part: Restricted Control Room Conditions (but modified a bit for this condition)

A restricted control room condition requires the following:

- All visitors out (following accepted security protocols, of course).
- All scheduled personnel must be in attendance: if assigned supervision personnel are not available, predetermined backup individuals must step in.
- All essential personnel at their designated station(s).
- All operations are conducted with heightened attention.
- All non-vital activity in the field, including maintenance, construction, survey, etc., is terminated (ensuring that the work is safe and workers are out but available on standby for emergency use).
- Where access to special or detailed documentation information is needed that is not normally readily available, a specific dedicated individual will be responsible for opening access and its use.

Second Part: Sterile Control Room Conditions

In addition, add these sterile control room conditions:

- All background activities are stopped: food is placed on temporary hold; music off; non-essential telephone calls and radio calls ended and all future ones screened.
- All non-vital activity in the field, including maintenance, construction, survey, etc., is terminated (rendering them safe). Unless maintenance personnel are required to ensure the safety of the holding of field activities, all are removed from the operating area. They are to remain at the ready in case of need.
- Any operators in training or newly assigned must have their fully trained and qualified backup take over their activities.
- Operators whose duties are split between the control room and the field must have both the control room and field manned by separate qualified individuals.

- Operator and supervisor relief handovers due to shift change are delayed pending a staffing determination. However, relief personnel should report and remain present for backup.
- All non-essential movement within the control room ended.
- All non-essential conversation ended.
- All essential conversation is conducted in a muted fashion designed for clear understanding and minimal distraction to others. The recipient confirms each exchange via voice-back repeat.

Termination of Sterile Control Room Condition

It is important to note that the mere ending of any specific situation or condition that caused the control room to initiate sterile operation is not sufficient to end sterile operating conditions without further care and attention. Care must be taken to ensure that all relevant aspects of operation within the control room and the enterprise are sufficiently understood and determined to be within reasonable or normal ranges. Unless there is confirmed reason to believe that all initiators of a restricted control room are absent, a transition to a restricted control room would be advisable. Supervisory personnel would normally determine this.

31.3 Control Room Codes

There are specific impactful situations where it is important to communicate a special operational status to everyone. In a control room, it can be used to quickly change operating modes in a controlled and effective manner due to an abnormal operating condition. This is so that everyone in the control room will know rapidly and without ambiguity that things as usual have changed in a significant way. During a code, everything you consider or do must explicitly ensure that everything must be thought about differently according to the situation identified by the code.

This section documents a new control room concept borrowed from the proven protocols for emergency medical operations used today in most all hospitals and trauma centers [31.1]

A medical example of this change in operating modes is a *code blue.* A code blue means an individual (patient, visitor, or otherwise) at a specified location in the hospital has stopped breathing (hence turning blue).

The pre-planned code blue response is as follows:

- All pre-assigned responding individuals immediately move to the specified location.
- One or more of these individuals finds and delivers the required predefined and prepared cart of equipment and other equipment to the specified location.
- The pre-assigned individuals now function as a team, each performing required activities under the pre-assigned team leader, until the event is over (announced by a "code clear").

When the immediate need for mental concentration is high and risk appears to be all around, it is extremely difficult for a highly focused team to be able and willing to realize that their world has undergone a significant and risky change (reached a tipping point) that requires a dramatic transition from normal protocols. [31.1]

For your enterprise, the appropriate specific process and procedures will need to be added to each code requirements. The specific control room codes are explained below:

Code Outside

Code outside is a situation where a significant threat exists or is anticipated, to the integrity of a facility that comes from **outside** the enterprise [**Q13a**]. In this situation the focus for concern is at the enterprise boundary. Prior to the threat, the enterprise is operating properly and faces little or no operational threats. The new threat is announced by "code outside." A code outside risk is one that could potentially cause a significant change in operations. This includes likely curtailment, disruption, or danger. It must be recognized and operations modified to accommodate it. Examples include severe, disruptive weather or natural disaster on the near horizon; fire or toxic release; civil unrest likely to directly affect operations; impending curtailment or loss of essential utilities such as electrical power, natural gas, sewer malfunction; impending significant loss of ability to store or ship product or dramatic increase for producing product; disease or risk of disease exposure that reduces availability of sufficient operations personnel; and more.

The specific details of the code outside will be set by the enterprise.

Code Inside

Code inside is the immediate recognition of significant trouble that is present or developing rapidly **within** the plant or facility itself [**Q13b**]. In this situation the focus for concern is within the enterprise. This internal threat would always be unexpected at the time. There is imminent and significant danger when it happens. Responses require deep understanding of the particular inner workings and function, and the ready availability of proper equipment and staff. Response activity requires careful coordination of personnel within the enterprise and effective communication with personnel outside.

Example initiator events would be outbreak of fire; explosion; loss of containment of flammable, toxic, or excessive materials; major malfunction or failure of equipment that exposes significant operational stress, including inadequate personnel in attendance; presence of unauthorized personnel potentially able to initiate dangerous operation or threaten safety of personnel; and more.

The specific details of the code inside will be set by the enterprise.

Ending Codes

Properly closing down a code is nearly as important as initiating one. Each code must have a clear ending procedure and process by which enterprise or unit operations are:

- restored to partial operation as a result of effective action; or
- returned to good operation by going through a carefully choreographed plan; or
- declared to be inoperative, so it is in a status of
  - "parked" or
  - shutdown.

31.4 Codes and Conditions

Control room codes and control room conditions may seem to be so similar to each other as not to matter. They are similar. Yet they are different enough to really matter. The best way to keep them in good focus is to understand that a control room **condition** is about changes made in who is allowed to be there, how all of the physical distractions inside are managed, and how the load on the personnel is controlled. A control room **code** is about what the immediate focus of everyone is, needed extra hands are in place, and what specialized roles individuals play, known and observed.

> A code is quite different from circumstances where these same situations would develop gradually. In those circumstances, everyone would be quite aware of how things were progressing and would have time and practice for managing them. They should be part of the enterprise strategic planning and implementation that would be a management initiative.
>
> Code actions must be pre-planned, of course, but to use them, the arrival of the situation would be largely unexpected and the necessary response tactical.

It could be perfectly normal for a situation to exist where the currently present control room personnel were fully able to manage, but the situation would be best managed if all other unnecessary aspects were put aside or out of the picture for the duration of the situation. The appropriate control room condition would do this nicely. On the other hand, it would be easy to see a situation where everyone in the control room needed to be immediately focused on the exact same vital situation so all "hands could pull their oars together." And, at the same time, any extra hands would be summoned, and all would know how to work together. For this, a control room code would be used also.

> Codes and conditions should be combined whenever necessary and covered by experience and plant protocol.

Yes, of course, some control room codes might be aided by the concurrent use of a control room condition. Both are designed to work together and complement each other's strengths and bridge needed differences or gaps. Neither is required with the other.

Module 32: Defensive Operating

The best way to handle trouble is by not getting too close to it.

32.1 Concept

Defensive operating is a unique and useful way to keep ahead of trouble by seeing it coming and working to avoid tight spots. Defensive operating is borrowed from the very successful and well-known principles and tools of "defensive driving." You might already be using those ideas.

> "The Smith System for defensive driving employs five basic principles. Each principle is designed to reduce the risks involved in driving by teaching drivers to anticipate dangerous situations. By driving defensively, traffic-related injuries are reduced, even in adverse weather conditions. An important rule in defensive driving is anticipating other drivers' errors, mistakes in judgment and/or carelessness." [32.1]

Your specific operator duties, tasks, and responsibilities, such as daily orders, shift handover instructions, and general duty lists of activities and responsibilities are already appropriately documented and part of your existing enterprise requirements. Defensive operating is supplementary to all other good operating practices. It provides a mindset for the activities you do. It represents another aspect of competency. It is a part of what you would do when monitoring things. It adds substance and preparation for your being ready. It is another useful tool in your box.

32.2 Key Behaviors

The policy and tools for defensive operating is the "elevator message," or mission statement of situation management. These five behaviors provide practical guidance for how you can maintain effectiveness. They are designed to reduce risks. They sharpen and focus your attention so that you are better at doing what you are tasked to do. They can be a key part of your professionalization as operator [**Q14a**].

1. **Aim High**—Develop and maintain an endpoint view for what is expected and/or required during and at shift end. What do you need to have done as opposed to just nice to do? Keep looking ahead to see how where you are now contributes to or hinders the current shift objectives.
   Do I need to prepare for or delay a scheduled repair?
   Will my operation be suspended because today is Friday and we are closing down for the weekend?

Are we running low on our monthly allocation of overtime?

Am I too tired to finish my shift and still be able to get home safely?

2. **Get the Big Picture**—Keep alert to the other aspects surrounding what is happening now. Check all of the operational limits. See if you are close to any or way too far from others.

   Is there anything about what is going on now that bothers you?

   Are there any production or other problems left over from previous shifts that need attention or follow-up?

   Is maintenance work going on in your unit?

   Are storms in the area?

   What do the production trends look like? Do they suggest something you need to check?

   Are lab values where they need to be and do they appear to be solid?

   Are other units around yours starting to prepare for significant changes in operation? Are they just recovering from them?

3. **Keep Your Eyes Moving**—Observe current operations to the point of knowing where things are both spot on or need closer attention. Account for all details, no matter how small.

   Where are things going smoothly?

   Where are the things that you need to pay attention to right now?

   Is anything unusual going on in your unit?

   Are any other units around you experiencing problems, e.g., small fire, partial planned shutdown in progress, deliveries, maintenance activities, etc.?

4. **Leave Yourself an Out**—Always, always have a backup plan. Respond to operation that might be too near to operational limits by taking steps to reduce the chances of limit crossing. Take advantage of operating space when there is still enough room and time that you can use.

   Do you need to change the operation of your overhead air coolers for this storm heading your way? What do you need to do to get ready?

   You are seeing pump efficiencies starting to drop. Should you consider cutting rate to ease up on their load until maintenance can have a look?

   The next operator position is taking 30% of your production but is experiencing a problem. Should you cut production in case he has to suddenly stop receiving?

   Is there room in a storage tank you can use to build inventory in case an unforeseen problem comes up?

5. **Make Sure They See You**—Ensure that the rest of the enterprise knows what you are doing and is reacting appropriately.

   If you are changing products or altering the operating situation, do the upstream and downstream operations know that?

   If you have an important maintenance operation scheduled soon, does the rest of the enterprise know? Are you sure?

32.3 Added Guidance

There are five additional guidelines to further clarify how you operate defensively. These five help inform all of your regular tasks and responsibilities. They are reminders to always keep in mind, even while doing things you do all of the time almost without consciously knowing you are doing them. This helps keep you sharp and "in the moment." Remember: what you miss seeing or doing will always find you. It might take time, but it will. Defensive operating can give you an opportunity to keep the bad things visible and hopefully away from harming you because you can see them coming in time to react [**Q14b**].

A. Use Extra Caution When Making Operational Changes

When you are maintaining operation, you are directly engaged in activities that you have already been doing for a while. You are aware of what is going on. This familiarity and operational history helps maintain a frame of what is normal and what might need attention. When you are making an operational change, all of this changes. Extra caution and a heightened sense of awareness are required.

Operational changes fall into two broad general classes:

- Maintaining current operation, including adjusting for load changes, managing disturbances and equipment malfunction, altering production rates, and the like
- Changing operation like starting up, shutting down, diverting product, transitioning from one product to another, and more

B. Leave Some Space—Operating at a functional limit line may seem to be efficient use of resources, but will not be when that very extra is needed to account for the unexpected.

- *This is not to say that everything needs to be saved in case it is needed. But there are resources that it would be prudent to keep in reserve. Chemical production that operates near a transition that could lead to exothermic reactions needs to have sufficient cooling at the ready in case it does. Network trunks need to have capacity or a plan to shed load in order to be prepared to handle unexpected yet vital communications.*

C. Pay Attention to Current Surroundings—Operators need to be in the moment as they do their work in the control room.

Personal telephone conversations or emailing takes attention and emotional room away from the present control room task. Shifting back is not seamless. Intervening history is not so easy to see looking back as it is watching in real time. A task half complete is harder to pick up again without missteps.

D. Trust but Verify—Confirm that what you expected is actually there.

Completed maintenance is not ready for service until you verify and prove that it is. An outside delivery of a chemical additive is not completed properly until you check it to be sure. A customer saying he is ready for delivery of more product is not ready until you watch him taking it and it works. Never assume.

E. Always Yield—Teamwork and safe operation is all about working as a team and doing it safely. Yield first and find out why later.

When coordinating operation with others, if your particular operation is in the lead, but you see that maintaining that lead might cause a problem, first back off, and only then, find out what is going on.

32.4 What to Watch Out For

Think of using defensive operating just like sunglasses. You put them on as soon as you get in the sun (take over your shift) and see everything through them. Operating with them on is going to be second nature.

Anytime you find yourself close to the edges, a bright red flag needs to start waving. This means any edge, not to mention two or more. That red flag says unless you do something explicit, it is time either to get help (if you have enough time) or otherwise move to a safe operating state. You have procedures that you have practiced before and you know that you can perform these properly this time. If needed, you will use *permission to operate* [**M35**] right now.

Module 33: Taking Action

Operators have two important duties: taking action when it is proper and needed, and not acting when it is neither proper nor needed.

33.1 What Taking Action Means

As an operator, almost everything you do during your shift falls into one of the following general categories:

- Be the "point man" for all that needs to flow into, out of, and needs to be managed by the control room.
- Understand the current status and likely future status of your unit.
- Take action where needed, and decline to take action where you judge action is not needed.
- Prepare required reports, issue necessary permits, and ensure all are proper and being followed.

> All actions you take must be followed by actively ensuring that the actions you took are having exactly the expected result you sought to achieve and all side effects have been recognized and properly managed.

This module visits when and how you take action, or why and how you decline. Always keep in mind that every action must always be fully appropriate and followed by actively ensuring that the action taken is having the expected result. If not, correction is required.

There are lots of tools and aids for you to reference and follow. They are discussed in other modules.

- Policies
- Procedures
- Protocols
- Operating instructions
- Training experience
- Checklists
- Alarm response sheets [**M19.4**]
- Monitoring and adjusting

Taking Action vs. Reacting

Action in a control room means you do something or specifically decide not to. When you do something, the cardinal rule to follow is to only act with sufficient reason, and act by diligently following all the rules, procedures, and protocols.

> You should take no action without sufficient reason, and then act (or decline to act) by diligently following the rules, procedures, protocols, and all other directives and requirements.

Reacting is a form of action, but it is done without sufficient analysis to properly convince you that you understand the situation. It is not enough that you do the reacting by faithfully following the rules, procedures, and protocols. You first need to have sufficient justification for considering that action! And when you either take the action or decide not to act, you need to ensure that what happens yields the required results.

The Surprise Effect

Once in a while, something significant happens completely unexpectedly, without warning. Basically, we are caught off guard. When it happens, the response can be so startling that it evokes something akin to a reflex response. Instead of pausing and taking a moment to figure things out, we will reach for the first remedy we can think of and push it hard. Not good in a control room.

"In an automated process the surprise effect, caused by unexpected changes, can affect or delay rational behavior. Upset prevention and recovery training aims [among other things] to reduce the negative effects caused by unexpected changes." [33.1]

The immediacy and shock of a sudden surprise is powerful but manageable. Control room protocols and training will make the difference between guessing and luck and a proper response. Each enterprise should lay out this situation for its operators. Take the time to find it and work out how you are going to react. Your reacting only needs one critical behavior: never do anything immediately (except run to safety if reacting to fire or other clear and present danger).

Always pause and follow the standard four situation management steps.

1. Define the problem that has surprised you? Identify it clearly.
2. Do what you know how to do to confirm or disconfirm that what you see is actually a situation needing immediate response.
3. If it is as important as you think, and you know what to do, do it (while factoring in anything else it may affect and staying in familiar territory).
4. Make sure that what you thought was the problem is still the problem and what you have done to manage it is working. If not, and you have enough time, start over.

At any moment where you feel it is not working and you are running out of time, use *permission to operate* [**M35**]. Being surprised is not a free pass around anything you need to do to keep safe and be responsible.

33.2 Alarm Response Manual

The *alarm system* [**M19**] contains everything you need to manage every active alarm. It specifies the steps to take to convince yourself that the alarm actually represents an abnormal situation and therefore will need your active intervention. It contains the specific actions you will consider and take, if needed, to respond to that situation. The actions are carefully laid out. All have been formally approved. Think of each alarm response sheet as a specialized collection of analysis and remediation information for exactly the current abnormal situation you are tasked to manage. It is a procedure that was custom-built following a policy-based protocol. It contains all of the four steps of proper situation management, but with a built-in advantage: everything has been carefully designed and spelled out for you to use.

33.3 Responsible Operation Issues

Duty to Manage

As the sitting operator, you understand the operational hazards associated with your job. You may have specific instructions from your enterprise as to what components constitute these hazards. You are charged with reporting any deviation from the norm when dealing with these hazards.

There may come a time when you notice a new hazard that is not on your "list." This could be something like the presence of a hazardous chemical in an analysis of exhaust gas. You know there are hazardous chemicals that might be present but that are deemed safe if kept below a predetermined level. However, this is something new. What do you do? If it is covered in an applicable procedure or protocol that applies to the stream or unit in question, follow it. If not, immediately report it. If precautionary actions are prudent, initiate them to provide sufficient staff, protect personnel, protect property, and manage environmental damage. If needed, get help while there is enough time.

Duty to Report

Let us compound the earlier situation. Even if reporting to proper authorities will result in a reportable incident or violation, putting the enterprise in an unfavorable situation, you still must report it. If you do report it to your supervisor and your supervisor chooses not to act upon it, you should follow established enterprise procedures for this. In the event that there are no workable or available enterprise escalation procedures, the prudent action would be to seek outside guidance in order to protect you, the enterprise, and the public.

33.4 Managing Problem Situations

There are times during your shift where things just do not seem to be going the way they need to. Some piece of equipment is not working the way it should. A laboratory report that you wanted is late or missing values. Your relief is late. You seem to be running behind, and it is beginning to worry you. You have visitors coming, but there is a spill (e.g., it is nothing consequential, but it looks awful and you are not able to find a clean-up crew).

Your backstop is time. If you have it, and you know that you do, relax. Not too much, because you need to get help right away. You should become very familiar with the ways to get help [**M34**]. It will enable you to figure out what you will do to get it. Remember, if one way does not seem to be available or to work, there are lots of others. Persist until it is done or you are running short on time. Before you run out of time, invoke *permission to operate* [**M35**] to get things safe.

There are other times, rare though they are, where you are getting really desperate. Something is going on and you feel as if a tsunami is heading directly for you. You can see your cool evaporating. Or you feel sick. Or you become injured enough that it gets in your way. Be responsible: *escalate* or invoke *permission to operate*. Always.

33.5 Problem Requests from Supervision

This topic is delicate. It has difficult ethical and perhaps legal ramifications. What follows are suggested recommended procedures to be approved by your enterprise prior to your following them. No one can do more than advise you of things to consider when you are confronted with it. Please, in advance and with the cooperation of your supervisor, ask your enterprise to provide the policy you follow.

> This topic is delicate. It has difficult ethical and perhaps legal ramifications. No one can do more than advise you of things to consider when you are confronted with it. Please, in advance and with the cooperation of your supervisor, ask your enterprise to provide the policy you can follow.

The operational area boundaries between command and control in the control room are visited in the earlier module on *boundaries* of responsibility [**M30**] and a later module on *escalation* [**M34**]. Unless you explicitly agree and relinquish operational control to another, full operational responsibility remains with you. There are a few very special cases where the decision can be removed from you without requiring your agreement. One was where the supervisor (who must be in your direct chain of command) ordered you to leave or relieved you. The other was where you are functionally unable to understand or conduct duties due to incapacity or absence. Both are very clear exceptions.

> Every enterprise must identify the necessary situations and put policies and procedures in place for appropriately managing situations where the sitting operator is instructed (ordered, required, demanded) to perform operations that he would neither consider nor do based on his due diligence.

An important situation still remains. It is when you are in charge and given a difficult operational order by a supervisor in the direct chain of command. It is generally recognized that the sitting operator is the single individual responsible for all hands-on operational management. Hands-on means taking actions. It also includes responsibility for the consequences of not taking actions when actions should have been taken. If you feel that a specific operational order is contrary to legal, safe, and responsible operation, what is the proper course of action? To the authors' knowledge, there are no statutory requirements that fully address the situation (other than your refusing to violate the law, of course). What follows is an explanation of the circumstances and possible suggestions. Every enterprise must identify the necessary situations and put policies and procedures in place for appropriately managing them.

There are two situations. The first is where your supervisor is not formally qualified to be an independent operator. The second is where your supervisor is formally qualified.

Supervisor Not Fully Qualified to Be Operator

The concern here is that, while a supervisor is fully trained and qualified in a supervisory role, he is not fully trained and qualified to actually sit in the chair as an operator for the specific enterprise in question. We take a lead from the Department of Transportation (DOT) Pipeline and

Hazardous Materials Safety Administration (PHMSA) regulations. [33.2] [33.3] Only a fully qualified operator for a specific control room is allowed to make operational changes from a control room or to order others to make operational changes under his authority. Therefore, for all operations managed from a control room, no unqualified individual may make, or order you to make, any operational change, or to prohibit you from making an operational change that you deem necessary. They may suggest changes, but may neither expect nor require the changes to be followed. This prohibition extends from the lowest level supervisor to the highest enterprise authority level.

This would seem to be a prudent policy to adopt by all enterprises. The only exception, and this is an important exception, is the supervision's authority to safely and immediately order a cease of operations. This is retained by anyone in proper authority.

Supervisor Fully Qualified to Be Operator

This situation is when your supervisor is fully trained and fully qualified to operate the plant or enterprise for which he gives an operational order to you. What must you do if you feel that the operational order is unsafe or will result in unlawful operation, or likely lead to unsafe or unlawful operation? We cannot prescribe a remedy here, yet this situation is all too real. It must be adequately, formally, and fully covered in the enterprise policies and procedures.

A few suggestions might be useful for consideration. Once the order is given or you think the order is given, and the order is one that you would not on your own consider doing:

1. You restate the order to the supervisor and wait for confirmation from the supervisor that you got it right.
2. You (or anyone else on the operations team) invoke the principles and tools for crew resource management (CRM) [**M34**]. This means respectfully questioning the order and suggesting the reasons why following the order will or may result in unsafe, unlawful, or other significant damaging consequence. Suggest alternates that could be acceptable to both you and the supervisor.
3. You formally state that you object to following the order due to its being unsafe, unlawful, or will cause significant damage. Your objection should be witnessed and documented.
4. If the operation is unlawful, you have a legal responsibility to not follow it. No supervisor or enterprise manager has the authority to ask, much less order, anyone to commit a criminal act.
5. If the ordered operation is unsafe or damaging, but not to your knowledge illegal, you may:
   a. request a relief from operational responsibility (meaning that you would not be operating and that someone else would need to follow the order).
   b. follow the order, with the documented full understanding that it is done under protest and that the ordering individual accepts full and complete responsibility for all outcomes, including any errors made in good faith by you.

33.6 Bottom Line of Action

Your authority to take action comes with the job. It is expected. You take action almost without actually thinking of each as an action. But the "taking action" part comes with the critical components you already know. Be sure that they stick in the front of your mind as you work.

1. Before, during, and after you act: look for, recognize, and process any conflicting or cautionary issues raised by anyone working with you [**M34.3**].

2. The decision to act must be based upon sufficient justification.
3. The act (actions) must follow approved established processes.
4. The results of the action must be tracked and managed.

The critical bottom line is that you always ***factor in threats*** and ***stay in familiar territory***. In all cases where there is doubt, escalate, collaborate, or delegate when you are certain that you have time; otherwise, invoke ***permission to operate*** when you do not.

Module 34: Getting Help

34.1 When You Need Assistance

This module covers important ways for you to get assistance when you think that an extra view or additional pair of hands would be helpful. Methods such as delegation, escalation, and collaboration are not meant to be reserved for just special situations. If they are, you and the team will not be practiced enough to be good at them. And likely you will have waited too long to do them. This in no way suggests that you are not fully up to the job. Actually, it suggests maturity and responsibility on your part. Few of us have the collective experiences of everyone else with whom we work. None of us can know it all. When operating situations get extremely complex, confusing, or you would just be more comfortable with a second opinion, it is good to know that you can reach out to others. Here are ways it can work.

34.2 The In-Charge Rule

The in-charge rule applies when you are the scheduled, on-shift operator, obviously in charge. As you are aware, lots of things can be going on at the same time in your control room. What if something big is going on? What if others are present from management, or crisis response, or any other significant part of your enterprise? What if they are acting instead of you? We need to clarify exactly who has the bottom-line responsibility to make sure the individual seemingly in charge actually is responsibly in charge and everyone knows it. It is you. Let us call it the *in-charge rule.* This is what it is:

At all times when you, as assigned operator, are present in the control room, your overarching responsibility is to ensure that someone is always in charge. No matter what the status of escalation or delegation or any other activity that appears to or actually changes operational roles and responsibilities, you make sure that someone is in charge and knows that he is acting in that role. You have an affirmative duty to know who that is. You have an affirmative duty to make sure that that individual knows that he is. And you have an affirmative duty to make sure everyone else knows and understands what it means to them. If you see or suspect a problem, this duty would be met by respectfully reminding the individual in charge that he is exactly that. If that individual refuses or is unable to assert that role, then you must find a replacement or assume the role yourself.

34.3 Collaboration

Collaboration is a specialized communication event. During collaboration, the parties work together to clarify a status or situation about which neither alone may have sufficient knowledge. They work together. To do that, it is essential that they both understand the same problem and have the same goals. Here is a working communications checklist developed by Weick and Sutcliffe [34.1] for making sure that it works [**Q15**].

Five Steps for Collaboration/Coordination

1. *Situation*: Here's what I think we face.
2. *Task*: Here's what I think we should do.
3. *Intent*: Here's why I think that is what we should do.
4. *Concern*: Here's what we should keep our eye on because if that changes, we're in a whole new situation.
5. *Calibrate*: Now, talk to me. Tell me if you don't understand, cannot do it, or see something I do not.

   (Ask the following focus questions to round out Step 5)

   a. What have I missed?
   b. What do you know that I don't know?
   c. Where do we go from here?

You already know an extremely important special-case tool called *mindful conversations* [**M6.3**]. It is used during emergencies in situations requiring strong command and control. It is understood to be a part of *crew resource management*, which is covered next.

Collaboration must never start out to be a confirmation of an initial suspicion. Specifically, collaboration avoids anything to do with confirmation. The results of collaboration might (or might not) end up being a confirmation of an initial suspicion. But it must never start out to be. Ask, "What do you make of this observation?" Do not ask, "Is this really happening?" Seek assistance to understand. Then seek options to manage a confirmed situation. Seek extra eyes, ears, hands, knowledge, and experience. Listen. Compare. Understand. Work together, even though you have final authority.

34.4 Crew Resource Management

Crew resource management (CRM) is a collaborative process of decision support. It is also called Cockpit Resource Management by the United States Federal Aviation Administration (FAA) and Bridge Resource Management (BRM) by the United States Maritime Administration (MA). The FAA derived the need for CRM as a result of the investigation into the 1978 fatal crash of United Airlines Flight 173. While actually in the air, this flight "ran out of fuel while the flight crew were troubleshooting a landing gear problem." [34.2]

CRM recognizes that different members of a close-knit team can have different expertise, differing responsibilities, see different information, and reach different interpretations of shared situations. The power of CRM comes from its ability to interject critical information from subordinates into the command consideration process during the moments of greatest need.

> "[The FAA] began reappraising the traditional unwritten rule that the captain had ultimate authority on a flight and could not be questioned. From that point onward, first officers were encouraged to speak up if they believed a captain was making a mistake." [34.3]

In the Control Room

CRM in the control room enables all participants to actively engage in shared situation awareness, situation assessment, and decision management activities. Effective CRM does not undermine authority nor should it be interpreted to suggest any individual's lack of ability or inattention to duty. Just the opposite. CRM recognizes that the whole can be greater than the sum of its parts. You want to take advantage of this. Here are the key roles:

- Each **member** of an operations team has a duty not to be distracted by a singular problem that might interfere with their responsibility. They are to remain focused on the full scope of their specific responsibilities, using the full extent of their training, responsibility, and experience.

  Exception: There are critical times when delegation to an individual of a specific, narrowly focused task or responsibility is crucial to detecting or following something of vital importance. At this time, that individual has no other responsibility but to be hypervigilant to the assigned task and advise others of its status.
- Each **member** of an operations team has a duty to search for and identify:

  (a) any operational parameter or condition that may contribute to a problem, and

  (b) suggest when the present team's understanding might be inaccurate, incomplete, or incorrect.
- Any **subordinate member** of an operations team, having identified any potential abnormal operational aspect (see above bullet) has the affirmative responsibility to respectfully and accurately bring the identified operational aspect to the attention of the primary member of the operations team.
- The **primary member** of an operations team has the specific affirmative responsibility to hear, in the moment, all subordinate members' concerns and factor them into a proper operational frame.

Three Step CRM

To keep this basic and simple, assume just two parties: a *lead* and a *follower*. The *lead* could be a first level supervisor or just another operator (more senior or not) than the *follower*. The lead is involved in managing or otherwise in charge of some aspect of the plant. The follower is at the same plant aspect. Among other responsibilities, the follower has access to some, most, or all of what is going on in the parts of the plant that are being managed by the lead. The follower notices an abnormal situation that must be managed to avoid consequences. Effective CRM ensures that all parties to the situation share all known information, know that it is shared, and ensure proper action, or avoid taking action when unnecessary, to manage the abnormal situation.

In addition to the active engagement to ensure sharing of information and concerns between the working individuals, there is a specific CRM working procedure called the *challenge*. It is used to ensure that an abnormal situation does not remain unnoticed or unmanaged.**[Q20]**

The Challenge

Here is how the three-step *challenge* process plays out.

The follower notices something (termed the *facts*) in his area, or in the lead's area that might affect what the lead is doing. The follower is not sure that the lead knows what the follower knows because he has not communicated it or otherwise indicated it.

The steps are (1) Statement of Facts, (2) Challenge, and if it comes to it, (3) Take Action.

<u>(1) Statement of Facts</u>

The follower states the facts in order to confirm that the lead knows them, or advises the lead even when the lead knows, but has not indicated that he does. This is a succinct statement.

EXAMPLE. "The tower temperature is way off normal."

(2) Challenge

The follower restates the facts AND personalizes and quantifies it. This is designed to firmly refocus the lead away from any immediate distractions, or actions under the influence of his assumptions.

EXAMPLE. "[Use the actual name of the lead], are you going to continue the normal flow rate into the splitter tower? Check the low temperature at tray 22."

(3) Take Action

If the lead fails to exhibit direct engagement with the situation because the *challenge* goes unnoticed, or if noticed, goes unprocessed, or if processed and the remediation action is inappropriate, the follower must take immediate action. This action would be the least impactful one to ensure avoiding the abnormal situation consequences.

EXAMPLE. Follower, at his own initiative, immediately reduces the tower feed rate by 20% and monitors the tray 22 temperature. He tells the lead of his actions and states, "I will watch the tray 22 temperature until you relieve me."

> If the lead's actions are appropriate but do not appear to be effective, the follower's only supportive role is to immediately work to identify other actions to propose to the lead. Taking action on his own must not be done.

34.5 Triangulation

Another way to collaborate is *triangulation.* [34.4] Triangulation provides a second way to view and see something from a different perspective. The word triangulation comes from noting that a triangle is formed when an object is viewed from two vantage points (view one, view two, and the object) In situation management, these dual perspectives can expose differences that help provide a better understanding of what is seen, what is happening, and what to do about it.

Consider using triangulation when something is amiss and you need to know more. Again, this is not about confirmation. For triangulation, eight questions are asked. The answers (or lack of answers) for each of the eight will be reviewed and factored into your thinking. Remember the *knowledge fork* [**M11**]. Use only known information. Only when all information is in place, and there is enough time to work carefully through it, do you assemble a picture of what is going on and what the options are for dealing with it. If the situation still is not clear, escalate (discussed later in this module). You are not going to magically figure this out.

When do you use triangulation in the control room? You use it whenever you suspect something amiss in an area of your plant but cannot see or know for sure. If there are others available, you can ask for their assistance. There are two methods of applying this tool: (1) *job/task triangulation*, and (2) *experience triangulation*. Remember that you are specifically not asking anyone else to "see what I see." You are asking them, "What do you see?" Each method uses all of the triangulation questions. You should actually ask them!

Job or Task Triangulation

This approach takes advantage of situations where you and one or more other individuals are working on distinctly different jobs but in the same physical area. The jobs share general plant knowledge. There may be some, or even significant, job work overlap. When you find yourself with a problem or issue, you ask the other collaborating individual for his input. You, being responsible for the plant, are familiar with the other individual (field operator, maintenance technician, or such) and what he is doing. In this situation, you ask the eight triangulation questions below.

Experience Triangulation

In *experience triangulation*, you are not directly involved or working with the collaborator. An experience-based approach involves your actively seeking out someone with experience in your area of operation, but that person may or may not be actively working in or near your plant. It could be another operator in the control room (or available via electronic linking). If it is around shift change, it could be the relieving or the arriving operator. It might be the process engineer, safety engineer, trainer, or others. As before, you ask the eight triangulation questions, but it is understood that the collaborator will answer from his experience.

Triangulation Questions (to ask the other person) [**Q19**]

1. Do you see **anything unusual** or out of the ordinary from where you are?
2. Are you having **any problems** doing what you are doing?
3. What are the ways that what you are doing may affect my area of the plant?
   a. Are you **seeing everything** that you need to see to check that what you are working on is properly functioning?
   b. Are you seeing **all of the indications** that confirm that what you are doing is working properly?
   c. Do you see **anything amiss or unusual** that might suggest that something in your area might not be going as expected?
   d. Are you **unable to see** anything that could clarify or expose any of your concerns?
   e. Do you see anything that you **cannot fully and simply explain**?
4. Assume that you are doing my (the operator's) job:
   a. How could what I am doing **affect** what you are doing?
   b. Do you see **anything unusual** or out of the ordinary in my area?
   c. Do you see everything that you would **expect to see** to ensure that my area is properly functioning?
   d. Do you see anything that you would **not expect to see** going on in my area?
5. Are you aware of **anything going on in the larger plant** that might affect what you or I are doing?
6. Are you aware of anything planned to happen in the larger plant that **is not happening** or is not going according to plan?
7. What do you (operator asks the collaborator) know that I don't know?
8. What might I know (operator asks the collaborator) from my perspective that you do not?

Processing the Answers to the Triangulation Questions

Only if sufficient time is available, do you work out the answers to the triangulation questions. Use this new or confirmed knowledge to understand and manage. If it does not work, or you are too short on time, escalate or rely on *permission to operate*.

34.6 Worst Case First

Group discussion between informed, competent participants is an often-used process for gaining collective knowledge and concerns. *Worst case first* is a special procedure that works to overcome the preconceived notions that we all have. Worst case first shapes the thought process to reduce biases [**M16**]. It does so by starting with the problem and identifying not the most likely or most reasonable or most familiar, but the worst cases first. Notice that this is exactly what *forward extrapolation* does for weak signals [**M23**]. The group is asked to assume that the situation will play out in the most extreme (but not fantastical) way. It may seem to be leading with pessimism, but actually is far from it. We all hope for the best, but hoping for the best is not a plan. By not planning, you can actually be planning for disaster. Finding hints of that disaster early is what this is about.

To start, the group formally asks each participant to put forward what they can visualize as the most impactful result or outcome—their worst-case scenario. One rule: the visualization must be reasonable. Participants will need to be seriously honest, but stay "possible." Staying "possible" includes remote yet reasonable outcomes. In fact, these are the very situations that the method wants to bring to everyone's attention.

There is one additional important part of this tool to use. The most junior, least experienced member goes first. This way, the first voice does so without knowledge of what his peers and superiors are thinking. And so on up the chain. Working the discussion this way has two important benefits. First, experience, by its very nature, favors what is known or what is expected. This is not a problem when it comes to action where we need all available experience and competence. But when it comes to understanding possibilities, if done too soon, it can cloud the issue. The second benefit is that it avoids a whole host of biases. By not hearing from anyone with more experience or more insight or actual "stories" about their history, the junior participant is forced to rely on his own ability to think and visualize what might be possibly going on. No idea is examined until all participants have voiced their scenarios. If you are junior, have courage. If you are senior, be patient, open, and considerate.

34.7 10th Man Doctrine

The 10th *man doctrine* was developed by Israel after being blindsided by the attack during the 1973 Yom Kippur War.[34.5] This doctrine was put into practice via establishing a special branch of intelligence to take the devil's advocate position on everything. It reduces the untoward effects of groupthink in situations such as the one in the following illustration:

> *A group of individuals must come up with a collective decision. There are vitally important consequences for not getting it right. Each participant is an expert. After discussion and deliberation, the matter is polled and the entire group is of one mind. Every one, each of the participants, is ready to conclude that the group decision is the exact one they would support without reservation.*

Unanimity is always very comforting. It reinforces acceptance of the decision. It emphasizes the conclusion that the decision is right. However, it is the very uniformity that should be discomforting! No carefully considered decision is without doubt or is so certain. Uniformity can be achieved for difficult decisions, but it should not come readily.

Here is how the 10th man doctrine works. If a group of decision makers all reach the same conclusion, one experienced member of the group (against his personal convictions or not) is given the explicit task of going "outside of the box" to find all of the reasons why the chosen decision is not the right one, no matter how improbable the disagreement might seem. The results are brought back to the rest of the team. This enables more discussion, perhaps more investigation, and ultimately a better decision because all new issues are recognized and thoroughly worked through. The team now must reach a decision again. If it is the same as before, there is the additional reinforcement. If different, the decision now has an effective foundation to find out why the earlier one was not the best.

34.8 Red Teams

Red teams are a special type of collaboration. It is also a team form of the 10th man doctrine discussed above. Red teams are a very useful way to find gaps or holes in important thinking to ensure that it is well founded and without ambiguity. Ambiguity in situation awareness and assessment is a very difficult thing. When there is ambiguity, everyone works hard to ignore it so it will go away. Yet it is almost always a useful indicator of imperfect, closed, or biased thinking. The following quotation is from the *MIT Sloan Management Review*:

> *The British Armed forces and other organizations deploy so-called red teams to accomplish this [way to find holes]. The red team is a parallel task force, made up of senior leaders and support staff, whose only mission is to collect and synthesize information to prove that the current plan is wrong and needs to be changed. This team plays the role of the loyal opposition, in the spirit of Alexander the Great, who would periodically ask himself how much evidence it would take for him to abandon the current plan.* [34.6]

Using Red Teams

Red teams are useful for three situations: time-critical, decision-critical, and strategic. The complete procedure and process for using red teams would be spelled out by existing and approved site protocols and training. If none exist, it strongly suggests that any results would not be useful and, in fact, may likely be damaging.

Time-Critical Red Teams

Do not consider using them here! Where time is the critical factor, red teams can be started only when you know for sure that there is sufficient time for using any alternatives, including *permission to operate*. But, and this is a critical aspect, attempting to use the red-team approach where anyone on the red team is concerned about time, the press of time will almost always overrule any benefit. Unless there was an amazing degree of luck, any results cannot be trusted. Please, do not bet anything of value on luck!

Decision-Critical Red Teams

Where a concerning situation is recognized, and it is clear that any untoward effects are not coming soon, a tactical red team can be considered. In the control room, an ad hoc tactical red team can be assembled by anyone qualified except the *operator*. The operator cannot be a member. In every respect, the sitting operator must be completely outside of this activity. Yes, he will know that one is working. No, he will not interface or track or in any other way be aware of its activity. And yes, as part of their activity, the red team may ask the operator for specific observations and understanding. But no feedback can be given. Likewise, the operating supervisor must not be a member. His job is to maintain an overview. Participation in a red team will create an irresponsible distraction for him.

The composition of the tactical red team might include the following: arriving operator before his shift begins (even if it is delayed), the departing operator after his shift has ended (but he is kept over), area process engineering, technical specialists as it relates to the general problem area, any pre-assigned call-ins for red team duty, and any other available and relevant individuals. Keep the team small. The supervisor will charge the team only. The senior red team member will organize the team and manage its full conduct.

Strategic Red Teams

Red teams can have another important use for the control room. They can find operational vulnerabilities before you encounter them. The best way to manage abnormal situations is to not have them. In order to not have them, the ones you do not want need to be identified specifically, or the general situation that leads to them needs to be found. This is where the strategic part comes in.

Say that you and your fellow unit operators are experiencing situations that are not clear, yet are troublesome. You sense a vulnerability. But all of the usual suspects have alibis. Nothing appears to suggest a need to follow up. Doing what you did before either seemed to work or the situation went away on its own. No one seems to have enough time to really look into it. But the few who did look were not able to find anything useful. A red team here could be able to uncover what is going on.

Supervision forms a red team and charges it with finding operational vulnerabilities that might be lurking and on the verge of impacting the existing controls and safeguards. In this case, you may be a member of the team but only during your off-shift time.

34.9 Four Corners Tool

Not all plans or situations need to lead to action taken. Sometimes the best action is no action. The *four-corners tool* [34.7] is best used to balance consideration of a plan that allows either action (decision to do something) or inaction (explicit decision to do nothing). It does so in a way that sharply exposes the relevant choices and consequences. Deciding which to choose requires careful examination and consideration. Our normal inclination is to consider only the best outcome if we act. We tend to overlook what good thing might happen if we do not act. We really should know what might be good **or** bad for both the act of making a choice and the alternative of making no choice. This is a way to get the full picture for a responsible decision.

Pictorially, Figure 34.9.1 illustrates the framework to guide how you find a proper balance. A question is posed for dots in each corner. Understanding the answers for all four questions leads you to the best answer. Here is how it works.

If I Take Action Questions

If I **do** what I am thinking of doing (along the top edge of the square):

Q-1. What is the **best** that can be expected to happen (upper left corner)?

Q-2. What is the **worst** that can be expected to happen (upper right corner)?

If I Do Not Act Questions

If I **do not do** what I am thinking about doing, meaning probably doing nothing (along the bottom edge of the square):

Q-3. What is the **best** that might be expected to happen (lower left corner)?

Q-4. What is the **worst** that might be expected to happen (lower right corner)?

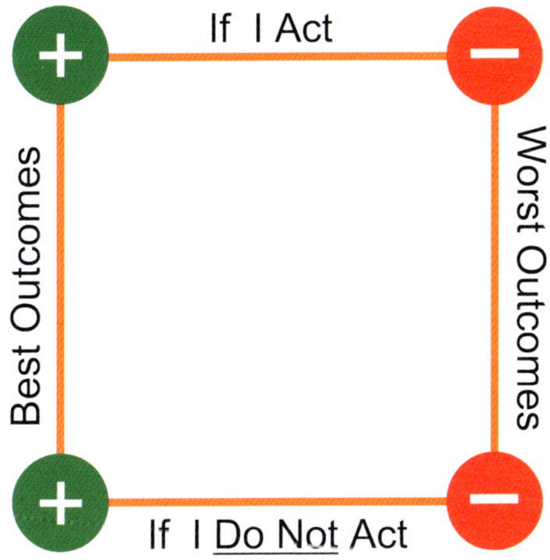

Figure 34.9.1. Four-corners decision framework that exposes needed information for a proper balance between what might be good or bad for both the decision to choose to act and the alternative of choosing to take no action (keep things as they are). Visit all corners. With this information in hand, an informed decision is more likely.

Result

Which corner is where the most favorable outcome is found? Is the value of that expected outcome for making the decision to take action better or worse than the expected value from the expected outcome if no action at all were taken? Simply, what is the best of the best? And are you protected if the worst of the worst outcome actually happens despite your choice?

Discussion

This is a simple, effective little tool. It looks at all four logical outcomes. Yet it goes a long way toward maintaining a balanced consideration. Going to all four corners tends to reduce bias or preconception. With the best of the best and worst of the worst known, it should be enough for you to weigh the value of the best against the damage of the worst. Notice that the best of the best is not necessarily found at any specific left corner. Nor is the worst of the worst at any specific right corner. There is no need for any probabilities for the outcomes, so please do not estimate or assume any. It will certainly defeat the whole objectivity of the process. Be comfortable with reasonable knowledge. Additionally, considering each part (corner) containing an alternative, one at a time, provides an excellent procedure for others to participate with both the outcome predictions and the selected decision.

There is an additional important benefit. By looking at both of the worst corners, unless you have made more than one miscalculation, you might get a good chance to see what might come to pass if you made an error in the selection of one of the "best" outcomes. You are looking for the best. But if the worst is bad, it might need to be a factor. If those "worst" outcome prices are really high, it suggests that it might be a good idea to do the decision-making much more carefully. In summary, the tool provides a compact and logical way to visit and document the process and decision.

34.10 Escalation

Escalation is an essential safeguard for responsible operation. It improves with use. Even fully qualified, well-trained, and engaged operators at times face situations that exceed their strengths. This is recognition of the sheer complexity of plants and operations, as well as the diverse ways operators can get into trouble. The fact is that some situations may be beyond any single person's ability. Control room protocols need to address how the enterprise expects these situations be handled.

Escalation should occur whenever you recognize that the current situation (abnormal or not) seems to you to be beyond your resources to manage. "Manage" includes understanding the situation, developing a plan for the situation, implementing the plan, understanding whether the entirety of that process is working satisfactorily, and making adjustment if needed. Escalation is a pre-trained and proceduralized activity for you to recognize a need to get help and get it. It is a formal process. It is not passing the responsibility to someone else. It is a responsible activity. Specific things happen when it is used as part of a planned program of getting help. It must be carefully laid out by the enterprise in policy and with procedures and training.

> *Remember the In-Charge Rule*: At all times where you as operator are present in the control room, your overarching responsibility is to ensure that **someone is always in charge**. No matter what the status is of escalation or delegation or any other activity that appears or actually changes operational roles and responsibilities, you must make sure that someone is in charge and knows that he is acting in that role.

There is a clear difference between seeking reactions (communicating) [**M6**], seeking advice and possibly assistance from others (collaboration), and escalation. The difference is that, before escalation, **you** maintain full and exclusive control over this problem and the rest of your responsibility. While the same individuals may be consulted or asked to monitor or "keep an eye on" something or another, it would not be considered escalation. Such activities are part and parcel of a cooperative workplace and should occur in a comfortable and effective way.

Depending upon the enterprise protocols, escalation may also be declared by someone other than you. Thus, an operationally qualified supervisor may intervene to invoke the escalation and take over or pass to another the processing of a specific situation. In even more rare situations, any operationally qualified individual other than a supervisor may intervene when it is clear and completely self-evident that you are unable to function. The specific recommended process for escalation is described in *Escalation Teams* later in this module. They are what you will use to supplement existing enterprise protocols. As always in this handbook, your enterprise must agree with this process.

Escalation vs. Transfer of Responsibility

Only specific problems or highly ambiguous situations are escalated. An escalation must be as clear as your knowledge and awareness allows. Some examples follow of what escalation is for:

- Get significant help to discover what might be going on
- Get serious advice for how best to confirm and/or test facts or issues
- Obtain confirmed experience as to what should be considered and eventually acted upon to remedy the problem

When you identify an escalation situation, you still retain full operational authority and responsibility unless it is explicitly transferred. All intervention activities recommended by the escalated person should either be executed by you or executed with your specific knowledge and permission (in the role of a *long arm of the operator* [**M4.4**]). For example, it is not a good protocol for the escalating operator to give blanket permission to do whatever it takes to resolve the situation. This would effectively remove both knowledge and activity away from you. This lack of coordination would override any benefit.

Escalation Management and Consent

Though operational control is separate from escalation, there are rare escalations during which a transfer of control would be beneficial. They are discussed here.

- If the escalated-to person is fully oriented to the current situation *and operator qualified*:

  The person who is escalated to will formally acknowledge, "I accept the escalation. I am now in control." Or for another situation, "I accept the escalation. You remain in control; I will assist."

  This formality would continue until escalation is ended.
- If the escalated-to person is not oriented to the current situation but *is operator qualified*:

  The person who is escalated to would formally acknowledge, "I accept the escalation. You remain in control; I will assist." Later, as the situation and/or operating protocols permit, as operator, control may be transferred by declaring, "I am transferring control from me to you." This transfer must be acknowledged by the new recipient with, "I am now in control."

  This formality would continue until all escalation is ended.

The Process of Escalation

An illustration of the general flow process of escalation is shown by Figure 34.10.1. In your enterprise, a specific escalation process would identify and explain when, why, and how each step would be carried out. This is an example. When your enterprise has worked out the approved process, you will use that one. Begin at the top left shaded box.

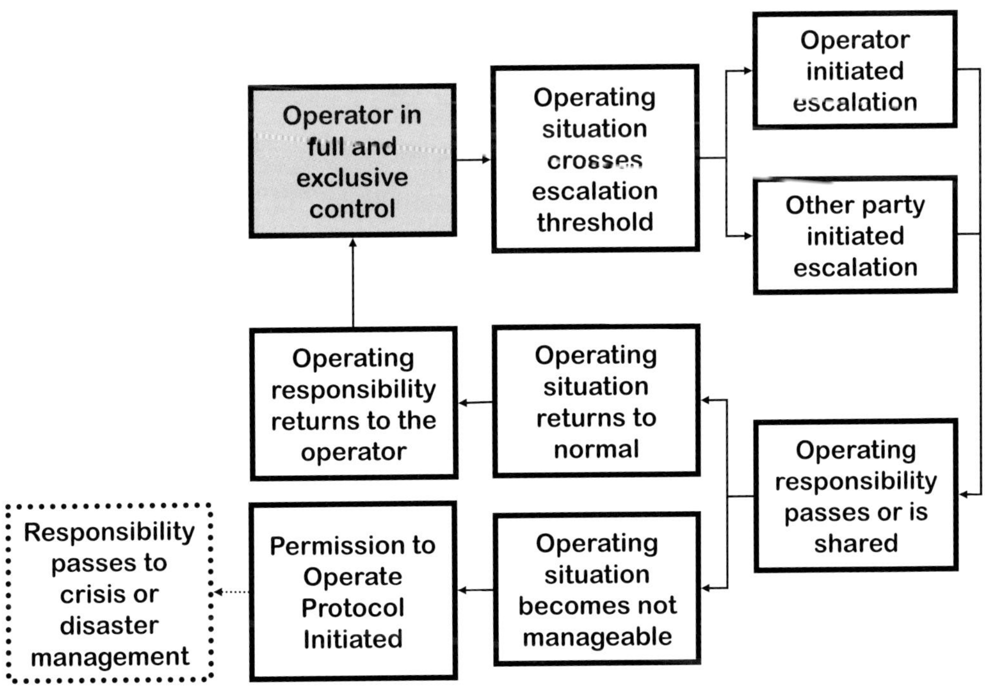

Figure 34.10.1. Begin at the shaded box at the top left. Escalation process framework, illustrating the initiation events and approved transfer of responsibility that can take place to ensure that the enterprise brings to bear the right skills for the right operating situation **[Q16a]**

When to Escalate

Enterprises must establish clear, unambiguous rules when escalation is used. There are times when escalation should be mandatory. And there will be times when you, without fault or blame, should be able to elect to escalate before a specific situation might require it.

Escalation should be available for supervisors to initiate. However, this action must not be reflexive or authoritative. There is no room for "I know better so I'm taking charge now."

Escalation should be available for others on the operational team to initiate. There may be times where the individual closest to the front line may be so engaged that a larger picture is missed. This is at the core of crew resource management discussed earlier in this module. When covered by enterprise policy, escalation may also be initiated by supervision.

Escalation can be an important tool in the following example situations.

Operational Area Boundaries Are Crossed

Escalation would be necessary (unless different protocols are set out for you) when the effects of abnormal operation cross an operational area boundary [**M30**]. An operational area boundary cross may be:

- from your operation area to another operator area (or from another operator's area to yours).
- from your operation area to the enterprise.
- from your operational area to outside the boundaries of the enterprise.

In all of these situations, the effects of operation are beyond your ability and authority to manage. Explicit site-approved protocols must be used.

Current Situation Is Over-Demanding

Operating at times of stress or unusually abnormal situations can be daunting. The enterprise should have specific and clear policy developed for ensuring that escalation is considered for these situations. In addition, *Situation Management* [**M29**] and *Managing the Control Room* [**M31**] cover how to transition the control room and the operations team for response to extreme situations. At all times, when time or resources are running out, *permission to operate* [**M35**] specifically applies. *Escalation* would not be considered or used.

Escalation Resources

Escalation means that an additional person or people will be directly involved in the current operational situation. They bring the following to the scene:

- Additional manpower for examining documents, assistance to follow protocols, tracking external effects, guidance, very temporary relief, etc.
- Broader or more in-depth experience
- Added authority to manage or track operational risk
- Extra set of hands (only for operationally qualified personnel, or under the explicit direction and control of the operator)

When an enterprise wishes to establish a policy to use escalation, it must be clear to all what is being added to the operational situation and who will do what. Specifically, there must be clear command and control authority at every step of the way.

Direction of Escalation

For situation management during escalation, we can look for added resources upward, sideways (peer-to peer), or downward in the organization. Therefore, when you find yourself in a situation where you feel escalation must be utilized, you should identify the resource and initiate the process.

The illustrations that follow should help clarify the intent:

(a) Another operator observed you having what may reasonably appear as a significant enough medical, psychological, or other impairment issue.

(b) A peer or subordinate (if qualified) may request escalation if the operator is cognizant enough to understand and agree—or may declare the situation escalated if the operator lacks sufficient awareness. If the peer or subordinate were not sufficiently qualified, then he would follow procedures to escalate on behalf of the operator only to someone who was available and qualified.

None of this discussion is meant to modify, supplant, or contravene any operating protocols for fire, explosion, insurrection, or other crisis or disaster situations.

First Duty of Escalation

Immediately when you consider or activate escalation, it must be verified that sufficient time is left to manage the current situation, once escalated. The escalating operator must do this before activating escalation, or the escalated-to individual before attempting any managing activities. There must be no realistic opportunity for things to go horribly wrong because the decision to continue operating by employing escalation was made instead of outright invoking *permission to operate.*

Escalation in Perspective

Escalation is a tool, not an end result. Using it can permit an enterprise to bring additional resources for managing. All existing enterprise policies, procedures, safeguards, requirements, and public responsibilities remain the same before, during, and after escalation.

34.11 Escalation Teams

Your enterprise has emergency, crisis, and disaster management teams because management understands that some situations can become serious enough to require a certain level of expertise to manage that is way beyond any individual operator's control. This leads us to the desirability for your enterprise to have escalation teams at the ready, fully trained. They just might be able to stall off a serious problem before it becomes unmanageable. This section introduces the concept of *escalation teams* (abnormal situation management escalation teams). Their function is to augment or take over your current operating responsibility with enough experience and resources to prevent an abnormal situation from becoming a full-blown emergency. Where an emergency cannot be avoided, then their efforts would reduce the seriousness of the impacts. The appropriate manning of these teams and their roles during operation will be left up to each specific enterprise. Here, we are introducing the team as a concept.

The following definitions are useful for understanding Figure 34.11.1 below:

> **Abnormal Situation**—"An abnormal situation is one in which it is no longer possible to continue operations (production, manufacturing, etc.) using the normal procedures. If left unmanaged, it can become an emergency that can endanger safety, financial integrity, and environmental containment." [34.8]
>
> **Emergency**—"An emergency is a situation that requires immediate attention, but is usually small [or limited] in scale." "An emergency can turn into a disaster if left unchecked. However, not all disasters are preceded by an emergency." [34.8]

The escalation team takes over whenever the operational situation is within the orange dotted box (abnormal operational mode) in Figure 34.11.1. This should be considered the next level of situation management. It is unlikely that any operational change could fully restore the plant to normal operation. Left unmanaged, the plant will likely suffer heavy losses. The specific timing of when the escalation team begins operation is covered under the topic *The Escalation Activity* later in this module.

The role of the escalation teams directly relates to the basic purpose of this handbook: the ability of an enterprise to provide the effective infrastructure. Enterprises will readily deploy escalation teams as often as needed for safe and effective operations. The ability to rapidly deploy competent escalation teams to the control room can make the difference between an orderly transition to a safe operating state with the clear ability to recognize incipient danger, and failure via inappropriate risk-taking leading to dangerous operation.

Operational Modes:	Plant States:	Critical Systems:	Operational Goals:	Plant Activities:
Emergency	Disaster Accident	Area Emergency Response System Site Emergency Response System	Minimize Impact	Fire Fighting First Aid Rescue
Abnormal	Out of Control	Physical and Mechanical Containment System Safety Shutdown, Protective Systems,	Bring to Safe State	Evacuation
	Abnormal	Hardwired Emergency Alarms DCS Alarm System	Return to Normal	Manual Control & Troubleshooting
Normal	Normal	Decision Support System Process Equipment, DCS, Automatic Controls Plant Management Systems	Keep Normal	Preventative Monitoring & Testing

Figure 34.11.1. Plant state vs. operational modes and goals identifying the situation management escalation region (orange dotted box) in which the operator changes the objective from trying to return to normal toward bringing the plant to a safe state (perhaps shut down if that is the only appropriate goal). When sufficient time is available, a purpose-defined escalation team can best manage this mode of transition **[Q16b]**.

Escalation teams are manned from a predetermined list. As soon as an escalation is declared, the team assembles. They may use special rooms outfitted to support this situation, or portions of existing facilities as needed. The rooms may have other uses for emergency, crisis, or disaster management. The team functions as long as needed and performs activities in support of predetermined plans governed by the particulars of the current situation. Once the abnormal situation has passed, the team disbands. Although there may be reports and follow-up investigations done by certain members of the team, for the most part, the team responsibility ends with the end of the event.

They may also have important roles in addition to managing abnormal situations. See Section 34.12.

Escalation Team Composition

An escalation team is composed of individuals with specific capabilities. And those special capabilities are extensive. The members are highly experienced in the design, operation, and risks of misoperation of the enterprise they support. There are experts in the detailed equipment design and operation, in the handling of extreme changes in operating environment that cause or contribute to the emergency, and in understanding human limitations during severe conditions. They have the ability to engage the efforts needed for success. A team is composed of an appropriate number (from 4 to 6) of the following:

- Senior (not specifically top) enterprise management
- Operating superintendents and senior operations leaders
- Equipment experts in design and operation
- Process experts in production behavior and limits of operation
- Maintenance supervisors and senior members of the maintenance team
- Senior logistics personnel
- Senior safety, health, and environmental personnel

The Escalation Activity

It is important to understand when operational control should transfer from you to the escalation team. You are always there. The escalation team is not. Assembling the team will take some time. We want to call them out when there is a reasonable likelihood that we should. On the other hand, we do not want to be caught unaware by not having them at the ready to do their job when needed. Each enterprise should lay down explicit guidance for this to happen properly. Too early is much better than even a bit too late! Some of the triggering events include the following:

- Activation of any single highest priority process alarm
- Activation of two or more independent second-highest priority processes alarms
- The moment an operator decides to initiate a shutdown of a significant portion of his operational responsibility to manage an out-of-control abnormal situation, but maintains operation of the rest
- Operator requests the team (likely where the operator has significant concern about the operational conditions being improper, but shutting down has its own significant risks)
- Supervisor requests the team (where the operational situation is reasonably likely to undergo significant stresses, such as severe weather or significant operational demands that threaten the integrity of the plant or impact essential product delivery—or there is sufficient uncertainty in his mind about what is happening or likely to happen)

Enough lead time must be allowed after notification so that they can actually be assembled into a physical (or virtual) functioning team. Once the escalation team has been called, it is the responsibility of the team to actually decide if and when to take some or all operational control from you. Usually, you join the team with specific functions under the direction of the team leader. In addition, the act of assembling the team should provide sufficient notice to the enterprise to alert the emergency, crisis, or disaster management teams to the situation and prepare to interact with them, as needed and prescribed by enterprise protocols.

34.12 Additional Roles for Escalation Teams

Front-Line Coaching and Mentoring Role

An activated escalation team provides you the perfect opportunity to be coached and mentored. Before that can be done, you must be reassured by the team that your decision to call them out was responsible and the right thing to do. It should in no way be considered a defect in your actions or skill. The fact that you or someone else was concerned enough to call the team is due to the overriding desire for safe operation and protection of the community. It is for that reason that the escalation team needs to reassure you that this safeguard is the right thing to do.

As the escalation team prepares or takes charge, they would be actively engaged in thinking out loud. Each member of the team shares his thoughts and plans with you. Where the team acts collectively, someone interprets their actions for you and the group. Each question, each concern, each search for status and other information, is accompanied by introduction before it is done. Afterwards, an explanation is given as to what was found and how it fits or not. Everything unexpected is voiced as it happens. All anomalies are discussed, explaining what they might mean and what will be done next. "What if" questions work well. Using this process also ensures that the team functions as a unified team.

Exposing you to the entire process builds experience and confidence. At the end of the day, the enterprise has the benefit of the best operating support they can have, and you have the opportunity to learn, first-hand, how it is all done.

Preparedness Evaluation Role

Escalation teams have the ideal makeup to also plan for and conduct *preparedness evaluations,* not only for themselves as a team, but also for the enterprise as a whole. Readiness evaluation allows you and your management to know how prepared you and your support team are for an emergency before it appears. The better you are at your jobs, the less likely an emergency is to appear. And if and when one does, you are better prepared to recognize one and support the management teams when they take over. Preparedness evaluation might study enterprise effectiveness in two areas:

1. The design of the enterprise for operation, the control room, the HMI, maintenance, and support
2. Operator performance and effectiveness, through fatigue assessments, the operating procedures, permitting, reporting, and shift handovers

Training and Coaching Role

Escalation teams have an excellent understanding of the operation, of all equipment, and all process elements in the enterprise. When called, they will partner with you. Who better than they to "invent" operational irregularities and abnormal situations? Who better than they to test and coach you with weak signal processing? Good weak signal processing requires in-depth understanding in order to do the following:

1. Recognize weak signals from operating data and displays
2. Connect each weak signal to its worst likely situation (forward extrapolation)
3. List the likely early indicators of the worst likely situation, assuming it to exist (backwards projection)
4. Find all early indicators of the worst likely situation that are actually present (evidence)

Who better than the members of the emergency management team or a special escalation team to take near hits (near misses) and use them for effective training scenarios for coaching and training operators, shift supervisors, and other mid-level management and support leaders?

34.13 Delegation of Responsibility

Delegation of responsibility, also called delegation of authority, is a powerful operational tool. The concept model in this handbook differs in one small respect from common practice. Yet this difference makes all the difference for safe and reliable operations.

Delegation is considered when the primary responsible individual is unavailable or unable to perform the duties or actions for which he is responsible. Delegation allows an organization to continue activities without undue delay caused by an unavailable responsible person. We are conditioned to view delegation as being done **down** the chain of organization. If a team leader were on vacation, he would usually choose a senior individual working under him on the team to stand in. But the act of delegation does not at all suggest such a direction. The best delegations are up, not down. It is the choice of direction that determines everything!

Why "up?" Three strong reasons:

1. The individual up the chain of authority has equivalent or more experience than the delegator.
2. The individual up the chain of authority has sufficient "authority" to back any evaluation or decision he makes in the role of delegate.
3. In order to step into the role of delegation, the individual up the chain of authority must be fully engaged in following and understanding the work of those responsible to him and for whom he may be called to step into the role of delegate; this reinforces active engagement with subordinates.

Module 35: Permission to Operate

Permission-to-operate is the most important people managed operational safeguard there is.

35.1 Why We Need the Idea of Permission

The default operational goal for almost every industrial plant is to keep running. Only a very few (such as a nuclear reactor) have a default state of not operating. *Permission to operate* works to ensure that any default thinking or predisposition will not unduly get in the way of safe and effective operation. It is an essential safeguard for you to know when and how to take appropriate action to protect the enterprise, its personnel, and the surrounding community at large. This module explains what it is and how it is used.

In the press of upset, the desire to manage "the problem" often works very hard to crowd out the more important decision of whether there is a proper chance to succeed. By the time that realization sets in, it is usually too late. It is too late either because there is not enough time to do what was necessary to return to a safe operating mode or because you cannot determine what is wrong or what to do. In all cases, the result is the same: equipment damage, production losses, personal injury, environmental degradation, or worse. We manage this problem by managing your authority to operate. Once operating authority is withdrawn, you must cease all attempts to remedy or remediate. Chevron's "Tenets of Operation" [35.1] captures it all: *if you do not know what is going on, stop*. Always: when you do not know, you have lost permission to operate.

Before this discussion goes any further, it is vital to state that "withdrawing" or "losing" permission to operate is an operation safeguard. Moving to a state when permission no longer exists has nothing to do with your skill, training, actions, or anything else. It means that there is not sufficient evidence that continuing operation is the responsible action. And the defining procedures and protocols to make that decision have been approved at all levels of enterprise management.

The operating situation that defines where permission to operate applies is when a part of the enterprise is abnormal and you must decide clearly and cleanly whether it is just abnormal or if it is really out of control. Figure 35.1.1 depicts this region as the yellow shaded area boxed with a solid red line.

The permission-to-operate question can be asked this way: Is the actual operational status (the red box around the yellow area labeled "abnormal") of your enterprise **above** or **below** the dashed red line within the solid red box in the figure? If it is **below** the dashed red line, and you are confident that it is, your operational mission is to use your best efforts to restore normal operation. However, if your actual operation is **above** the dashed red line, you must immediately change operational goals entirely. No longer is it permissible to work to restore normal operations. Now you must put all of your efforts into transitioning the enterprise to a safe state. This is not optional. Your entire management and supervision personnel and rules demand it. This module covers how you determine on which side of the dashed red line you are. Are you "abnormal" and just "abnormal" (below the red dashed line), or are you "abnormal" and "out of control" (above the red dashed line)?

Operational Modes:	Plant States:	Critical Systems:	Operational Goals:	Plant Activities:
Emergency	Disaster	Area Emergency Response System	Minimize Impact	Fire Fighting
	Accident	Site Emergency Response System		First Aid Rescue
Abnormal	Out of Control	Physical and Mechanical Containment System Safety Shutdown, Protective Systems,	Bring to Safe State	Evacuation
	Abnormal	Hardwired Emergency Alarms DCS Alarm System	Return to Normal	Manual Control & Troubleshooting
Normal	Normal	Decision Support System Process Equipment, DCS, Automatic Controls Plant Management Systems	Keep Normal	Preventative Monitoring & Testing

Figure 35.1.1. Enterprise operational modes, plant states, critical systems, operational goals, and plant activities. Note the red-boxed area surrounding the yellow "abnormal" states, and the red dashed line defining when permission to operate is either retained (below it) or lost (above it) **[Q17b]**.

35.2 Defining Permission

Permission to operate is one of the most important safeguards there are for operators. When you understand how it works and master the tools for using it, you will be able to confidently manage everything coming your way. You will be able to clearly identify the limits to your experience and the effectiveness of your actions. Let us take a closer look.

When you have permission to operate, it means that your senior enterprise management empowers you to do everything proper and in your power to keep things operating correctly. However, as soon as you cross the line of no longer being able to understand and manage, you must immediately stop and move everything to a safe state. Your enterprise will have specific protocols, procedures, and guidelines in place for this. You will have been trained how to use them. You will likely have been a part of actual exercises to test and evaluate your skills. You will know when and you will know how. It is that important.

35.3 How Permission Works

Please keep in the forefront that each individual plant or enterprise will be shaping specific guidelines and policy for you. You will use those instead of anything here that conflicts or differs from those requirements. What follows is an explanation of the concept and lots of ways it can be designed to work.

You retain the full duty and responsibility to operate your plant so long as you fully understand everything that is going on within your area of responsibility. As long as you retain the effective duty and responsibility to manage the situation, you may continue to do so. However, once that responsibility is lost (permission to operate is lost), you must immediately cease all attempts at operation and move all affected areas to shutdown or safe state.

> You retain the full duty and responsibility to operate your plant so long as you fully understand everything that is going on within your area of responsibility.
>
> As long as you retain the effective duty and responsibility to manage the situation, you may continue to do so.
>
> However, once that responsibility is lost (permission to operate is lost), you must immediately cease all attempts at remediation and other operation and move all affected areas to shutdown or safe state.

Permission to operate has only two states: (1) Permission, or (2) Lost Permission. When you have permission to operate, you are empowered to operate the unit(s) under your care and responsibility by the authority of your management. Until permission is lost, all staff will employ their best efforts, utilizing plant procedures and training, to ensure safe, productive, and responsible operation. If permission is lost or withdrawn, all of your current efforts to keep the plant running must cease; all future efforts will be directed to placing the plant in a safe state.

In order to maintain permission to operate, you must at all times have a firm and accurate understanding of the operation. You must be making planned, appropriate, and positive remediation or control progress. No additional threats are appearing.

You have permission to operate so long as you **fully understand** what the process is doing. Once lost, you have an affirmative duty (you must act) to immediately, and with the direct purpose of safety, cease operation and move to a safe state. When permission to operate is lost, and neither the proper shutdown nor effective safe park has been attained, operations will presumptively be in a crisis or disaster management mode. Someone else will be managing things now. Your management will specify which of the following methods for having permission to operate should be used. Or they will have developed ones of their own for you. Make sure you understand your specific rules and procedures.

You lose permission when, for example, you are responding to a significant alarm and you have completed all documented steps and all appropriate procedures specified in the alarm response sheet, but the abnormal situation has not effectively resolved.

35.4 Methods for Having Permission

As long as you have *permission to operate*, you are to use all proper means at your disposal to keep things running. Your management will specify which of the following methods for having permission to operate should be used. Or they will have developed ones of their own for you. Make sure you understand your specific rules and procedures.

- *Full authority*. Permission to operate is retained if, and only if, you are reasonably sure that you **fully** understand what is going on and are quite confident of your ability to manage. It is up to you to properly decide.

- *In usual upset.* Permission to operate is retained if, and only if, you are reasonably certain that the plant is in a usual upset state, regardless of whether or not you understand specifically what the current upset is, and you are confident you will be able to manage it successfully. If anything about this upset is unusual, or not fully understood, permission to operate is lost. It is a bit more subjective than full authority, but still clear and unambiguous.
- *Manageable unusual upset.* Permission to operate is retained if, and only if, you are reasonably certain that, even though the plant might be upset, you have known and clear reasons to know or expect the upset to be manageable. The moment you determine or begin to suspect that the upset might not be manageable, permission to operate is lost. The only manageable upsets are those that are routinely seen and for which procedures exist that are expected to work. This case is quite a bit more subjective and requires substantial competence on your part to employ it.
- *Operating limits challenged, reached, or exceeded.* As soon as any part of your operation threatens or comes against or exceeds any operating limit, no matter how insignificant it may seem at the moment, permission to operate is lost.

Again, just to be clear, your enterprise will determine whether to use the *methods* above, the *situations* below, a combination, or something different.

35.5 Operational Situations for Maintaining Permission

We next visit specific ways for you to identify additional situations that may clarify how a clear decision can be made. We provide four important situations to help you decide whether or not you have permission to operate. Any item from any situation below that is true withdraws your permission. You have lost permission to operate. Please keep in mind that these are important guidance aids. However, those of your plant always supercede anything here. If you do not know them or have them at hand, it will be all too easy for you to put this aside until later in the press of operational difficulties. The purpose of the ones below is to provide clear, understandable alternatives for you to understand in their absence. It is fine to use as many of the types as needed, unless your enterprise develops new ones.

Situation 1: Qualifying Abnormal

Let us assume that the plant is now abnormal. The following situations will identify what becoming abnormal means. If any one of these exists, then you must conclude that the plant now is actually abnormal, **regardless of what you might think.** Permission to operate is lost.

- Additional alarms activate in the same or a closely related plant area.
- Few, if any, existing alarms in the same or a related plant area have returned to normal within a reasonable time.
- There is one or more new items of equipment (valve, pump, sensor, or anything else non-trivial) that then fail or are newly discovered as having failed earlier.

Situation 2: No Help at Hand

Difficult abnormal situations where there is not a second experienced plant person at your "left hand" in the control room (or virtually, with dependable technology) should automatically suggest that you are either already in trouble or soon to be. Consider this to be sufficient to escalate. If you cannot escalate or do not have enough time, you lose permission to operate.

Situation 3: Operating Limits Challenged

Your plant usually has a wide range of operation. When operations threaten any single normal, usual, or customary operating limit, it means that whatever you are doing or not doing is not working well enough. The threat may be either by coming too close to abnormal, or approaching it at too high a rate of change. Unless you specifically know what to do, it is sufficiently proceduralized, and you have enough time to comfortably do it all, permission to operate is lost.

Situation 4: Do You Understand

The answer to each of the following six questions will determine whether or not your understanding is enough to retain permission. Any single "discontinue" always means that permission to operate is lost.

An Imaginary Operator Is Asking You:

- *Question 1.* Have you tried more than one approach to turn the abnormal situation around and it is still quite abnormal?

  YES—Discontinue operation NO—Continue operation
- *Question 2.* Can you clearly explain what is likely to happen to the plant in the next ten to twenty minutes?

  YES—Continue to operate NO—Discontinue operation
- *Question 3.* Are you using a predetermined documented approach to work the abnormal situation?

  YES—Continue to operate NO—Discontinue operation
- *Question 4.* Has the situation remained quite abnormal (even if it does not appear to be gravely so) for longer than a reasonable time?

  YES—Discontinue operation NO—Continue operation

As the Operator, You Ask Yourself:

- *Question 5.* Do I feel too harried to take a careful, methodical approach to understand, evaluate, and manage the situation?

  YES—Discontinue operation NO—Continue operation
- *Question 6.* Do I feel that there is enough time to slow down and think things out?

  YES—Continue to operate NO—Discontinue operation

35.6 Lost Permission

The moment you lose any single (or more) criteria from any of the defining ones you are authorized to use for continuing operation, *permission to operate* is lost. You must directly execute the preplanned process of moving the operation sufficiently away from the unsafe or unmanageable condition. This requires your ability to change the operating mode from operating to safe state. It must be an integral part of the plant design and your training. In addition to proper protocols and procedures, controls and equipment must be in place for you to command the change, monitor the progress, and remediate any defects along the way. The choice of either moving to a shutdown state or to a safe-park state is an operational one.

35.7 Putting It All Together

We ask you, as the operator, to learn, understand, and use the rules you have. Fully understand them. Properly use them. The bottom line is whenever there is uncertainty—either explicitly, by your not knowing or understanding, or indirectly, from what is happening to you—you should close things down. Figure 35.7.1 illustrates our decision process as a flow chart. It is suggestive. At every decision point, you should be able to rely upon your enterprise rules. Always keep track of time. If it gets even a little short, you are not going to succeed. Use the little time left to become safe.

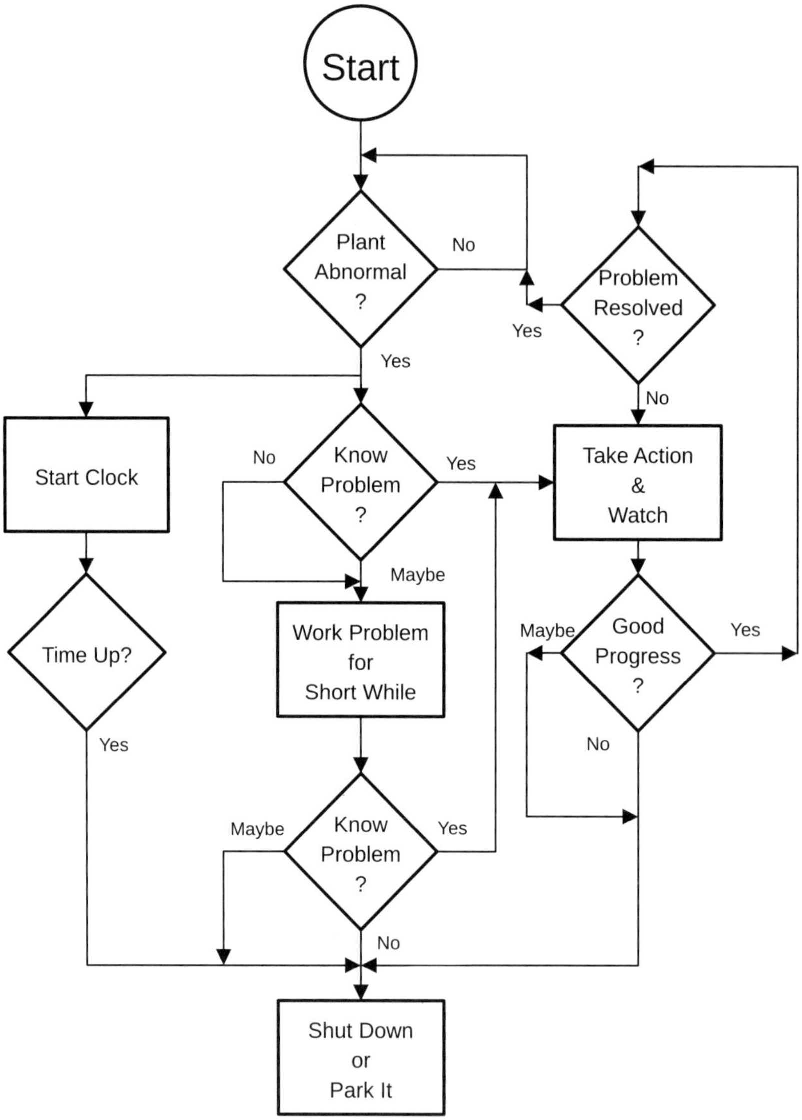

Figure 35.7.1. Permission to operate diagram depicted as a formal flow chart. Note the parallel "clock" that ensures that working on a problem resolution does not take too long without a change in tactics **[Q17a]**.

35.8 Time vs. Permission

Note the parallel "clock" in Figure 35.7.1 Its single and vital purpose is to ensure that working on a problem resolution does not take too long without a change in tactics. This means that you must ALWAYS respect that everything takes time. You must always have enough time to do everything you need to do to operate properly. No matter how confident you are, no matter how close you may seem to be to a solution, as soon as you do not know for sure that you have enough time for what you are doing, you cannot continue. When the clock is running out, it means unequivocally that you are not going to be successful without a miracle. Miracles do not make safety. You have lost permission to operate. Shut down or safe park.

Module 36: Operator Redeployment

This topic is new. As you understand why it is important, it should change the way you face significant operational problems. Having it available for use means that your enterprise recognizes that multitasking does not work. They do not expect you to use it.

Major situations can be extremely impactful. They risk extensive enterprise damage, personnel injury, and environmental degradation. Thankfully, they come very infrequently. But when they do, things are never the same. They will almost always overwhelm the capabilities and resources of even a most experienced operator. Because these operational situations are so difficult and require complex concentrated activities, it is important that the normal personnel roles be modified to permit a focused and tightly integrated team to do the work together. Call this situation-based, dynamic operator role deployment. It carefully shifts all operational roles.

36.1 Operator Burden in Major Situations

Operator redeployment is an effective way to tactically adjust operational roles to form a highly trained, effective critical situation response team. This new team will exert a profound positive effect for effective action. In this situation the operator has not yet lost *permission to operate* [**M35**]. However, the situation is extremely likely to develop in ways that exceed his ability to keep on top of it. Too many parts are moving all at once.

Think of operator redeployment as a specialized form for an emergency management *escalation team* [**M34**]. It is an actual working procedure for US nuclear power plants to manage upsets. It has been used there with considerable success. They have it proceduralized and practiced. It is part of the fabric of their operations. As a result of their success, many leading industrial enterprises use it. If your enterprise has provided for this capability, here is how it might work. If they do not have anything in place now, you might bring it to the attention of your supervisors to consider adding it to the situation management kit.

A note about terminology before we begin. We will not be referring to "you" in the following discussion. As the operator, you may or may not be acting in the explicit role as *console operator* during redeployments as described in this module. It would be easier to read this if you were to assume that you might be acting in any number of operational roles for this protocol.

36.2 Operator Positions

Single Operator

Figure 36.2.1 illustrates a single control room operator in a single "chair." He is the console operator. He is located at his operating console that happens (for this schematic illustration) to

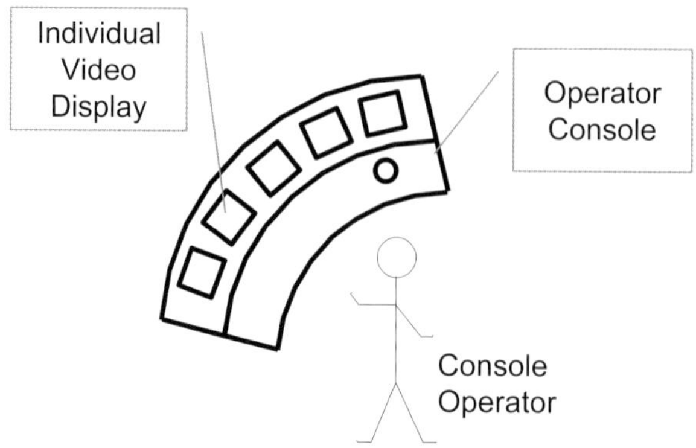

Figure 36.2.1. A single control room operating "chair," showing the operator console and five typical video displays

have five video display units. He also has the usual accompaniments of telephone, enterprise radio, and such, assumed present but not shown here.

Additional Operators in the Control Room

Put our single console operator into a central control room with three others and a *master operator* (supervisor). Figure 36.2.2 shows a typical enterprise arrangement. The basic and primary materials flow is shown progressing from the upper-right operator counterclockwise through all four stations. The primary exporting is done from the enterprise managed by the operator of the lower-right station. Each operator has his own area of responsibility with separate activities. Their interaction with each other is normally limited to product coordination, though they might also relieve one another for short breaks. In some enterprises, these operators may rotate responsibilities on a scheduled basis to ensure cross training, if that is needed.

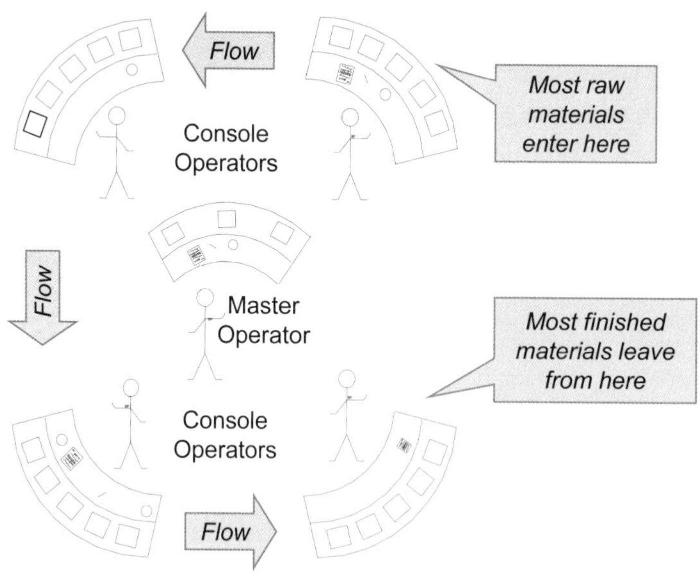

Figure 36.2.2. Control room with four console operators and master operator (supervisor), illustrating an arrangement where each operator, in turn, manages a part of the plant before sending product downstream to the next operator.

Significant Operational Problem

Suppose that the second operator (the upper-left position in Figure 36.2.2) experiences a problem in a part of his responsibility, depicted in Figure 36.2.3 as a red screen on the display at the left end of his console. We know that screens and displays are not dedicated to specific enterprise equipment, but let us just assume this case for pictorial illustration convenience.

The current problem is significant and involves about 20% of his enterprise. It is significant, because the left console shows it as red. It is 20% in this illustration because only one of the five displays is involved (1 in 5 is 20%). This is just an illustration. In reaction to this significant situation, several important operational events will happen in quick, preplanned, and proceduralized steps. Each is discussed, step by step, in Section 36.3. The first step is shown in Figure 36.3.1. The last is shown in Figure 36.3.2.

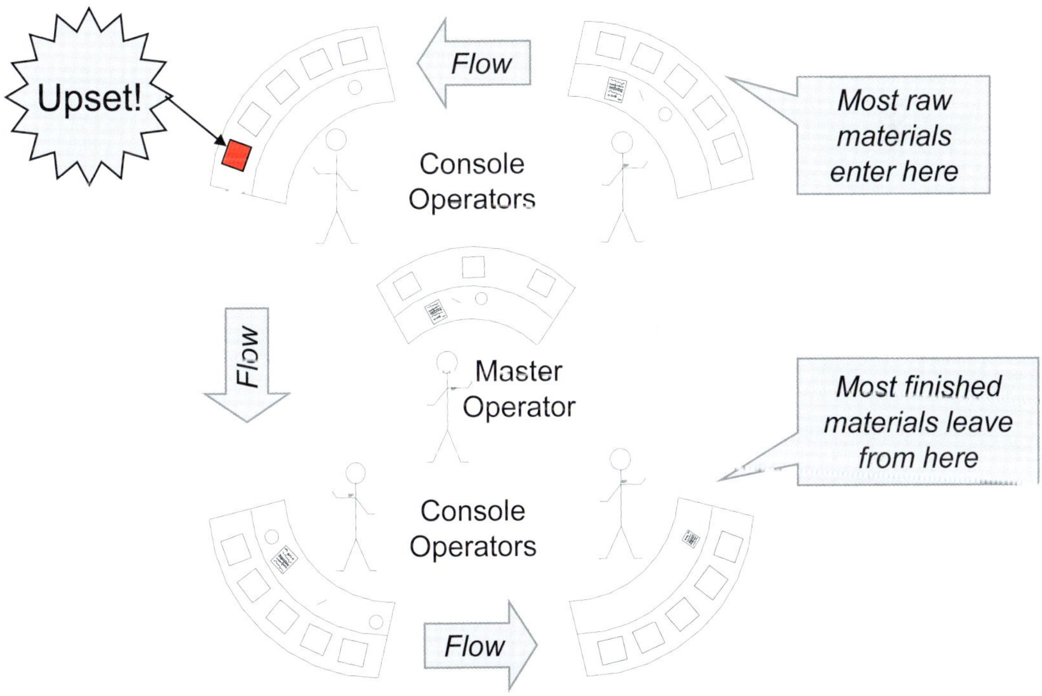

Figure 36.2.3. Control room with four console operators and master operator, depicting a serious upset at the second (in the flow of production) operator position. The upset is depicted as a single red console display screen.

36.3 Redeployment, Step by Step

The following is the four-step process for setting the plan in motion. Please keep in mind that all steps are done in quick succession. It is not until all are in place that the effective work of managing begins. So, as you read this, understand that the detailed activity happening for each step is ongoing. No one will wait for the completion of responsibility for one step before moving on to the next. Everything is going on in a controlled way at the same time as the role changes evolve and are then complete, with everyone in place and functioning.

Step 1: Affected Operator's Load Reassigned

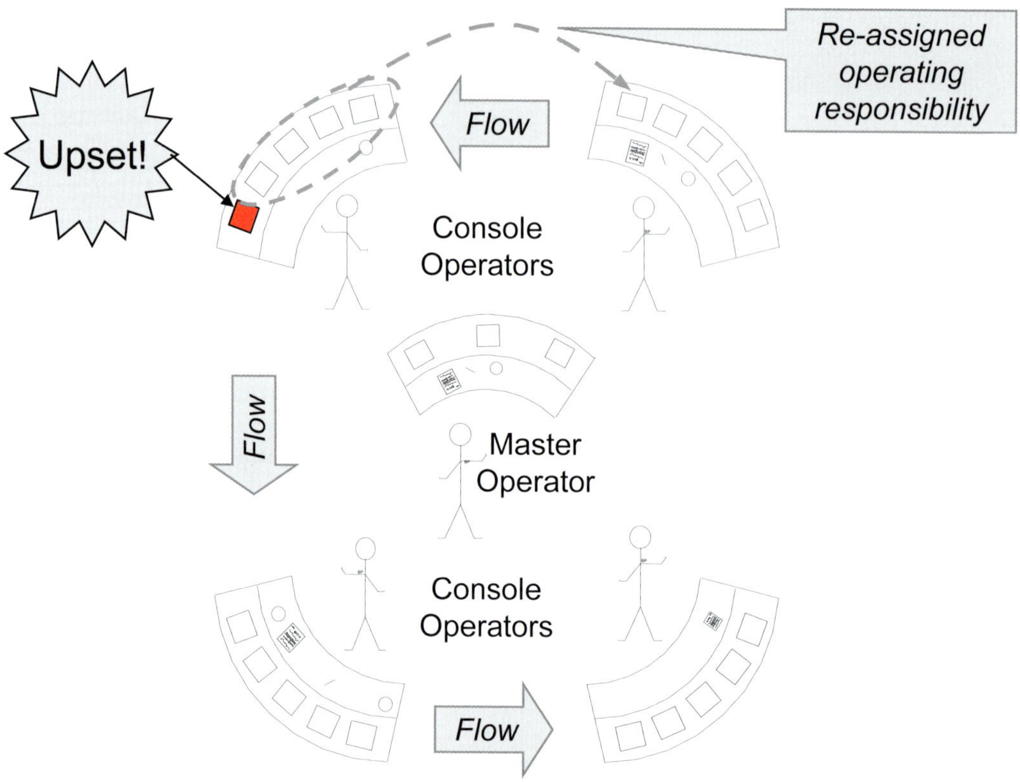

Figure 36.3.1. Control room with four console operators and master operator, depicting a serious upset at the second (in the flow of production) operator position—now showing the (temporary) assignment of his operating responsibility for his non-upset units over to the first upstream operator.

All of the (approximately) remaining 80% of the affected operator's operational responsibilities are passed to the control and management of the nearest upstream operator. In our case, his unaffected 80% is transferred to the closest operator upstream in the material flow path—upper right. It is passed there for two reasons. First, this operator will be the least influenced if the current problem is not managed effectively. Second, this operator may be able to make changes in his operation to better enable the affected operator to manage. If there were not an upstream operator available, the responsibilities are passed to the operator furthest downstream.

Step 2: Narrowed Focus for Affected Operator

The affected operator remains on station. However, his focus is dramatically modified. Now, it is narrowed to include only the limited portion of his plant that is upset. For the moment, his only task is to locate the appropriate displays and control handles that might be needed for situation management. He will take no independent actions. Only at the direction of the managing operator (explained in Step 3) will he perform clear and distinct hands-on control and other changes to plant equipment that are possible from the control room. His counterpart in the field will do the same for field equipment. They will take only the actions requested by the *managing operator* and only according to his schedule and scope.

Step 3: Assignment of Managing Operator

Another position, that of the managing operator (Figure 36.3.2), steps in. This post may be filled by the operating supervisor (potentially a master operator) for the shift, by a senior floating operator, or by any other pre-qualified operations individual. The managing operator will, for the entire course of the upset, be responsible for the minute-by-minute operational tactics to be used to manage this situation. The managing operator will direct the console operator to perform single tasks, one at a time. Only the current task will be told to the console operator. When that task is complete and confirmed, he will be told the next task. This activity is continued as long as needed.

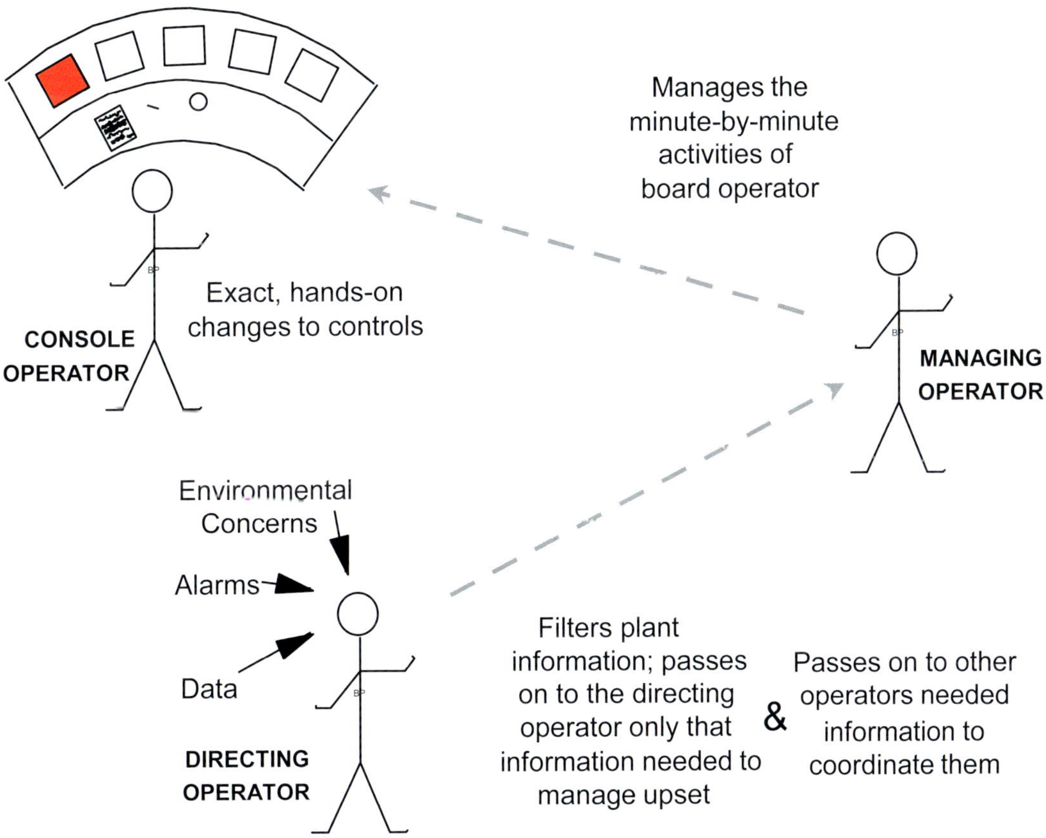

Figure 36.3.2. The upset operating situation, showing the assigning of responsibilities for the escalation team. (For simplicity, the other three area operators are not shown.) The console operator is now only responsible for all hands-on changes and eyes-on monitoring of the upset unit. The managing operator is responsible for planning the proper operator responses and then asking the console operator to execute each step, one at a time. The directing operator is responsible for filtering out all plant situations except for what is needed to understand the current operating risks.

Remember, only one task at a time is on the table for the console operator. It is up to the *managing operator* to decide what to do and in what order. It is the sole job of the managing operator to understand everything in the current abnormal situation (keep situation awareness), decide what is amiss (perform situation assessment), and decide what to do and how it shall be done (step-by-step situation management). His primary focus is the entire affected 20%—though he must factor in everything about the rest of the operations (relying on the *directing operator* to keep him informed).

The *managing operator* instructs the *console operator* to perform tasks in sequence, one at a time, each announced individually by the managing operator to the console operator. Note that the console operator will confirm each instruction to the managing operator. The managing operator will confirm each observation back to the console operator [**M6.3**]. At each step and for every step, unless there is a modification to the order, the managing operator will issue no next order until the console operator has confirmed the instruction, confirmed the action, and confirmed the result back to the managing operator for this order. The following are example instructions from the managing operator to the console operator:

1. Put the reflux flow controller in manual and set the valve output to a bit less than 25% flow.
2. Stop the pump on the rundown tank.
3. Reduce the steam flow to the reboiler to keep the temperature in the upper tray area around 165°F.
4. (And so on until the current specific tactical plan of the managing operator has been carried out.)

Each instruction by the managing operator to the console operator will be confirmed back by the console operator to the managing operator. This way, no instruction should be missed or misunderstood.

Step 4: Assignment of Directing Operator

A *directing operator* (Figure 36.3.2) steps in to assist the *managing operator* by providing all external situational information needed to fully understand the rest of the enterprise and its state. If the shift relief operator has reported but not assumed duties, he may assume this role. If the *shift operator* has been relieved but is still present, he may assume the role. Otherwise a knowledgeable individual must be located to fulfill this role. That individual need not be "operator qualified" in the formal sense. His job is to be aware of the entire rest of the enterprise and only pass on to the managing operator what is needed to manage the immediate and specific threat posed by the upset event.

36.4 Recap

Are you thinking to yourself that all of this sounds like some imagined, over-the-top complicated thing to manage, when just plowing ahead should work just fine? It is not. It is not, because each of us has a limit point of being so stretched that we have real trouble keeping focused. It is why some of us sit down to think about a complicated or troubling problem. Others stand up to do it. All of us are searching for a way to snip the horrendous problem into manageable bite-sized pieces. We are looking for any way we can find to separate that web of seemingly endless lists of pieces all working against your working things out. At that moment, and for as long as the heavy load hangs around, we do not do a good enough job at slowing down, picking up a piece, playing with the piece; and simultaneously keeping all eyes out for likely disaster lurking everywhere. We cannot always keep everything straight, separate, and attentively work on all of it. We miss important things. We overly exaggerate trivial or inconsequential things. We make mistakes. We miss seeing that we made a mistake or left out something. It is the key reason why this redeployement works. Please do not underestimate its value.

Module 37: Operational Safety Nets

37.1 Safety Nets

Every enterprise works diligently to ensure good and safe operation. Their operators—you—do their best to make sure that everything goes well and according to plan. Yet things go wrong. Even the most carefully imagined and built industrial control room infrastructure may sometimes come face to face with circumstances that can challenge even best efforts to properly and efficiently manage. At that very moment, lives may be at risk; equipment faces destruction; loss of financial integrity becomes a reality; and damage to land, air, and water are exposed. When this occurs, it is important to be able to limit the damaging effects. Safety nets is a collection of important ways to do just that. They help you manage abnormal situations to keep them from getting even worse, if such management is at all possible. They are pre-planned and pre-designed.

> Whatever proper situation management is reasonable and likely, it can be utilized so long as the problem at hand is manageable and there is enough time.

This final module on *operational safety nets* provides important and useful tools to significantly improve your management of operations to maintain safety and manage losses. The basic safety nets are: *permission to operate* (covered fully in the previous module), *safe park*, *getting help*, and managed shutdowns. Please become comfortable with them all.

37.2 Tenets of Operation

The permission to operate [**M35**] concept began with Chevron's "Tenets of Operation." Tenets of Operation is a set of very high-level rules that allows for minimal operational drift. Chevron Oil developed them after they experienced a costly operational mishap that really should never have happened. They set about to fix it. A careful investigation showed that, although they had all of the rules in place to prevent the very situation they experienced, there were too many of them. It was too hard to know which were critical and which were just instructive. They were confusing and difficult to follow. Little by little, the employees slipped a bit here and there. Eventually, they became used to slipping and stopped watching for it. Then something awful and expensive happened. It should not have. If all of the old rules had been followed, it would not have. Clearly, there was a problem with a lot of rules, even very good ones. Chevron's solution was to set a few very high-level rules that allowed no drift. These are the rules for following all the rest of the rules. They called them "Tenets of Operation." [37.1] This type of philosophy is also referred to as "rules to live by." No slippage is allowed. Adhering to them provides a very reliable safety net for the big things. Of course, all of the rest of the rules are still in place. Slippage is not encouraged, but where slippage does occur, it will not be so costly.

Ask your enterprise to visit this protocol list and implement as much as possible for operators. Here are the basic Tenets of Operation that Chevron developed. The first three are key principles and are the foundation for everything.

- Do it safely or not at all.
- There is always time to do it right.
- If you do not know what is going on, stop.

The next rules are still basic and high-level but add more substance. The tenets address a wide range of behaviors. The key word in the tenets is **Always.**

1. Always operate within design and environmental limits.
2. Always operate in a safe and controlled condition.
3. Always ensure all safety devices are in place and functioning.
4. Always follow safe work practices and procedures.
5. Always meet or exceed customers' requirements.
6. Always maintain integrity of dedicated systems.

> Most enterprises have something very similar. If not, suggest that your enterprise consider establishing rules like them, so that you might avoid falling prey to unintended operational protocol drift. You need a good, reliable, approved backstop!

37.3 Alternate Safe Operating Modes

Alternate safe operating modes are practical ways for you to move to place your operation in a safe condition or state, whenever permission to operate is lost. This is how to implement your vital need to buy enough time to avoid serious problems. You buy that time by either shutting down or safe parking.

Operator-Initiated Shutdown

The conventional emergency shutdown (ESD) is familiar to all current industrial manufacturing plants. OSHA mandates their presence. They are currently designed as a last-resort safe-operation control measure. Operator-initiated shutdown is one small but significant step back from that. This is an option for you to consider when you have lost permission to operate. For this, care is required to do things in an orderly, controlled, but direct manner. It is a pre-designed, pre-implemented capability that you initiate using the PCS for a sequence of operations that shut down the unit. This is important. We depend on this solution not to place unnecessary risk on the enterprise. Implementing this capability usually requires the replacement of most manually operated field components with ones that can be operated remotely. A few manual ones can be left, but the "people action" must be carefully coordinated. When implemented correctly, this will ensure that the enterprise will be in a much less damaged shutdown state than if the ESD system had done it. This should eliminate or significantly reduce equipment damage. As a result, it should facilitate a faster restoration of production when the enterprise is ready to restart. Of course, those situations that could not be properly managed by this controlled shutdown will be caught by the ESD, thus ensuring process safety integrity foremost.

Automated Shutdown

An automated shutdown fully automates a reliable shutdown using instrumentation and controls equipment. It is much "kinder" on the equipment than an ESD. As soon as the enterprise is unusually abnormal, rather than attempting to work out what might be amiss and try and fix things, the existing equipment initiates the automatic shutdown process. The enterprise will quickly and properly shut down. This activity often utilizes some of the same equipment and enterprise infrastructure that the operator-initiated shutdown uses, but will initiate without operator action. This flexibility is not ordinarily economical for smaller enterprises.

37.4 Safe Park

Shutting down can be harsh. Even when done carefully, it is time-consuming, expensive, and fraught with opportunity for more trouble, with equipment damage and poor product. But it does the job. *Safe park* is an option to avoid shutdowns. Safe park is a methodology that designs appropriate ways for production units to be moved to a much safer and more reliable operational mode—to buy time and, at the same time, to avoid the problems of an extensive shutdown. It is used only if it can be done safely and effectively. The recurring principle is to not do something without having a procedure for it that has been effectively learned and practiced. Operating is not inventing. You will need a few basic guidelines. In order to use this tool, your plant will have all of the specifics set out for you to follow. Everything will have been pre-planned, pre-approved, and supported by effective procedures and supporting HMI display screens.

Safe Park Design

For each intended safe park process unit, a carefully worked-out scenario is developed that permits it to be parked instead of shut down. Simple examples are placing distillation towers in total recycle, putting pumps on low power with sufficient kickback to avoid operational problems, and placing a furnace on low fire but in a way that controls thermal stresses and inventoried product problems. Other examples from transportation systems would be to sidetrack a train off of the main track when another train in an unknown position may be attempting to share it. Air traffic controllers will use this (to airplane travelers' frustration) to place all airplanes in a holding pattern around a potential airfield until a weather problem or another emergency is sorted out. Note that "parking" cannot be done forever. It is a tool, not a solution.

The particular unit will govern the specific parking details, of course. Any additional needed instrumentation and controls are added. Operating procedures should be expressly designed for this activity, and the corresponding protocols need be in place.

Safe Park Illustration

The following illustrates the concept. You may exercise a safe park operation only if all of the following are true:

1. Current extent and character of the abnormal situation is clearly and fully understood.
2. All external factors are either not limiting or accounted for.
3. The proposed solution (in this order).
    a. has less or equivalent operational risk than any other safe choice.
    b. will not violate existing specific or implied safe operating requirements.
    c. has a better chance of being effective than any other safe choice.

4. After implementing the proposed solution, the overall operating situation will not expose the enterprise to increased operating risk or operational stress later on.

Full automation for startup and shutdown is normally not considered for a complex processing unit. A safe park option is designed to provide additional time for you to work problems. Safe park can greatly reduce exposure to the danger and damage of continuing to operate. It reduces the risk of unwanted consequences and permits a smoother and more efficient return to normal operation when it is okay to do so.

37.5 Keeping Perspective

If you ever get to a place where it is too hard to understand what your process is doing—no matter what the reason—at that very moment you do not actually know what is happening. Using the permission-to-operate protocol, you are obligated to cease operations and move to safe or shutdown state. The more effective you are at earlier managing, the better job you can do at foreseeing the abnormal and managing it before things get out of hand. But when you run out of time, you run out of options. Those we covered are the ones of last resort.

37.6 This Is Not an Ending

You are reading the last section of the last module in your handbook. You have been diligently exploring most, if not all, of this material. Or, having picked this up and without any thought whatsoever, you jumped to the end. Or you have been looking here and there, and are now reading this. No matter how you came to be here reading, the authors Beverly and Douglas are gratified that you are. It is our wish and hope that the ideas, suggestions, and experiences in this handbook might be in some beneficial ways valuable to you.

We have a few more words. They are our last. Yet they are far from an ending. There is no end to your handbook. Everything connects back to everything else. Nothing you learn here is a linear path to traverse. Now that you are reading this, you are so much **better able to read again what came before. Experiences during your reading, in ways small or surprisingly impactful, will bolster your courage and calm your uncertainties.**

Be proud. Be successful. Be as safe as you can. Enjoy your career.

ADDITIONAL MATERIAL

In this part of this handbook are items that complement and expand to provide added depth and breadth to the coverage in the specific modules.

Material that has been discovered or better stated by others and included in the modules are identified in the Endnotes. If you would like to know more, here is where to look.

About the Authors gives you a bit of background on us.

In the Credits we acknowledge the help from others who read drafts, recommended approaches, and topics, and generally challenged and assisted us to prepare the handbook and make it the best we could.

Additional Reading is for those who find specific topics interesting and would like to know more.

The Appendix includes an interesting discussion of how documents printed on paper can be used differently from documents viewable on a display screen; examples of "soft" weak signals; and the complete list of cue cards (to provide a permanent list of those cards supplied separately).

Finally, the Glossary and Index so you can find anything specific you might want.

ADDITIONAL MATERIAL

Endnotes

[1.1] Douglas H. Rothenberg, *Situation Management for Process Control: Situation Awareness and Decision Making for Operators in Industrial Control Rooms and Operation Centers*, (Research Triangle Park, NC: International Society of Automation [ISA], 2019).

[1.2] "Occam's razor," *Wikipedia*, accessed Dec 4, 2016, https://en.wikipedia.org/wiki/Occam%27s_razor

[1.3] Richard Swinburne, *Simplicity as Evidence of Truth* (Milwaukee: Marquette University Press, 1997); Thomas Aquinas (Italy: 1225–1274 ACE).

[2.1] Lasky, Rick and John Salka, *Five Alarm Leadership: From Firehouse to Fireground*, (Tulsa OK: PennWell, 2013), https://www.pennwellbooks.com/.

[2.2] Alan Deutschman, "How Authentic Leaders 'Walk the Walk,'" *Bloomberg Business Week*, September 18, 2009, https://www.bloomberg.com/news/articles/2009-09-18/how-authentic-leaders-walk-the-walk.

[3.1] "Safety," *Wikipedia*, last modified May 15, 2018, http://en.wikipedia.org/wiki/Safety

[3.2] Lord W. D. Cullen, *The Public Inquiry into the Piper Alpha Disaster*, Volumes 1–2 (London: Her Majesty's Stationery Office, 1990).

[3.3] American Institute of Chemical Engineers, *Guidelines for Safe Automation of Chemical Processes*, (New York: Center for Chemical Process Safety, 1993), xx, xxi.

[3.4] Don, Joel (ed.), *AutoQuiz: What Are Multiple Independent Protection Layers in Relation to Industrial Safety?*, International Society of Automation, ISA Interchange, Automation.isa.org, accessed January 3, 2020, https://automation.isa.org/autoquiz-multiple-independent-protection-layers-isa84-industrial-safety/

[3.5] ibid, Don, Joel.

[3.6] 29 CFR 1910.119, *Process Safety Management of Highly Hazardous Chemicals* (Washington, DC: Occupational Safety and Health Administration [OSHA]).

[3.7] "Duty of care," *Collins Dictionary of Law*, accessed May 8, 2016, http://legal-dictionary.thefreedictionary.com/duty+of+care.

[3.8] "Swiss cheese model," *Wikipedia*, accessed November 27, 2016, https://en.wikipedia.org/wiki/Swiss_cheese_model.

[3.9] http://jamanetwork.com/data/Journals/SURG/926092/srv120009f1.png

[4.1] International Society of Automation (ISA), *The Automation, Systems, and Instrumentation Dictionary*, 4th ed. (Research Triangle Park, NC: ISA, 2003).

[4.2] ISA 82.02.01-1999, *Safety Standard for Electrical and Electronic Test, Measuring, Controlling, and Related Equipment: General Requirements* (Research Triangle Park, NC: ISA, 1999).

[4.3] Pipeline and Hazardous Materials Safety Administration (PHMSA). Hazardous Liquid Integrity Management Enforcement Guidance Sections 195.450 and 452. (Washington, DC: PHMSA), Accessed September 4, 2018, https://www.phmsa.dot.gov/sites/phmsa.dot.gov/files/docs/Hazardous_Liquid_IM_Enforcement_Guidance_12_7_2015.pdf

[4.4] Coy, Byron, "Control Room Management," Presentation to the Southern Gas Association, Jacksonville, July 2011. {ref USDOT/PHMSA, 07-12-2011}

[4.5] "Maslow's Hierarchy of Needs," *Wikipedia*, last modified June 4, 2018, http://en.wikipedia.org/wiki/Maslow%27s_hierarchy_of_needs

[5.1] "Impairment," *Dictionary.com*, accessed Dec 1, 2019, https://www.dictionary.com/browse/impairment?s=t

[5.2] Martin Moore-Ede, "The Definition of Human Fatigue" (white paper, Stoneham, MA: Circadian, 2009).

[5.3] ibid, Martin-Moore-Ede, white paper.

[5.4] Martin Moore-Ede, "Working Nights: Health & Safety Guide," (Stoneham, MA: Circadian, 2008).

[5.5] Douglas H. Rothenberg, *Situation Management for Process Control: Situation Awareness and Decision Making for Operators in Industrial Control Rooms and Operation Centers*, (Research Triangle Park, NC: ISA, 2019), 166.

[5.6] Ibid, Martin-Moore-Ede, white paper.

[5.7] Douglas H. Rothenberg, *Alarm Management for Process Control: A Best Practice Guide for Design, Implementation, and Use of Industrial Alarm Systems*, 2nd ed. (New York: Momentum Press, 2018).

[6.1] "Communication," *Wikipedia*, last modified June 20, 2018, http://en.wikipedia.org/wiki/Communication

[6.2] Carrillo, Rosa Antonia, and Neil Samuels, "Safety Conversations: Catching Drift & Weak Signals," *Professional Safety*, (Park Ridge, IL: American Society of Safety Engineers, January 2015), 22-32, http://www.asse.org/

[6.3] Mindful Conversations, https://www.lifehack.org/articles/communication/have-mindful-conversation-9-easy-steps.html

[6.4] Leadership Now Blog "Problems with Virtual Communication & What to Do About It," December 27, 2018, https://www.leadershipnow.com/leadingblog/2018/12/five_problems_with_virtual_com.html, accessed 6/25/2020.

[6.5] Morgan, Nick, "5 Fatal Flaws With Virtual Meetings," *Forbes*, October 2, 2012, https://www.forbes.com/sites/nickmorgan/2012/10/02/5-fatal-flaws-with-virtual-meetings/#6999533f704f, accessed 6/25/2020.

[6.6] Peccolo, Gigi, "45 Texting Statistics That Prove Businesses Need to Take SMS Seriously," CustomerThink, September 14, 2015, accessed June 25, 2020, https://customerthink.com/45-texting-statistics-that-prove-businesses-need-to-take-sms-seriously/

[6.7] Edmonds, Jane, ed., *Human Factors in the Chemical and Process Industries*, (Amsterdam, Netherlands: Elsevier, 2016).

[7.1] "Watch This!" *The Brain Games*, season 1, disc 1, games 7 and 8, directed by Jeremiah Crowell, released October 9, 2011 (National Geographic Channel, 2011), DVD.

[7.2] Crenshaw, Dave. *The Myth of Multitasking: How Doing it All Gets Nothing Done.* (San Francisco: Jossey-Bass, 2008).

[7.3] Lohr, Steve, "Slow Down, Brave Multitasker, and Don't Read This in Traffic," *The New York Times*, March 25, 2007, accessed Feb 14, 2014, https://www.nytimes.com/2007/03/25/business/25multi.html

[7.4] ibid, *The New York Times*, article on multitasking.

[7.5] David E. Meyer, Cognitive Scientist and Director of the Brain, Cognition and Action Laboratory, University of Michigan, Ann Arbor.

[7.6] ibid, *The New York Times*, article on multitasking.

[8.1] Pacific Crest Group, "Getting Right with Policies and Procedures," San Rafael, CA, https://www.pcg-services.com/getting-right-with-policies-and-procedures/, accessed June 24, 2020.

[8.2] PowerDMS Blog, "Following Policies and Procedures and Why It Is Important: Making sure your staff knows how to follow procedures," May 2, 2018, https://www.powerdms.com/blog/following-policies-and-procedures-why-its-important/ , accessed June 24, 2020.

[8.3] Ibid, Pacific Crest Group.

[8.4] DeLuca, Rob, "Compliance: 4 Steps to Ensure Employees Follow Procedures," HR Insights 7 Min, July 27, 2015, updated October 31, 2019, accessed 6-24-2020, https://www.bamboohr.com/blog/procedures-employees/

[8.5] Ibid, Pacific Crest Group.

[8.6] Working in Partnership Programme, "Using protocols, standards, policies and guidelines to enhance confidence and career development," September 2006, NHS, UK, accessed June 24, 2020, http://www

.wales.nhs.uk/sitesplus/documents/861/Wipp%20Using%20Protocols%2Cstandards%2C%20policies%20and%20guidelines.pdf .

[8.7] "Flight Safety, Discipline and the Importance of Checklists," BAA Training Aviation Academy, Avia Solutions Group, Vilnius, Lithuania, 9/25/2017, accessed June 24, 2020, https://www.baatraining.com/flight-safety-discipline-and-importance-of-checklists/.

[8.8] Lateef, Fatima, "Simulation based learning: Just like the real thing," Journal of Emergencies, Trauma, Shock, 3(4), Oct-Dec 2010, p 348-352,Wolters Kluwer – Medknow Publications, accessed June 24, 2020, https://www.ncbi.nlm.nih.gov/pmc/articles/PMC2966567/.

[8.9] Shirts, R. Garry, "Ten Secrets of Successful Simulations," Simulation Training Systems, Del Mar, CA, https://www.simulationtrainingsystems.com/ten-secrets-successful-simulations/

[8.10] "Logbooks," from *Wikipedia*, the free encyclopedia, accessed June 24, 2020, https://en.wikipedia.org/wiki/Logbook.

[8.11] Oswald, Nick, Dr., BiteSize Bio, "Top Tips to Keeping Your Logbook in Shipshape," July 9, 2016, accessed June 24, 2020, https://bitesizebio.com/24852/top-tips-to-keeping-your-logbook-in-shipshape/

[8.12] Ibid, BiteSize Bio.

[8.13] Writing skills, "What makes a good report," University of Leicester, UK, accessed June 24, 2020, https://www.le.ac.uk/oerresources/ssds/writingskills/page_75.htm

[8.14] SiSense, What is Operational Reporting?, https://www.sisense.com/glossary/operational-reporting/

[8.15] NDMU Nursing, "How to Perfect Your End of Shift Report," February 12, 2018, Notre Dame of Maryland University, Baltimore, Maryland, accessed June 14, 2020, https://online.ndm.edu/news/nursing/perfect-end-shift-report/

[10.1] "Skill," Merriam-Webster, accessed July 14, 2018, https://www.merriam-webster.com/dictionary/skill

[10.2] Ibid, "Skill," Merriam Webster.

[10.3] Technology Training Systems, "Abnormal Situation Management Training for Process Operators: A Competency-based Approach to Sustained Performance Improvement," (student handout notes, Denver, CO: Technology Training Systems, 1998).

[10.4] Technology Training Systems, "Abnormal Situation Management Training for Process Operators."

[11.1] "Watch This!" *The Brain Games*, Dir. Jeremiah Crowell. Perf. Lera Boroditsky segment. DVD. National Geographic Channel. Disk 1 Illusions of Speech. 0:36:05 +. 2011.

[11.2] Socrates, Greek philosopher (470–399 BCE).

[11.3] Ed. Goodman: see http://www.edagoodman.com/the-value-of-knowing-what-you-dont-know-you-dont-know/

[14.1] FAA (Federal Aviation Administration), Appendix G to System Safety Handbook (Washington, DC: Government Printing Office, 2008).

[14.2] "Definition of Key Terms," AirSafe, last modified July 2, 2017, http://www.airsafe.com/events/define.htm; "NTSB Accident and Incident Reporting Requirements," NBAA (National Business Aviation Administration), accessed June 18, 2015, http://www.nbaa.org/ops/safety/ntsb/

[14.3] "Accident," The Free Dictionary by Farlex, accessed June 15, 2015, http://legal-dictionary.thefreedictionary.com/Accident

[14.4] "What Is Accident?" Black's Law Dictionary, accessed June 18, 2015, http://thelawdictionary.org/accident/

[14.5] "Disaster," Merriam-Webster, accessed June 8, 2015, http://www.merriam-webster.com/dictionary/disaster

[15.1] Baldoni, John, "Rafael Nadal Shows Us How to Use Doubt to Our Advantage," *Forbes*, January 16, 2020, accessed July 17, 2020, https://www.forbes.com/sites/johnbaldoni/2020/01/16/rafael-nadal-shows-us-how-to-use-doubt-to-our-advantage/#4264f68731e5

[16.1] List adapted from the table of contents from Joseph T. Hallinan, *Why We Make Mistakes* (New York: Broadway Books, 2009).

[16.2] Amagasaki Rail Crash. 2005. http://en.wikipedia.org/wiki/Amagasaki_rail_crash

[16.3] "Retraining sessions at JR West." The Japan Times Online, (japantimes.co.jp), 2007-06-29, http://www.japantimes.co.jp/news/2005/07/23/national/1182-retraining-sessions-at-jr-west/ - .VbT1hfl3poM

[16.4] Michael Shermer, "The Mind's Compartments Create Conflicting Beliefs: How Our Modular Brains Lead Us to Deny and Distort Evidence," *Scientific American*, January 1, 2013, https://www.scientificamerican.com/article/the-minds-compartments-create-conflicting-beliefs/

[16.5] F. John Reh, "Understanding Pareto's Principle: The 80-20 Rule," accessed August 8, 2018, https://www.thebalancecareers.com/paretos-principle-the-80-20-rule-2275148

[16.6] "Cognitive Bias Mitigation," *Wikipedia*, last modified May 29, 2018, https://en.wikipedia.org/wiki/Cognitive_bias_mitigation

[16.7] "Confirmation Bias," *Wikipedia*, last modified May 31, 2018, https://en.wikipedia.org/wiki/Confirmation_bias

[16.8] Ibid, "Confirmation Bias," *Wikipedia*.

[16.9] Michael Shermer, *Skepticism 101: How to Think Like a Scientist* (Chantilly, VA: Great Courses, 2013).

[16.10] Arien Mack and Irvin Rock, *Inattention Bias* (Cambridge, MA: MIT Press, Bradford Books Series in Cognitive Psychology, 2000).

[16.11] Richard E. Nesbitt, *The Geography of Thought: How Asians and Westerners Think Differently . . . and Why* (New York: Free Press, 2003).

[16.12] *Providence Journal-Bulletin*, review of *The Geography of Thought* by Richard E. Nisbett.

[16.13] Douglas H. Rothenberg, *Situation Management for Process Control: Situation Awareness and Decision Making for Operators in Industrial Control Rooms and Operation Centers*, (Research Triangle Park, NC: ISA, 2019).

[16.14] Schwartz, Barry. "Selling Coca Cola in Israel: Israel Marketing," *Barry Blog*, 7April 2009 http://www.cartoonbarry.com/2009/04/selling_coca_cola_in_israel_is.html.11/23/2011 (23 November 2011).

[16.15] David Bernard Guralnik, ed., *New World Dictionary of the American Language*, 2nd ed. (New York: Simon & Schuster, 1980).

[16.16] Daniel Stacey, "In Asia's Skies, Mistakes Go Unreported," *Wall Street Journal*, July 14, 2015.

[18.1] What Is Human-Machine Interface (HMI), Technopedia, accessed June 26, 2019, https://www.techopedia.com/definition/12829/human-machine-interface-hmi

[18.2] ISO 11064 Parts 1-7, *Ergonomic Design of Control Centers*, British Standard, BS EN ISO 11064.

[18.3] Few, Stephen. *Information Dashboard Design: The Effective Visual Communication of Data*. Sebastopol, CA: O'Reilly Media, 2006.

[18.4] Hollifield, B., D. Oliver, I. Nimmo, and E. Habibi, *The High Performance HMI Handbook*, Houston, TX: PAS, 2008. http://www.pas.com/

[18.5] Edward Tufte, "Sparkline Theory and Practice," Edward Tufte forum, last modified May 27, 2004, http://www.edwardtufte.com/bboard/q-and-a-fetch-msg?msg_id=0001OR&topic_id=1

[18.6] "Fan Chart (Time Series)," *Wikipedia*, last modified June 12, 2014, http://en.wikipedia.org/wiki/Fan_chart_(time_series)/

[19.1] Tyron Vardy, "Can Luck Cover for a Lack of Alarm Management?" Plant Engineering, August 26, 2014, https://www.plantengineering.com/single-article/can-luck-cover-for-a-lack-of-alarm-management

[19.2] Rothenberg, Douglas H., *Alarm Management for Process Control*, 2nd Edition—A Best Practice Guide for Design, Implementation, and Use of Industrial Alarm Systems. New York: Momentum Press, 2018. http://www.momentumpress.net/

[20.1] Drexel University, "Model Reference Adaptive Control (MRAC)," blog post, accessed July 19, 2015, http://www.pages.drexel.edu/~kws23/tutorials/MRAC/MRAC.html.

[20.2] Steven Obermann, "Measuring Multivariable Controller Performance: How to Tell if Your Multivariable Controller Is Doing a Good Job," *InTech* 62, no. 3 (2015): 16–20.

[20.3] Douglas H. Rothenberg, *Situation Management for Process Control: Situation Awareness and Decision Making for Operators in Industrial Control Rooms and Operation Centers*, (Research Triangle Park, NC: ISA, 2019).

[20.4] ISA-106 Series, *Procedural Automation for Continuous Process Operations* (Research Triangle Park, NC: ISA).

[20.5] Maurice Wilkins, "Improving Operations in Continuous Processes through Procedural Automation," *InTech* 58, no. 6 (2011).

[22.1] Center for Chemical Process Safety (CCPS), *Conduct of Operations and Operational Discipline: For Improving Process Safety in Industry* (New York: Wiley, 2011), http://www.aiche.org/ccps/publications/books/conduct-operations-and-operational-discipline-improving-process-safety

[23.1] Paul J. H. Schoemaker and George S. Day, "How to Make Sense of Weak Signals," *MIT Sloan Management Review* 50, no. 3 (2009), http://sloanreview.mit.edu/article/how-to-make-sense-of-weak-signals/

[23.2] International Association of Oil & Gas Producers (OGP), "Cognitive Issues Associated with Process Safety and Environmental Incidents," Report No. 460 (London: OGP, July 2012).

[23.3] Malcolm Gladwell, *The Tipping Point* (London: Little, Brown and Company, 2000).

[23.4] *Tipping point*, www.Dictionary.com

[23.5] Bryan S. Coffman, *Weak Signal Research—Part I: Introduction* (Louisville, KY: MG Taylor Corporation, 1997).

[23.6] Sir Arthur Ignatius Conan Doyle, *A Study in Scarlet*, (London: Ward Lock & Company, 1886). [presently Orion Publishing Group], (20 November 1886).

[24.1] Paul J. H. Schoemaker and George S. Day, "How to Make Sense of Weak Signals," *MIT Sloan Management Review* 50, no. 3 (2009), http://sloanreview.mit.edu/article/how-to-make-sense-of-weak-signals/

[24.2] Amy Webb, *The Signals Are Talking: Why Today's Fringe Is Tomorrow's Mainstream* (Philadelphia: Public Affairs, 2016).

[24.3] ibid, Webb, *The Signals Are Talking: Why Today's Fringe Is Tomorrow's Mainstream* (Philadelphia: Public Affairs, 2016).

[24.4] *Inflection point*, accessed June 9, 2020, https://www.thefreedictionary.com/inflection+point

[24.5] Malcolm Gladwell, *The Tipping Point* (London: Little, Brown and Company, 2000) https://www.littlebrown.co.uk/

[27.1] "Situation Awareness," *Wikipedia*.

[27.2] Sidney Dekker, *The Field Guide to Understanding Human Error*, 3rd edition, Boca Raton, FL: CRC Press, Taylor and Francis, 2014.

[27.3] Susan Milius, "Bold, Incorrect Spiders Mislead Groups: Wrongheaded Notions Prove Dangerous to a Colony's Health," *Science News*, July 11, 2015.

[27.4] Amy Chua and Jed Rubenfeld, *The Triple Package: How Three Unlikely Traits Explain the Rise and Fall of Cultural Groups in America*, (New York: Penguin Press, 2014).

[27.5] Weick, Karl E., and Kathleen M. Sutcliffe, *Managing the Unexpected: Resilient Performance in an Age of Uncertainty*, (San Francisco: Jossey-Bass, John Wiley & Sons, 2007), 154.

[28.1] Douglas H. Rothenberg, *Alarm Management for Process Control: A Best Practice Guide for Design, Implementation, and Use of Industrial Alarm Systems*, 2nd ed. (New York: Momentum Press, 2018).

[29.1] Center for Chemical Process Safety, *Conduct of Operations and Operational Discipline for Improving Process Safety in Industry*, American Institute of Chemical Engineers (AIChE), Center for Chemical Process Safety (CCPS), (New York: Wiley, 2011), 10, 19, 54, 109, 125.

[31.1] "Hospital Emergency Codes," *Wikipedia*, last modified June 16, 2018, https://en.wikipedia.org/wiki/Hospital_emergency_codes

[32.1] Andrine Redsteer, "5 Principles of Defensive Driving," *It Still Runs*, accessed August 26, 2014, http://www.ehow.com/info_8003043_list-five-principles-defensive-driving.html

[33.1] "Accident in Oajevágge, Norrbotten County, Sweden on 8 January 2016 involving the aeroplane SE-DUX of the model CL-600-2B19, operated by West Atlantic Sweden AB." Swedish Accident Investigation Authority; Final Report RL 2016:11e. File no. L-01/16. 12/12/2016. http://www.havkom.se/assets/reports/RL-2016_11e.pdf

[33.2] 49 CFR 192.631, Control Room Management, Transportation of Natural and Other Gas by Pipeline; Minimal Federal Safety Standards (Washington, DC: PHMSA [Pipeline and Hazardous Materials Safety Administration]).

[33.3] 49 CFR 195.446, Control Room Management, Transportation of Hazardous Liquids by Pipeline; Minimal Federal Safety Standards (Washington, DC: PHMSA [Pipeline and Hazardous Materials Safety Administration]).

[34.1] Weick, Karl E., and Kathleen M. Sutcliffe, *Managing the Unexpected: Resilient Performance in an Age of Uncertainty*, (San Francisco: Jossey-Bass, John Wiley & Sons, 2007).

[34.2] "Crew resource management," *Wikipedia*, last modified March 3, 2018, http://en.wikipedia.org/wiki/Crew_Resource_Management

[34.3] "Air Florida Flight 90," *Wikipedia*.

[34.4] R. Yin, *Case Study Research: Design and Methods*, 2nd ed. (Thousand Oaks, CA: Sage Publications, 1994).

[34.5] "World War Z (2013 movie): Do the Israelis really have a 10th man doctrine?" *Quora*, accessed October 10, 2014, http://www.quora.com/World-War-Z-2013-movie/Do-the-Israelis-really-have-a-10th-man-doctrine

[34.6] Paul J. H. Schoemaker and George S. Day, "How to Make Sense of Weak Signals," *MIT Sloan Management Review* 50, no. 3 (Spring 2009), http://sloanreview.mit.edu/article/how-to-make-sense-of-weak-signals/

[34.7] Ben Carson, radio broadcast interview, National Public Radio, ca. 2009. Author's note: I have been unable to locate even a semblance of a reference follow-up for this tool. I include it because it is quite valuable; without it, something is lost. My apology to the original source.

[34.8] "The Difference Between a Crisis, Emergency, and Disaster," Lighthouse Readiness Group, 2016, http://lighthousereadiness.com/lrg/difference-crisis-emergency-disaster/

[35.1] "Tenets of Operation," Chevron Corporation, accessed March 22, 2013, http://www.chevron.com/about/operationalexcellence/tenetsofoperation/

[37.1] "Tenets of Operation," Chevron Corporation, accessed March 22, 2013, http://www.chevron.com/about/operationalexcellence/tenetsofoperation/

[A2.1] Andrew Dillon, Cliff McKnight, and John Richardson, "Reading from Paper Versus Reading from Screens," *The Computer Journal* 31 (5), 457–464.

About the Authors

Douglas H. Rothenberg

Douglas H. Rothenberg is the president and principal consultant of D-RoTH, Inc., a technology consulting firm that provides innovative technology and services for industry. D-RoTH, Inc., currently specializes in alarm management, control room management, fit-for-purpose product design, and innovation development for new products and services. Rothenberg is a leading authority in the field and provides professional consulting and services worldwide. He is the author of *Alarm Management for Process Control* (2009, 2018), a defining reference book in alarm management, and *Situation Management for Process Control* (2019), a leading resource for engineers and managers of industrial processes. Rothenberg was a faculty member in the Systems Engineering Department of Case Western Reserve University and spent over twenty years with Standard Oil, BP Oil, and BP Amoco, where he was responsible for identifying and incorporating new state-of-the-art technology to support advanced manufacturing solutions.

Rothenberg has a PhD in systems and control engineering from Case Western Reserve University, an MS in electrical engineering from Case Institute of Technology, and a BS in electrical engineering from Virginia Polytechnic Institute. He holds several patents in instrumentation and controls and alarm management. He is active in the International Society of Automation (ISA) and is a member of Sigma Xi. He is the recipient of the 2005 Educator-of-the-Year Award from the Cleveland Technical Societies Council in Cleveland, Ohio.

Beverly Biel Kuch

Beverly is currently a freelance technical writer, specializing in petroleum refining, catalysis, corrosion, and industrial specialty products. She worked as an engineer in research and development at Standard Oil and at BP Chemicals for thirteen years, where she developed and presented training material for operators, written operating procedures for process plants and laboratories, and worked with plant operators on a daily basis.

Beverly was director of the Fenn College of Engineering Cooperative Education Program and a Career Pathways Specialist at the Cuyahoga Valley Career Center, where she authored "Developing and Implementing a High School Simulation Course to Provide Rigor and Relevance to the Curriculum," published in and presented at the symposium of the 2007 International Winter Simulation Conference of the Institute of Electrical and Electronics Engineers. Beverly founded HMK Enterprises, Inc., dba Water Station Plus, a retail water purification business. She has a BS in chemical engineering from Fenn College of Engineering at Cleveland State University.

Credits

The authors wish to credit the following individuals and organizations for providing source material that was included in this book.

Momentum Press engineering series, for permission to include materials previously published in *Alarm Management for Process Control*, 2nd edition, 2018.

International Society of Automation (ISA), for permission to include materials previously published in *Situation Management for Process Control*, 2019.

Honeywell, for permission to include graphics developed by the Abnormal Situation Management Consortium (ASM).

PAS, for permission to include an excerpt from a dashboard figure.

Additional Reading

The following items provide additional information, clarification, and discussion for the wide range of topics and ideas suggested throughout the handbook. You may find them helpful.

American Petroleum Institute. Recommended Practices "RP 1167 Pipeline Alarm Management." (2009) http://www.api.org/

Atherton, John, Frederic Gil, and AIChe. *Incidents that Define Process Safety*. New York: Center for Chemical Process Safety, 2008

Center for Chemical Process Safety (CCPS). *Recognizing Catastrophic Incident Warning Signs in the Process Industries*. New York: John Wiley, 2012.

Endsley, Mica R., Betty Bolté, and Debra G. Jones. *Designing for Situation Awareness: An Approach to User-Centered Design*. Boca Raton, FL: Taylor and Francis, 2003.

Engineering Equipment Materials Users' Association (EEMUA). *Alarm Systems: A Guide to Design, Management and Procurement*. EEMUA Publication No. 191, London: EEMUA, 2007. http://www.eemua.co.uk/

Flin, Rhona, Paul O'Connor, and Margaret Crichton. *Safety at the Sharp End: A Guide to Non-Technical Skills*. Burlington, VT: Ashgate Publishing, 2008.

Friedhoff, Richard Mark. *Visualization: The Second Computer Revolution*. New York: Harry N. Abrams, 1989.

Gawande, Atul. *The Checklist Manifesto: How to Get Things Right*. New York: Picador, 2011.

"Hazardous Liquid Integrity Management Enforcement Guidance Sections 195.450 and 452," Pipeline Hazardous Materials Safety Administration (PHMSA), Washington, DC: US Department of Transportation. Guidance for internal PHMSA enforcement staff use.

Hollifield, Bill R., and Eddie Habibi, *Alarm Management: A Comprehensive Guide, Second Edition*, International Society of Automation (ISA), Research Triangle Park, NC: ISA, 2011. http://www.isa.org/

International Society of Automation (ISA), *Management of Alarm Systems for the Process Industries*, ISA Standard 18.02, Research Triangle Park, NC: ISA, 2008. http://www.isa.org/

ISO 11064-5:2008. *Ergonomic Design of Control Centers, Part 5: Displays and Controls*. London: British Standards Institution (BSI), 2008.

ISO 9241-12:1998. *Ergonomic Requirements for Office Work with Visual Display Terminals (VDTs), Part 12: Presentation of Information*. Geneva, Switzerland: International Standards Organization, 1998.

Klein, Gary. *Seeing What Others Don't: The Remarkable Ways We Gain Insights*. New York: Public Affairs (imprint of Perseus Books), 2013.

Kletz, Trevor. *What Went Wrong: Case Histories of Process Plant Disasters, Third Edition*. Houston, TX: Gulf Publishing, 1994.

Marguglio, B. W. (Ben). *Human Error Prevention.* Cold Springs, NY: Bookinars, 2009.

Recognizing Catastrophic Incident Warning Signs in the Process Industries, Center for Chemical Process Safety, AIChE, New York: John Wiley, 2012.

Roberto, Michael A. *Know What You Don't Know: How Great Leaders Prevent Problems Before They Happen.* Upper Saddle River, NJ: Wharton School Publishing, 2009.

Shermer, Michael, *The Believing Brain: From Ghosts and Gods to Politics and Conspiracies—How We Construct Beliefs and Reinforce Them as Truths,* New York: Henry Holt, 2011.

Shirley, Richard S. "A Fog Index for CRT Displays: How Good Is Your CRT Display?" *Advances in Instrumentation,* v 40 part 2, proceedings of the ISA/85 Conference and Exhibit, Philadelphia, October 21–24, 1985.

Tufte, Edward R. *Envisioning Information.* Cheshire, CT: Graphics Press, 1990.

A1: Paper vs. Electronic Screen

A1.1 Paper vs. Electronic Screen

Electronic screens are the current norm in today's control room. Everything is becoming electronically based—controls, reports, checklists, procedures, shift handovers, schedules, and other odds and ends like work orders. Most of this movement away from paper has been driven by real and important reasons. It is a current view that electronic is better. It is assumed that paper is fraught with problems and issues and lost trees. This discussion matters because sometimes reality can be different from trends toward change. To be sure, there remain some drawbacks to electronic forms as well. Studies have shown that reading from electronic display devices is "slower, less accurate, more fatiguing, [and] decreases comprehension."[A2.1] Unless we carefully understand what this means to the individual operator or reader, we might be assuming our way into unrecognized problems—problems that can materially affect quality and outcome. Let us look together.

Pros and Cons

Pros: These are some of the benefits of electronic forms:

- Are readily available (not tucked away in some manila folder, cardboard box, or 3-ring binder)
- Are always up to date as approved
- Can be linked to specific tasks or display screens so context is preserved
- Can be specialized so that they are designed to apply only to specific situations at hand
- Can be immediately (electronically) shared with others

Cons: Electronic formats also have limitations and are not suited to all users or uses.

- An operator cannot find an electronic form if it is not where it should be. (Most operator search engines are nonexistent or primitive at best; a Google-like search capability is not the norm.)
- It is difficult to see a readable overview (either in document text view or quick whole-page-at-a-single-view).
- It is difficult to annotate (either directly or via Post-it notes) and archive the working copy "as worked" and flagged with a "push" to those who need to do more work on it or approve it.
- Electronic formats are tied to electronic viewers (portability, pocketability, and viewability).
- They are less tolerant of weather conditions and to viewing conditions such as sun or darkness.
- It is not easy to pass them on directly to another individual or to his device close at hand.

- They are difficult to "cut and paste" while in progress.
- They can inadvertently be sent to unauthorized others (unless specific controls are in place).

A1.2 Electronic Document Characteristics

Naturalness

Two general age groups form the staff in control rooms: the twenty- to thirty-something-year-olds and everyone else. The younger group has been raised with electronics. They take mobile communications, personal music, and small screens for granted, at times even preferring them to conventional displays. Email and text messaging are important and integral to their primary interacting skills. Give them a query task, and a Google search is an almost reflexive response. They are comfortable with interlinked views of almost anything—"click here for more information" is almost a "litmus test" requirement for acceptable content flow management. If it does not have a mouse and a touch screen and, more recently, voice recognition, there is no use for it. All in all, this group is ready, trained, and primed for electronic information.

On the other hand, the everybody-else group is different. They may be familiar with electronic tools—and exploit their use for advantage. They generally know how to interact with the devices and are comfortable doing so. But they are also comfortable with the predecessor mediums. They are more likely to search a long electronic document for a specific item and then pick up the printed version to look into it. They hope for a physical touch interaction when no other means is available. They are less likely to expect a touch screen-type of interaction when they see a mouse. They expect much more format flexibility (paper and other fixed-form documents) than their younger counterparts.

The bottom line is that it is important for designers to understand these differences and not force one mode over another as the required information access mode. They should also provide appropriate alternates for screen and for paper.

Readability

The screen electronic and presentation design must be high resolution and without functional artifacts and visibility distractions (glare, contrast, etc.). Here is a list of these and the rest of the static items requirements for electronic display:

- Sufficient pixel resolution to render all displayed items perfectly clear
- Sufficient display size to show needed information in an appropriate size for viewing and enough space for the depiction of related content without need to switch pages
- Comparable aspect ratio to content and everything properly rotated for viewing
- Comfortable fonts and graphic conventions
- Character and symbol spacing, line spacing, and format enhancements (underlining, italics, etc.) that are perfectly clear, conventional, and comfortable
- Color rendering, background controls, and contrasts that minimize reader fatigue

A1.3 Using Electronic Documents

This is an overview of the various characteristics and ways of using these documents. It is provided for you in order to assist you to use them, as well as for you to suggest improvements and modifications to your management and engineering if you are experiencing difficulty in their use.

Following the Thread

Electronic forms by their nature are flexible. Portable document format (PDF) documents are fixed in format but easily permit linking into and out of them. Other documents and screen content usually contain lots of navigation targets and real-time refreshed content. This flexibility can be important and useful for some tasks, but not all. Electronic documents provide a significant likelihood for out-of-order content, out-of-context content, stale content, and missing content. Clear construction and use controls need to be in place to prevent this from happening. You need to know how to use them all effectively and efficiently.

Procedures

Procedures require performance in strict order, with nothing left out and nothing added. During the execution, it may be necessary for you to access other information and perform specific controls manipulations. Access to these other items must be provided in a way that does not interrupt the procedure information presentation or flow. Both order and context must be preserved. They must support delay, leaving, and return and do it seamlessly for concurrent or successive users. You need tools and training in their use.

Documents

Documents are meant to convey information. This information is conveyed by the content (the words used, word order, and word juxtaposition with other words). It is also conveyed by the order of presentation: paragraph order and document section order. Therefore, the reader must be aware of the expected reading flow path through the document and diligently follow it. Any "click-outs" must be directly relevant and necessary for clarification or remedial understanding. They should be there for necessity, not for convenience or subject completeness. All click-outs must support direct return of the reader to the exact place he left. Long documents (more than three or four pages) should contain clear content lists and labels so that the reader is constantly aware of where in the document flow he is and where he might want to view next, all the while preserving the intended thread flow.

Content presentation should be placed into context with the larger whole. Anything snipped out should be referenced-lined back to where it came from (much like how a small area blowup of a graphic has a boundary line around it and a line connecting it to a much smaller boundary line showing its actual size and location in the source). Content should be displayed with sufficient surrounding content to clearly denote how it fits into the document. This suggests that paragraphs show at least the one before and the one after (or as much of each as reasonable given the size of the focus paragraph).

Content Change Management

This is not intended to be a discussion of change management in the traditional sense of management of change (MOC). Rather, it is literally the control of content changes. Documents in this category are commonly referred to as *live documents*. The most familiar would be Wikipedia. The most common control room example is operating displays with up-to-the-moment equipment status indications, operating values, and alarm indications. We expect this sort of freshness and clarity. What are less expected are procedures that may be updated during use. However, appropriate posting controls can handle any conflicts.

The issue is the document controls that the operator (or viewer) of content has available during explicit document use. There is a need to be able to fix the content of a display or a related group of them. Fixed content might be needed for reporting (status documentation, problem identification, etc.). It might be needed for understanding (freeze a display so that all the parts and information on it can be viewed at the same time and with the same values). When fixed, that status must be evident and audit trailed.

Related Content

Very little information is used in isolation. Where content refers to, is related to, or is needed to manage other things, those other things need to be visible to the operator at the same time. Where existing display screen real estate and resolution permit it, the other items may be shared in other pages or windows to be visible simultaneously. If it cannot be visible on a shared display screen, other displays are needed. They must be configured and available. If the other displays cannot be dedicated for the term of the need, then they must be capable of departing and returning to the same configuration as before departure. The end result is for you to be able to see at the same time enough of the content that is needed so you can understand and manage without navigating around to find it.

Personalization and Annotation

Personalization means the reader's ability to take his personal copy of the approved content and move it onto his virtual desk for use. On his desk, he has a full set of markup tools. A representative list follows:

- Underlining
- Highlighting
- Striking out (words and phrases, and whole paragraphs or pages)
- Adding marginal and in-text notes and diagrams and other annotation
- Replicating and editing parts of existing annotations for reuse, instead of needing to completely re-enter items
- Linking of text (e.g., circling one area of text or a diagram and drawing a line [visible electronic link] to another place or circled area in the document to where it is related or linked)
- Adding Post-it notes or similar
- Adding other differently formatted documents (pasting or inserting whole pages) so that they can be utilized as an integral part
- Adding labeled bookmarks so that the user can return to parts as needed
- Being able to contemporaneously share the personal document with others, as modified, but clearly identifying it as a working markup from a specific individual (useful for collaboration or approval)

Of course, these personal documents are for the operator who uses them. However, there is a need to preserve them, both temporarily and often permanently, as part of the operating record for later recall and understanding. This capability must be part of the embedded operating protocol for document control.

A1.4 Comparisons to Think About

We have briefly looked at some important aspects of electronically displayed content. This discussion concludes by returning to the strengths of paper content to further appreciate the important things electronic information display should master. Paper is clearly ordered in flow (page numbers, stapling or binding, etc.) from start to finish. It is much easier to position a paper document in the hands to find a comfortable viewing angle (devoid of reflections and glare), with appropriate lighting, at the right viewing distance, in a comfortable and convenient spatial location (desk with supplies, chair that is close to the kitchen), in a less distracting environment, and on a to-do pile. Reading and examining complex or lengthy content in paper form is less tiring, and there is a tactile connection that we humans seem to be sensitive to and like.

This discussion is not provided to convince anyone to adopt or abandon electronically displayed content. Rather, it is to advance understanding of some of the use realities and afford an opportunity to both appreciate them and plan to maximize their advantages and minimize negative effects.

A2: Soft Weak Signals

A2.1 Opener

Module 23: *Introducing Weak Signals* introduced and covered the new concept of weak signals. Most of the material there touched on tangible observations of small indications. For example, they included sensor readings that might seem a bit unusual. This appendix provides another facet of them. Here, we introduce small indications that in some sense are soft from a technical and equipment perspective and so relate more to infrastructure or people issues.

A2.2 Examples

Please note that the examples that follow are not designed to be complete or all-inclusive. They are intended only to give you an introduction into how it might work and suggest to you the broad scope of other possibilities for weak signals.

While attending to your regular duties, you just might happen to observe any of the following:

Observation (*weak signal*)	Possible Meaning (*forward extrapolation*)
Control valve position is near the ends of its range	Process is operating in a situation that is near or outside design expectations Valve itself is not working properly Measuring instrument is not operating properly Control loop is not working properly Process design is faulty
Lots of action (valve movements) made by the feedback controller	Control loop is working just as expected Possible tuning problems with the control loop Possibly too much noise in the measurement Excessive process noise Interfering electrical signals Missing or misadjusted deadbands Potentially sticking valve or leak in the pneumatic tubing

Observation (*weak signal*)	Possible Meaning (*forward extrapolation*)
Manufacturing or production efficiencies are changing more than 5 percent	Process is operating in a situation that is near or outside design expectations Process design is weak or faulty Measuring instrument is not operating properly Efficiency calculation is not functioning properly Process is operating according to design, but operational choices are affecting the efficiency (either improving when improvement was not being sought, or worsening)
Additives usage is changing more than 5 percent	Process is operating in a situation that is near or outside design expectations Process design is weak or faulty Measuring instrument is not operating properly Additive composition is changed or changing Additive addition calculation is not functioning properly Process is operating according to design, but operational choices are affecting the usage of additives more than expected
Normal variations of process measurements seems to have changed	Process measurement system is operating unreliably Faults in telemetry or data flow Some aspects of the operation have changed beyond normal variable limits or expectations Operator interface faults or failures, including not receiving updated information, or receiving updated information but failing to display it properly
Warning indicators are present for situations that are considered normal	Process is operating in a situation that is near or outside design expectations Instrument faults or installation errors Process design is weak or faulty Measuring instrument is not operating properly Operating situation is actually normal, but operator is misreading or misinterpreting the indicators as being abnormal

Observation (*weak signal*)	Possible Meaning (*forward extrapolation*)
Late laboratory reports and/or Missing laboratory values	Reports have been misdirected (either electronically or otherwise)
	Unusually high demand on laboratory resources (equipment and/or personnel)
	Difficulties in processing laboratory samples due to laboratory equipment issues
	Difficulties in processing laboratory samples due to the unusual nature of the samples
	Administrative or other personnel problems or issues at the laboratory
	Weak laboratory protocol
	Laboratory priorities are not aligned with the manufacturing requirements
Personnel inside the plant operating area who are not "checked in"	Weak or unenforced access control protocols for operating area access
	Unauthorized or unknown maintenance or cleanup activities in progress
	Unauthorized or unknown operations activities in progress
Open electrical boxes in the field	Unauthorized or unknown maintenance activities in progress, unless they are in a known maintenance area
	Poor housekeeping or weak maintenance and inspection protocols
	Physical damage preventing proper closure
Leaks on the ground or elsewhere	Leaks representing spills or events unrelated to operational problems or issues and thus represent sloppy work and/or poor housekeeping
	Leaks representing actual weakening and/or failure of designed containment of operating equipment
	Leaks representing equipment operated outside of design limits, causing temporary or permanent containment breach

Observation (*weak signal*)	Possible Meaning (*forward extrapolation*)
Changes in noise or smell in the unit	Changes represent events unrelated to operational problems or issues
	Chance observations that reflect transient operating conditions which are not indicative of any operational or equipment issue
	Changes representing equipment operated outside of design limits, causing temporary or permanent containment breach or damage to mechanical integrity of equipment
	Changes representing actual weakening and/or failure of designed containment, or stresses to the mechanical integrity of equipment

A3: Cue Cards

A3.1 Introduction

Cue Cards are a special part of this handbook. This appendix contains copies of the Cue Cards that are provided separately. You can pull them out, and use the information on each one to keep a particular activity focused and on course. Later on, perhaps your first day back on a rotation, pick a few out and use them as a handy review. This way, you can both ease into the shift and brush up on what those cards cover. Cue Card material is pulled directly from the text, where it is explained in the context with the situation it covers. That specific handbook location is identified on each Cue Card using the same reference types used throughout the handbook. For example, Cue Card 4a [**Q4a**] material has been taken from Module 6 section 3; e.g., [**M6.3**]. A similar referencing is used to refer to other Cue Cards [**Qx**] where the "x" is the specific number of the Cue Card. The color used for the corner identification of each card is matched to the color of the relevant handbook part. Hence, the Cue Card 4 label would be in green since it came from Part 2.

A3.2 List of Cue Cards [Q1b]

Here is the list of all 20 cards.

Card Number	Description
1	*Handbook overviews* FRONT: Handbook Map BACK: List of cue cards
2	Long-arm-of-the-operator checklist
3	*Situation management* FRONT: Basic 4 situation management components BACK: Control the Situation Management Work Process
4	Safe communications FRONT: Safe Communications Set-up BACK: Steps for Safe Communications
5	SUDA (See, Understand, Decide, & Act) FRONT: SUDA Diagram BACK: SUDA in Everyday Situations SUDA in Alarm Situations
6	Managing uncertainty

Card Number	Description
7	HMI Navigation Checklist
8	*Alarm response* Front: Alarm response sheet Back: 7 Steps for using alarm response sheet
9	Shift handover tasks and sequencing
10	*Maintenance handovers* Front: Maintenance handover Back: Maintenance handback
11	*Weak signals* Front: Weak signals diagram Back: Weak signals 7-step procedure with brief explanation
12	*Control room conditions* Front: Restricted control room Back: Sterile control room
13	Control room codes Front: Code Outside (emergency outside plant) Back: Code Inside (emergency inside plant)
14	Defensive operating Front: Defensive Operating Behaviors Back: Defensive Operating Added Guidance
15	Collaboration
16	*Escalation* Front: Escalation flow diagram Back: Escalation situation diagram
17	*Permission to operate* Front: Permission to operate flow diagram Back: Permission to operate situation diagram
18	Weak Signal Flags
19	Triangulation Questions
20	Crew Resource Management

A3.3 Cue Cards

See next pages. Where the card is multi-sided, the front will be first followed by the remaining cards in order.

Cue Card 1a *Control Room Operators' Handbook* Handbook Content Map

Control Room Operators' Handbook **Cue Cards**

Cue Card **1b**

Here is the list of all 20 cards.

Card Number	Description
1	*Handbook overviews* FRONT: Handbook Map BACK: List of cue cards
2	Long-arm-of-the-operator checklist
3	*Situation management* FRONT: Basic 4 situation management components BACK: Control the Situation Management Work Process
4	Safe communications FRONT: Safe Communications Set-up BACK: Steps for Safe Communications
5	SUDA (See, Understand, Decide, & Act) FRONT: SUDA Diagram BACK: SUDA in Everyday Situations SUDA in Alarm Situations
6	Managing uncertainty
7	HMI Navigation Checklist
8	*Alarm response* FRONT: Alarm response sheet BACK: 7 Steps for using alarm response sheet
9	Shift handover tasks and sequencing
10	*Maintenance handovers* FRONT: Maintenance handover BACK: Maintenance handback
11	*Weak signals* Front: Weak signals diagram Back: Weak signals 7-step procedure with brief explanation
12	*Control room conditions* FRONT: Restricted control room BACK: Sterile control room
13	Control room codes FRONT: Code Outside (emergency outside plant) BACK: Code Inside (emergency inside plant)
14	Defensive operating Front: Defensive Operating Behaviors Back: Defensive Operating Added Guidance
15	Collaboration
16	*Escalation* FRONT: Escalation flow diagram BACK: Escalation situation diagram
17	*Permission to operate* FRONT: Permission to operate flow diagram BACK: Permission to operate situation diagram
18	Weak Signal Flags
19	Triangulation Questions
20	Crew Resource Management

Cue Card 2 *Control Room Operators' Handbook* "Long Arm" Checklist

(Coordination between Control Room and Remote Site Personnel)

1. **Operator** identifies all equipment to be observed or adjusted by remote site personnel
2. **Operator** reconfigures and adjusts current operations as needed to ensure safe and appropriate observations and adjustments by the remote site personnel
3. **Remote site personnel** verify using independent means that they have properly identified the equipment to be observed and adjusted, and it is ready for their activities
4. **Remote site personnel** take the proper steps to observe and adjust the equipment.
5. **Operator** verifies step-by-step that the equipment can actually be taken back for normal operation and is fully ready for service; or is sidelined for other work. This is an active process where the operator and remote site personnel check, test, coordinate, and otherwise ensure that the appropriate adjustments are done and equipment is operationally ready or suitably set aside.

[M4.4]

Cue Card 3a

Control Room Operators' Handbook

Four Essential Activities for Effecting Situation Management

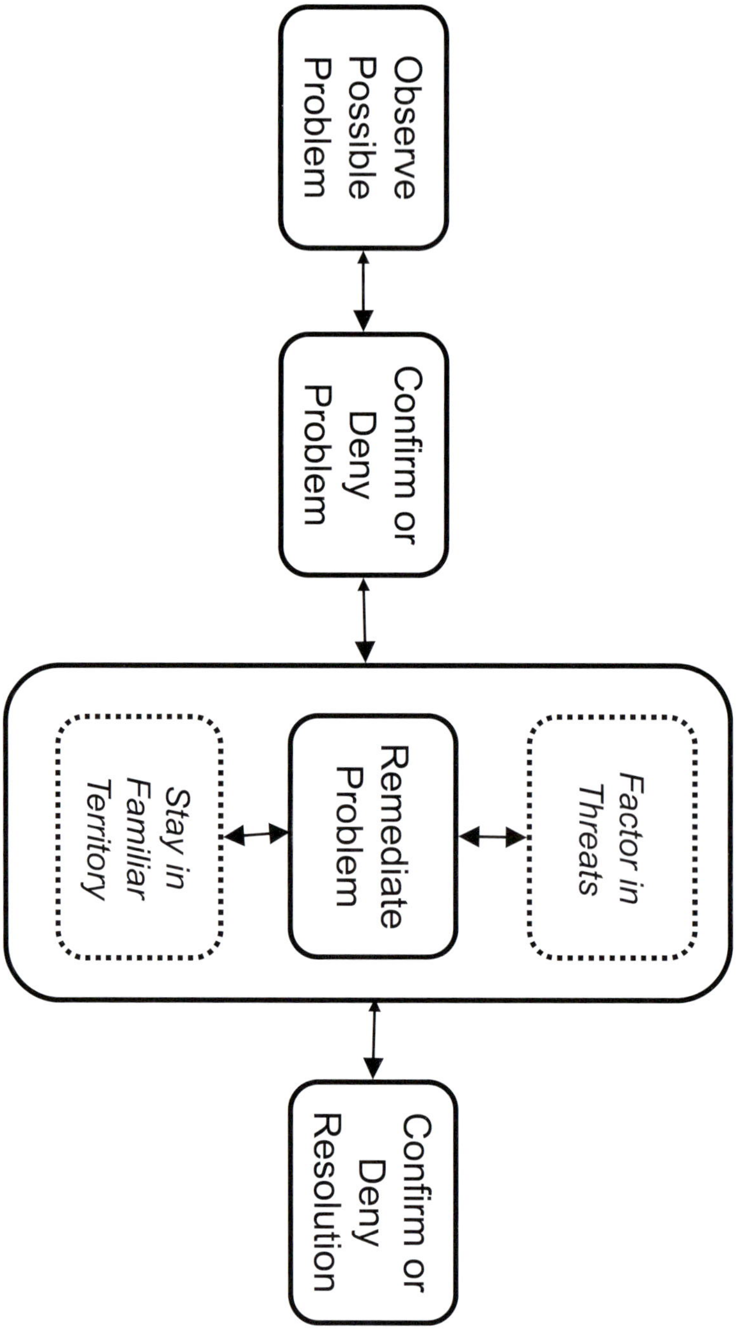

[M4.6]

Cue Card 3b *Control Room Operators' Handbook*

Situation Management Process

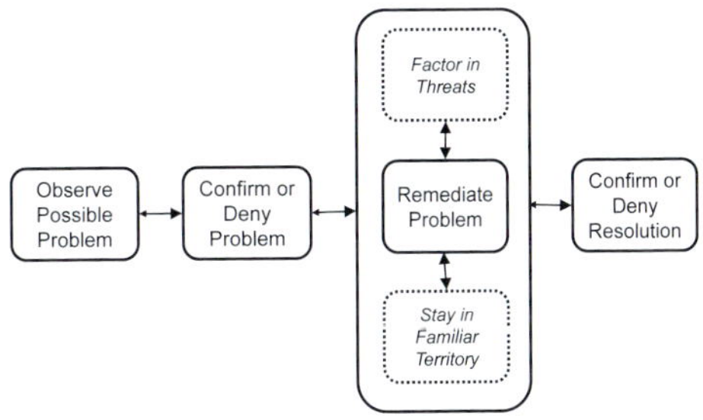

1. OBSERVE all potential problems.
2. CONFIRM the actual existence of all observed potential problems.
3. REMEDIATE all confirmed problems
   - Factor in all potential threats.
   - Never operate where you have no experience or explicit guidance.
4. CONFIRM resolution; otherwise start over

[M29.3]

Control the Situation Management Work Process

1. Use the effective operating protocols for situation awareness provided for you.
2. Ensure that there will be enough time for situation assessment and remediation activities; otherwise *permission to operate* **[Q17]** will be in effect.
3. Locate and use sufficient operational and directional information to permit successful management of the abnormal situation.
4. Utilize all safeguards and support mechanisms **[Q15]** & **[Q16]** & **[Q17]** in case your intervention is not obviously working as needed; or not going to be successful in time.

[M16.4]

Cue Card 4a

Control Room Operators' Handbook

Mindful Conversations

(Set-up for Safe Communications)

1. **Turn Off All Devices. Focus** on only individuals necessary to conduct the conversation.
2. **Use Your Eyes** to make contact. Pick up on all needed clues about how the other person is reacting about communicating with you.
3. **Listen** to the tone of voice to pick up on what might be below the surface, hidden, or hard to say. Hear the words said, the way they are spoken, and the care and feeling they convey.
4. **Say "no" to thoughts** that have nothing to do with the conversation. Get them out of the way.
5. **Attempt to truly understand** the communication. Hear without personal judgment, critique, or your own discomfort.
6. **Watch out for external conditions** unrelated to the conversation (e.g., clothing, gender, etc). These can lead you to presuppose a message.
7. **Respond instead of react.** Give yourself the time to process what you hear. You are in the listening pathway now. Listen only. Don't plan your response before its time.

[M6.3]

Cue Card 4b

Control Room Operators' Handbook

Steps for Safe Communications

1. **STATE WHAT IS KNOWN** and how it is known.

   Say exactly what is going to be said and how it is known.

   Repeat back to the other party succinctly what you think you heard said, and ask if there is anything more.

   Confirm that each party heard and understood what the other said. Notice and clarify any actionables, challenges or doubt. Keep a (mental) list and refer to soon later on.

2. **STATE EXPLICITLY WHAT IS THE CONCERN**

   This sets the entire reason for the conversation (but not the outcome).

3. **STATE EXPLICITLY WHAT IS EXPECTED** of each party. This sets the working relationship between each.

   Is confirmation expected or other opinion needed?

   Is actual *operational assistance* being requested (escalation, hands-on aid, collaboration) or something else? State the *something else* if so.

[M6.3]

Cue Card 5a

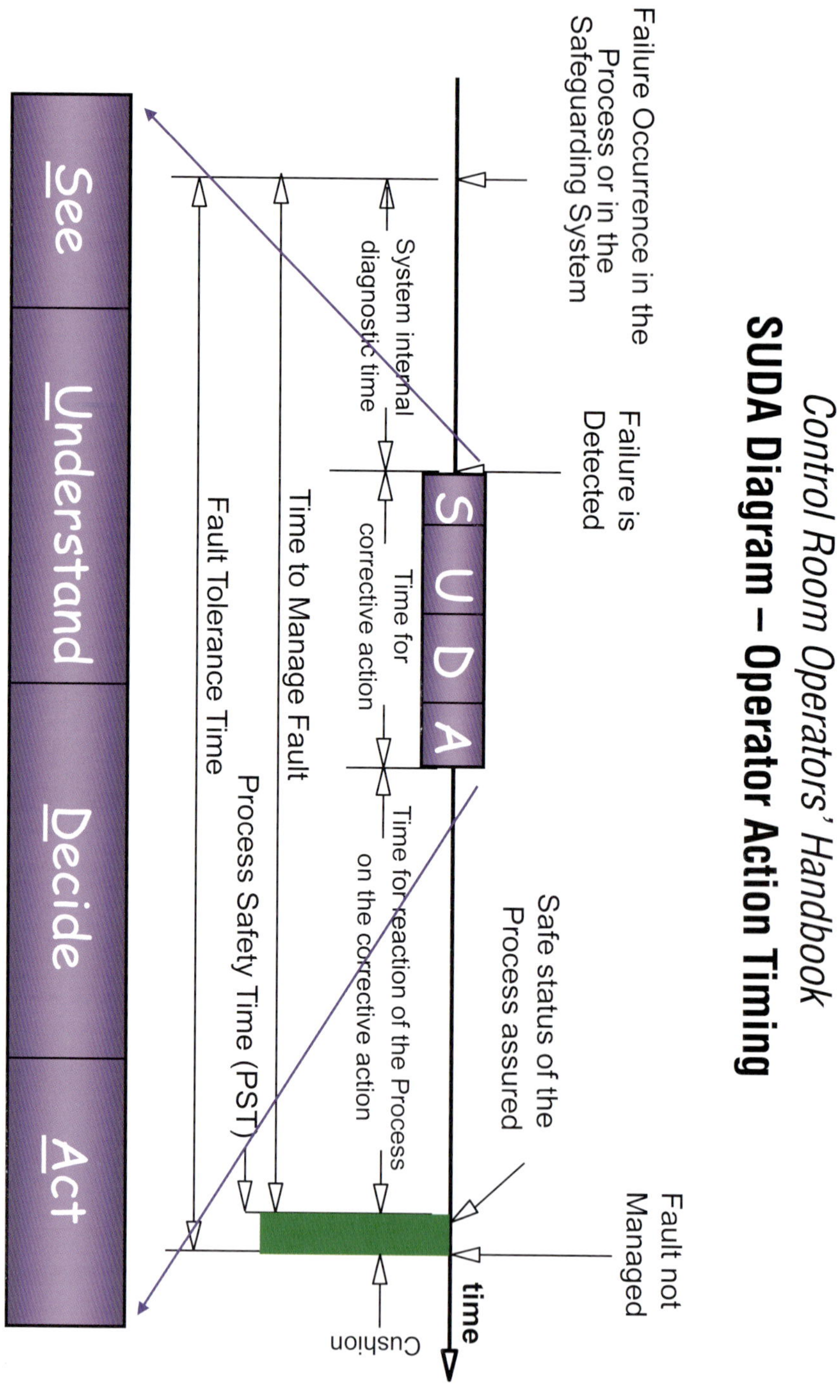

[M13.2] & [M19.2]

Cue Card 5b

Control Room Operators' Handbook

SUDA in Everyday Activities:

See: Be aware so you see that you have a problem

Understand: Understand the meaning of what you see and what the process will do if a remedy is not found or not found in time

Decide: Determine what to do to remedy the situation and decide how to do it

Act: Actually make all changes and other actions needed to implement your selected remedy

[M13.2]

Cue Card 5c

SUDA in Alarm Situations

For activated configured alarms (implemented in the Basic Process Control System):

1. **SEE**: see it (on HMI graphics and alarm summary screen)
2. **UNDERSTAND**: be able to sufficiently understand what the alarm means
3. **DECIDE**: decide what to do to respond to it
4. **ACT**: implement your selected response action. And make sure it is working

[M19.2]

Cue Card 6

Control Room Operators' Handbook

MANAGING UNCERTAINTY
TO UNDERSTAND, YOU NEED TO KNOW:

1. What are all SYMPTOMS of the concern?
2. What is KNOWN and what is the quality of that knowledge. Do not use what is unclear nor assumed unless and until properly verified.
3. What do you NOT KNOW but is essential to know and why do you need to know it?
4. What are the clearly perceived operational or other RISKS being faced?
5. What are the reasonable CHOICES and why?
6. What must be avoided?
7. What plan is the SAFEST and most likely to be successfully executed?
8. What is NECESSARY to execute that plan?
9. Is what is necessary to execute that plan AVAILABLE?
10. What is the BACKUP plan in case that plan cannot be selected or executed? Without a back up plan, your initial plan is unsafe.

[M15.2]

Cue Card 7

Control Room Operators' Handbook

HMI Display Screen Navigation Checklist

1. **Confirm current location** – is this specific navigation starting from the place you intended?
2. Before clicking, decide **what is needed next**. (This is the purpose of the navigation: exactly what it is that you are looking for).
3. **Decide where it is located** – decide where to find exactly what you are looking for. (You know the specific "location" where to look; or you have a good idea of what is near what you are looking for, so you can get close enough to find it.)
4. **Decide how to get there** using the existing tools and infrastructure; determine the exact, specific way to navigate (use a map; do a guided search; follow a **planned idea**)
5. **Mark where you are** and then get to where you want to go.
6. **Confirm where you went** is where you need to be. Did the navigation step(s) actually take you to the intended destination?
7. **Ascertain** that what you needed is where you went to find it.
8. **Get** the information or do the actions you need.
9. **Return** to step 2 and repeat as long as needed.

[M18.5]

Cue Card 8a *Control Room Operators' Handbook* ***Example* ALARM RESPONSE SHEET**

Tag ID: **FI-2009**
Point Descriptor: PB Feed Water Flow
Comments:
Setpoint: 155 Units: KLbHr
High Alarm: 185 (0 to 275 range)

Alarm Priority: High
Alarm Status: Enabled
Operating Group: 12
Process Area: PB
Trouble Point: 200
Rate of Change: 40

Abnormal Situation:
1. High steam consumption
2. Unusual boiler operation

Causes:
1. Failure of LVC/LIC-2011
2. Plugged flow element
3. Incorrect steam flow measurement
4. Instrument failure

Confirmatory Actions:
1. Check steam drum level LIC-2011
2. Check strip charts, check "periscope"
3. Check BFW header pressure FI-2078/2077, DEA Pressure PI-2098/2099
4. Check BFW pumps
5. Check flow control valve FCV-2009

Consequences of Not Acting:
1. Boiler could overheat and seriously damage tubes
2. Insufficient steam flow, high press.

Automatic Actions:
None

Time to Respond and Process to React:
90 seconds

Manual Corrective Actions:
1. If steam drum level and steam pressure are high, **put feedwater valve in manual and begin to close it.**
2. If steam drum level is low, **manually increase opening of feedwater valve**

Advanced Alarm Considerations:
1. Disable alarm when the boiler is shut down
2. Enable alarm when boiler is started up

Safety Related or Testing Requirements:
HaZop 12-4453; Yearly testing required

[M19.3]

Cue Card 8b *Control Room Operators' Handbook* USING THE ALARM RESPONSE SHEET

1. Compare the CURRENT VALUE of the process variable with the ALARM ACTIVATION POINT (setpoint) and the *Trouble Point* value.

   Compare the actual process variable with the *Nominal Rate of Change*. If considerably faster, then *escalate* [**M34**] or invoke *permission to operate* [**M35**] protocol.
2. Verify which of the ABNORMAL SITUATIONS is present and ensure that it is clearly. visible. If none, move quickly to step 3.
3. Work through each of the CONFIRMATORY ACTIONS to verify which are true and caused the alarm to activate. Check all! If non are true *escalate* or invoke *permission to operate*.
4. Review the CONSEQUENCES OF NOT ACTING so you will know where the trouble will lead.
5. Ensure that all AUTOMATIC ACTIONS have occurred and are working as expected. (Usually there are none, so no problem if that is the case.)
6. For each TRUE CONFIRMATORY ACTION, do the MANUAL CORRECTIVE ACTION and verify it is working. If not, *escalate* or invoke *permission to operate.*
7. If the alarm does not return to normal (clear) within a reasonable period, you must find the specific reason(s) why and remedy. If it still is not returning to normal, *escalate* or invoke *permission to operate.*

[M19.4]

Cue Card **9**

Control Room Operators' Handbook

Operator + Supervisor + Field Operator SHIFT HANDOVER Combined Timelines

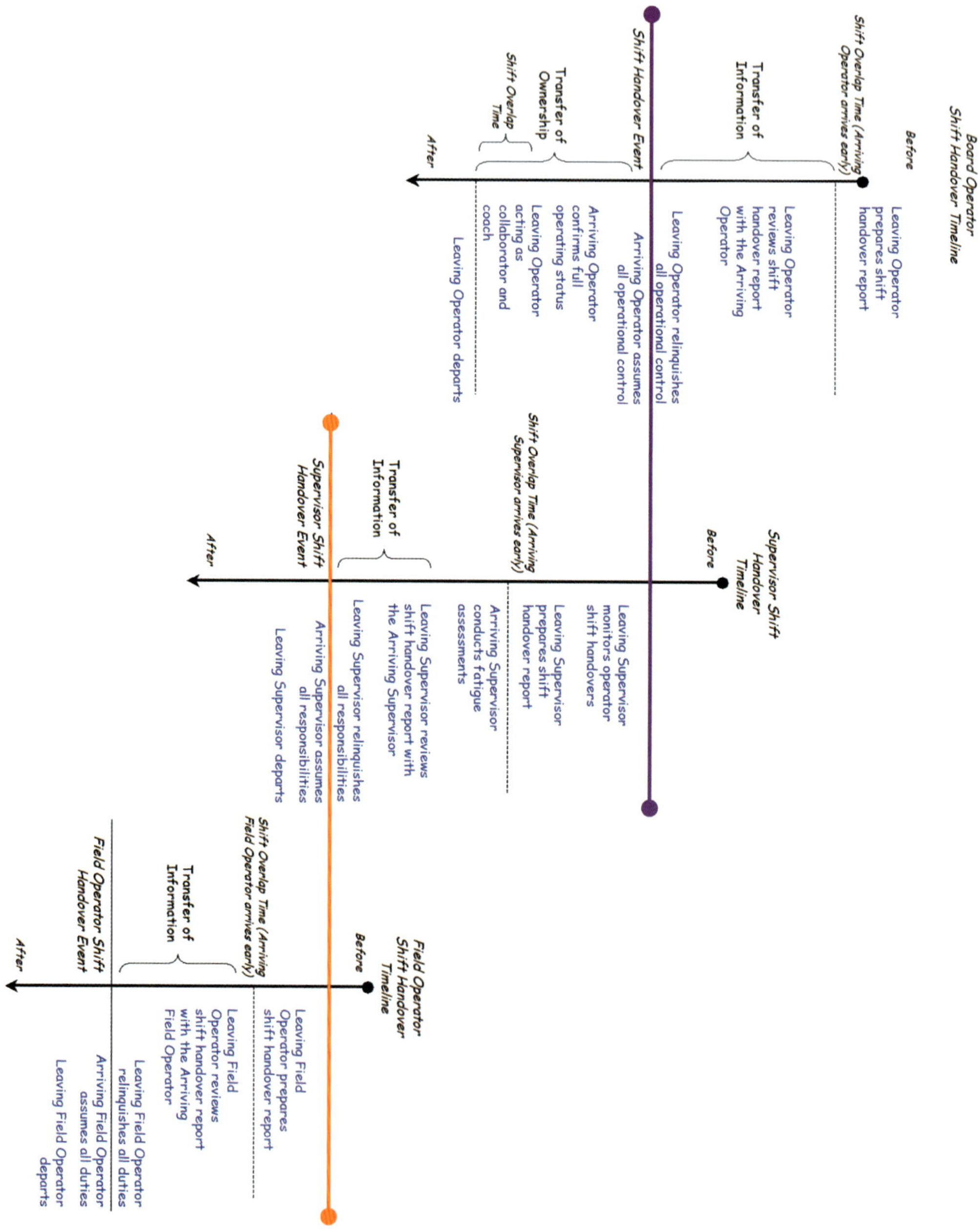

[M22.5]

Cue Card 10a

Control Room Operators' Handbook

Maintenance Handover

The board operator is the responsible individual since he must properly remove equipment from service before placing it in the hands of maintenance. This is a checklist of functionality, not a work procedure.

1. Board operator identifies all equipment to be taken by maintenance
2. Board operator reconfigures and adjusts current operations to remove the need for the equipment to be taken from production. Include all ancillary parts (like piping, instruments, etc.) in this action.
3. Maintenance verifies, by using independent means, that the equipment to be taken has been properly removed from production
4. Maintenance takes (temporary) ownership of equipment (and does their work/evaluation/inspection and anything else agreed to).
5. Maintenance takes the proper extra steps (not related to operations) to render the equipment safe and ready for maintenance activities; e.g., fitting of blinds, electrical isolation, erection of physical protection barriers, etc.

[M22.6]

Cue Card 10b

Control Room Operators' Handbook

Maintenance Handback

1. Maintenance identifies exactly what equipment is ready to be returned to service
2. Maintenance affirms that the equipment is properly repaired and/or evaluated and ready to be placed back in service; including removal of blinds, electrical restoration, clearing of physical protection barriers, etc.
3. Board operator verifies, step by step, that the equipment is actually ready for service
   a. This is an active process where the board operator and maintenance check and test and otherwise ensure that the equipment is ready to be back in service
   b. Note that this involves no verification of repairs; just that from the perspective of what the board operator can see and test, the repairs seem correct and the equipment should operate properly
4. Board operator takes back ownership of equipment
5. Maintenance leaves the area

[M22.6]

Cue Card 11a

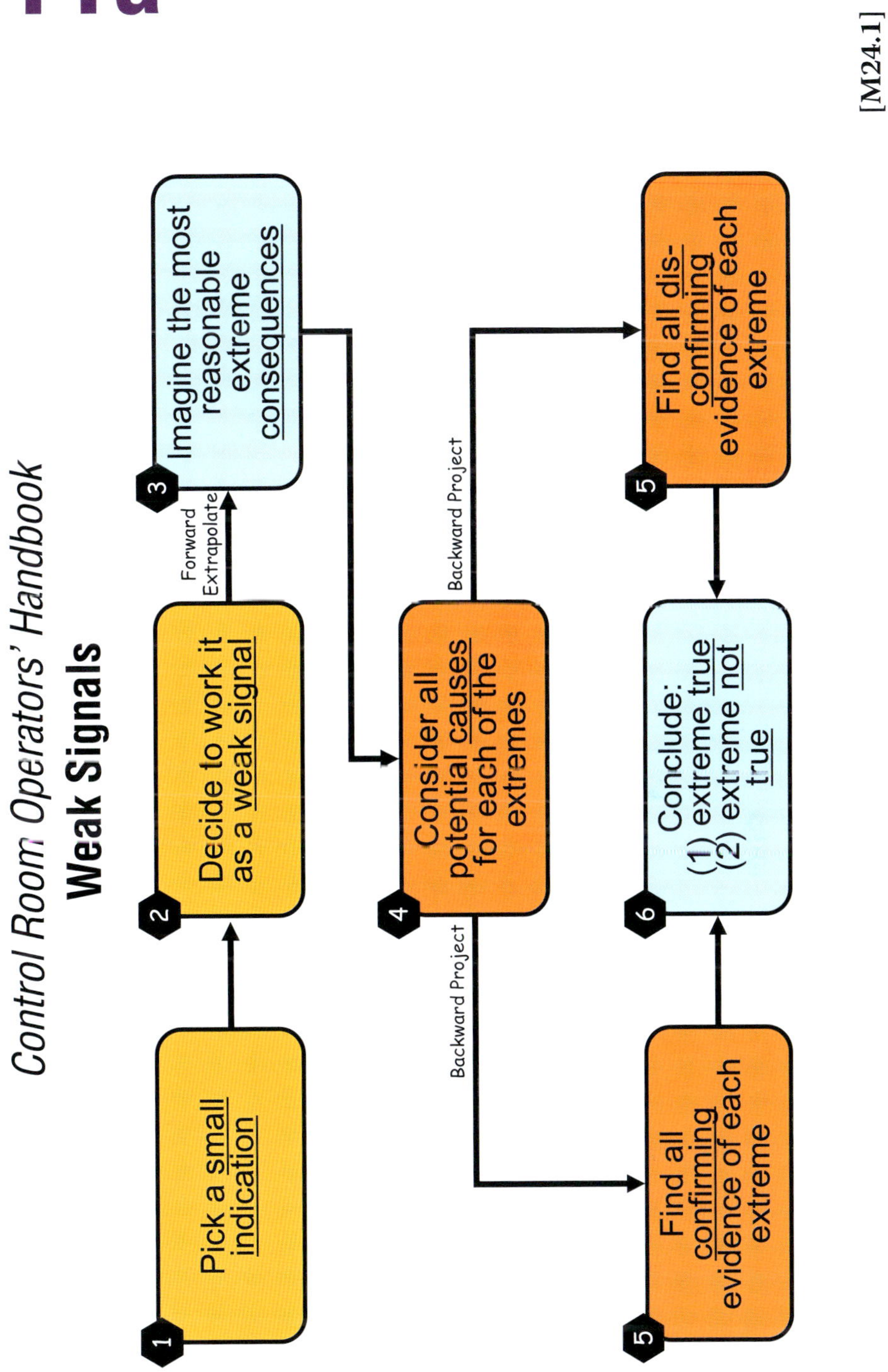
Control Room Operators' Handbook
Weak Signals
[M24.1]
1 Pick a small indication
2 Decide to work it as a weak signal
Forward Extrapolate
3 Imagine the most reasonable extreme consequences
4 Consider all potential causes for each of the extremes
Backward Project
Backward Project
5 Find all confirming evidence of each extreme
5 Find all dis-confirming evidence of each extreme
6 Conclude: (1) extreme true (2) extreme not true

Cue Card 11b *Control Room Operators' Handbook* Procedure for Working Weak Signals

Selection

1. Pick a small indication that you notice and might be useful.

You've seen or thought something and are curious or concerned enough about what it might mean to consider looking more.

2. Decide the small indication might be an indicator of something you want to know more about.

Now it is to be called a weak signal.Now it is a clue of something important affecting your operating area. From now on it is a weak signal.

Forward Extrapolation

3. Use your mind's process model to imagine the reasonable but worst that might be happening if this is actually a true early warning.

You are looking for the significant thing (quite bad; really unwanted) that might be happening if not fixed or worked on. If nothing is thought of, you are forgetting something about the enterprise operation. Rethink or get help.

Backwards Projection

4. For each worst thing that might happen (possible things that go wrong), make a list of all potential causes.

In steps 1 and 2, you noticed something unusual. In step 3 you imagined a potential important (bad) outcome. Now, in Step 4, list all things that might lead to that outcome.

5. For each potential cause on the list, look for everything that confirms (or denies) each supposed thing that could actually be happening.

Use the list from step 4 to tell you exactly what to look for and where to find evidence to confirm or deny.

Confirm Problem

6. For each confirmed weak signal in step 5, you have identified a confirmed problem you will need to resolve.

Confirmed means noticed, forward extrapolated to worst bad thing, backwards projected to look for evidence, and evidence found. So the original weak signal is probably real and in need of attention.

[M24.1]

Cue Card 12a *Control Room Operators' Handbook* RESTRICTED CONTROL ROOM

The *restricted control room* reduces extraneous distraction to a minimum by shifting the internal functioning of the control room to better respond to any significant operational situations present or coming.

Initiators:

- Severe weather now or coming
- Operations change requiring more than due care
- Medical emergency (in control room or field)
- Loss or failure to report for duty of minimum number of required personnel
- Observed or anticipated curtailment of any utility (electricity, water, outside fuels, waste or sewer)
- Below minimum needed operations personnel present
- By declaration of any authorized person

Conditions:

- All visitors out (to a safe, secure location, following accepted security protocols)
- All scheduled personnel must be in attendance: if assigned supervision personnel are not available, pre-determined back-ups must step in
- All essential personnel at their designed station(s)
- All operations are conducted with heightened attention
- All non-vital activity in field including maintenance, construction, survey, etc., are expedited for early completion, paused, or delayed (always in a safe manner).
- Where access to special or detailed documentation information is needed that is not normally readily available, a specific dedicated individual will be responsible for opening access and its use.

Termination:

- Authorized person declaring that the restricted control room is terminated. /or/
- By consensus of attending operators (where supervision is not in attendance; and initiating conditions are over)

[M31.2]

Cue Card 12b *Control Room Operators' Handbook* STERILE CONTROL ROOM

The *sterile control room* reduces extraneous distraction to a minimum by shifting the internal functioning of the control room to better respond to any significant operational situations, present or coming.

Initiators:

- Current situation has likely potential for loss of life or other significant dangers inside or outside of the enterprise
- Activation of any single highest priority alarm (or multiple second highest alarms)
- Significant loss of operational control
- Security (cyber or physical) breach
- By declaration of authorized person (either from a restricted status or directly to sterile)

Conditions:

- ***If not already set, invoke all restricted control room conditions*** (see **[Q12a]**). In addition:
- All background activities stopped: food on temporary hold; music off; non-essential telephone calls and radio calls ended and all future ones screened.
- All non-vital activity in the field is terminated (rendering them safe). Unless maintenance personnel are required to ensure safety, all are removed from the operating area. They are to remain at the ready in case of need.
- Operators in training or newly assigned must have their back up take over their activities.
- Control room and field manned by separate qualified individuals (no shared duties).
- All shift handovers are delayed pending staffing determination. Relief personnel should report and remain present for backup.
- All non-essential movement within the control room ended
- All non-essential conversation ended
- All essential conversation is muted designed for minimal distraction yet clear understanding. Each exchange is confirmed via voice-back repeat.

Termination:

- Authorized person declaring that the sterile control room is terminated.
  - ✓ AND room is transitioned to a restricted control room status
  - ✓ OR all unusual aspects no longer exist

[M31.2]

Cue Card **13a**

Control Room Operators' Handbook

CODE OUTSIDE (Outside Plant Fence)

In this situation the focus for concern is at the enterprise boundary

A significant threat exists or is anticipated from OUTSIDE the enterprise, likely to affect what you are doing or planning to do.

Initiators:

- Severe weather
- Civil unrest (likely to affect operations)
- Significant curtailment of any utility
- Requirement exists to isolate plant from the outside community (medical, political, or any other reason potentially impacting operations)
- Significant loss of ability to store/ship product, or dramatic production increase
- Disease or risk of disease exposure

Termination:

- Authorized person declaring **CODE OUTSIDE** is terminated.

[M31.3]

Cue Card 17a

Control Room Operators' Handbook

PERMISSION-TO-OPERATE

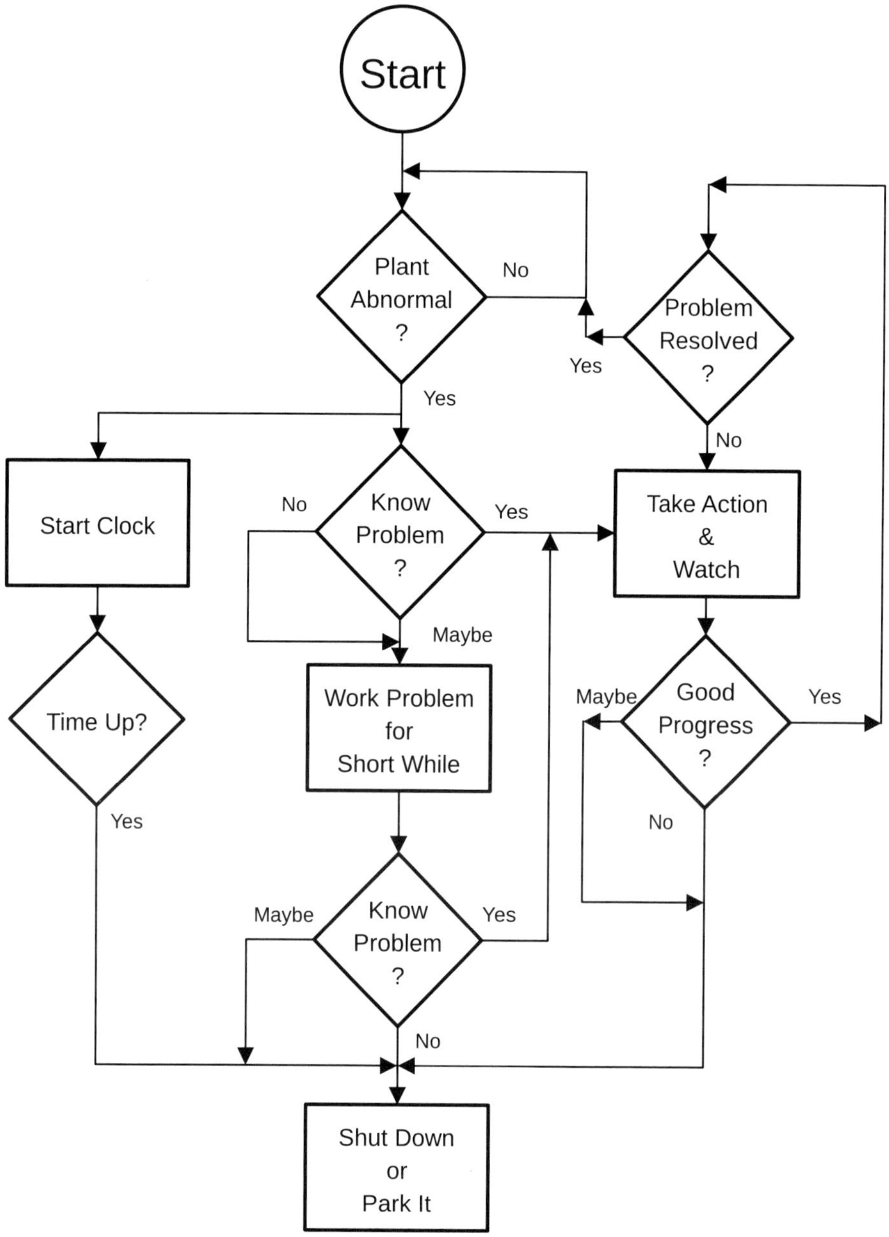

[M35.7]

Cue Card 17b

Control Room Operators' Handbook

PERMISSION-TO-OPERATE

Operational Modes:	Plant States:	Critical Systems:	Operational Goals:	Plant Activities:
Emergency	Disaster	Area Emergency Response System	Minimize Impact	Fire Fighting First Aid Rescue
	Accident	Site Emergency Response System		
Abnormal	Out of Control	Physical and Mechanical Containment System Safety Shutdown, Protective Systems,	Bring to Safe State	Evacuation
	Abnormal	Hardwired Emergency Alarms DCS Alarm System	Return to Normal	Manual Control & Troubleshooting
Normal	Normal	Decision Support System Process Equipment, DCS, Automatic Controls Plant Management Systems	Keep Normal	Preventative Monitoring & Testing

If operating within the red box and in the operational situation of being ABOVE the dotted red line, you have immediately lost permission to operate. Shut down or safe park..

[M35.1]

Cue Card 18

Operators' Companion Guide

Working Weak Signal FLAGS

1. INFORM supervisor whenever the number of confirmed weak signals become three or more.
2. Do FLAGS appear related
   (a) Yes– examine carefully for a new problem different from current supposed one.
   (b) No—work each problem; escalate as needed.
3. RECOMFIRM that current problem is correct.
   (a) Yes—carefully continue so long as it works.
   (b) No—for sure the current problem is wrong.
4. Current solution still working, but slowly?
   (a) Yes—Carefully look for any new problems
   (b) No—Problem is getting worse. Escalate.
5. Are ALL indicators for successful remediation still present?
   (a) Yes—Current solution needs improving
   (b) No—Wrong problem is being worked. Escalate immediately.

[M25.3]

Cue Card 19 — *Operators' Companion Guide* TRIANGULATION QUESTIONS

1. Do you (asking the other person) see **anything unusual** or out of the ordinary?
2. Are you having any **problems** doing your work?.
3. What are the ways that anything you are doing might affect me?
   (a) Are you seeing everything you expect to see to ensure proper functioning in your area?
   (b) Does everything in your area confirm that what you are doing is proper and working?
   (c) Do you see anything amiss in your area suggestion something is not as expected?
   (d) Are you unable to see anything you need to confirm your concerns?
   (e) Do you see anything you cannot fully explain (and understand)?
4. Assuming you are doing my job:
   (a) How could what I am doing affect you?
   (b) Do you see anything unusual in my area?
   (c) Do you see everything you expect to see in my area to ensure proper function?
   (d) Do you see anything in my area that is unexpected?

(continued on reverse)

Cue Card **19**

Operators' Companion Guide

TRIANGULATION QUESTIONS

(Continued)

5. Are you aware of anything going on in the larger plant that might unduly affect your or my current operations?
6. Are you aware of anything planned in the larger plant that is not happening or not going according to plan?
7. What do you know that I do not?
8. What might I know that you do not?

Processing Answers to the Triangulation Questions:

Work down the list of 8 items only so long as you know FOR SURE that both of you have enough time to completely work all problems fully. Use this new or confirmed knowledge to understand and manage.

Otherwise ALWAYS *escalate* (if time permits) otherwise invoke *permission to operate* [**M35**].

[M34.5]

Cue Card 20

Operators' Companion Guide

CRM – Crew Resource Management

This is the control room equivalent of "see something, say something."

The person in authority is the LEADER; the other junior individual is the FOLLOWER.

1. The FOLLOWER states the facts.
   (a) Here exactly is what I see or think.
   (b) Example: The tower temperature is way off normal.
2. If the LEADER does not affirmatively indicate hearing and understanding the facts:
   FOLLOWER issues the Challenge.
   (a) FOLLOWER restates the facts with more context and personalizes the message.
   (b) Example: [Use the name of the LEADER], are you going to continue the normal flow rate into the splitter tower? Check the low temperature at tray 22.
3. If the LEADER fails to exhibit full engagement with the situation and act appropriately:
   (a) FOLLOWER takes least impactful but adequate action on his own.
   (b) FOLLOWER identifies any other facts to LEADER but takes no further action.

[M34.4]

Glossary

Abnormal: The state of not being Normal; ordinarily in need of recognition and attention

See: Normal

Acknowledge: The first Operator action that indicates the recognition of an Alarm to the alarm managing portion of the Control System

Activate: The Alarm becomes "in alarm" by the monitored entity passing over the Alarm Activation Point from the Normal into the Abnormal condition.

Activation Point: The threshold value or discrete state of a process variable that when crossed triggers the alarm into the Active state

Alarm: An audible or visible means of indicating to the Operator an equipment or process malfunction or abnormal condition requiring a response

Alarm Acknowledge: See: Acknowledge

Alarm Activation Point: See: Activation Point

Alarm Flood: The situation when the number of alarm activations exceeds the Operator's ability to process them

Alarm Indication: All means of indicating to the Operator that an alarm has been activated

Alarm Limit: Archaic use

See: Alarm Activation Point

Alarm Log: The historical record of Alarm indications

Alarm Management: The processes and practices for determining, documenting, designing, monitoring, and maintaining alarm systems

Alarm Message: The text message that is normally used to convey identification information about an alarm when it Activates.

This is an important part of the Alarm Summary.

Alarm Philosophy: The guiding document for design or redesign of the entire alarm system

Alarm Point: See: Alarm Activation Point

See: Alarm

Alarm Priority: The attribute of an Alarm that defines a specific level of importance to the Alarm to be used by the Operator in deciding which alarms to work in which order

Alarm Rationalization: See: Rationalization

Alarm Response Manual: The set of all Alarm Response Sheets

Alarm Response Sheet: All information about a given Alarm that fully documents every operational aspect of the alarm. It is extremely useful to both the Operator and the alarm system designer.

The format may be printed forms, on-line printed forms, or dynamically assembled information relevant to the alarm and plant states.

Alarm Setpoint: Misleading term

See: Alarm Activation Point

Alarm Summary: A graphic Display that lists alarm indications over a period of time. List includes the status of all active alarms and all recently cleared alarms.

In general, the display is pre-configured by the BPCS manufacturer with only minor display customization allowed.

Alarm System: The entire construction that detects Alarm State, conveys that information and other supporting information to the Operator, manages the activation parameters, and archives all alarm information

See also: Alarm Summary

Alarm, Stale: Any alarm that has been active and not cleared for an extended period of time, well beyond any usefulness to the Operator (usually several shifts)

Alarm, Standing: Any alarm that has been active and not cleared for a given short period of time (usually a shift of eight to twelve hours)

Alert: A message or other visible and/or audible means of indicating to the Operator of a plant situation or equipment condition that does not require a response

Useful to replace old Alarms that were only for the purpose of notifying the Operator of conditions or status of equipment

BATNEEC: Best available technology not entailing excessive cost

Bias: Influences for action or decisions that interfere with proper consideration or understanding. Often they operate without our conscious awareness.

BPCS: Basic Process Control System

Bypass: To manually modify a function to prevent its activation (This term is not normally used to describe alarm readiness states.)

This is used to imply that some mechanical method (jumper wires, etc.) has been employed to ensure that the affected alarm will not activate.

See also: Disable

Clear: See: Cleared Alarm

Cleared Alarm: An Alarm that has returned to Normal

CM: Change Management

See: MOC

Configure: To modify the Control System parameters by entering appropriate values into a pre-structured database provided by the equipment manufacturer

Confirming: Evidence that directly confirms the reasonable truth of a worst, most impactful, or most extensive implication of a weak signal

Console: The interface for a single Operator to monitor the process, which may include multiple displays and other annunciators and communications equipment

See also: Display, Screen, Operator Area

Control System: See: BPCS

Controller (equipment): A device or program or alternate managing capability that operates automatically to regulate a controlled variable to a desired value or range of values

Controller (individual): The individual or individuals who operate equipment for its intended purpose by appropriate monitoring and intervention

Controller is an alternate term used in some industries and often in operation centers.

See also: Operator

DCS: Distributed Control System

Disable: To manually or automatically remove the alarm function from a process indication

Disconfirming: Evidence that directly refutes (disconfirms) the reasonable truth of a worst, most impactful, or most extensive implication of a weak signal

A failure to confirm cannot be assumed to disconfirm.

Display: A given rendition of a graphical depiction for showing on a Screen

See also: Screen

Enhanced Alarming: Any method of modifying alarm parameters or information to ensure that the alarm maintains its relevance for the current operating conditions. This also includes the provision of advice and information to the Operator regarding the interpretation and managing of alarms.

Enterprise: See: Plant

ESD: Emergency Shut Down

Evidence: Information that directly confirms or disconfirms the reasonable truth of a worst, most impactful, or most extensive implication of a weak signal

Extrapolation, Forward: Treating a chosen small indication as a weak signal, this is the activity of imagining the worst, most impactful, or most extensive implication for it.

Fault Tolerance Time: See: Process Safety Time

Four Corners Tool: The activity of examining a situation requiring a decision by examining four questions: best and worst result if the decision is made; and best and worst result if decision is not made

HMI: Human Machine Interface

Initiating event: A malfunction, failure or other condition that can cause an alarm indication

Intelligent Fault Detection: See: Enhanced Alarming

LO/TO: Lockout/Tagout is the coordinated action (according to enterprise rules) of physically locking out equipment (mechanical, electrical, etc.) using an actual mechanical lock; and affixing an appropriate physical tag notice. It is specifically intended to protect personnel from the unprotected effects of stored energy (electrical, mechanical, thermal, chemical, and radiological).

Major Event: An abnormal situation in which the plant or production unit is disturbed to the point of seriously jeopardizing its operational integrity

See also: Alarm Flood

Message: A text message that is delivered to the Operator or others meant to convey information. It is distinguished from an alarm so should not be used for situations requiring alarms.

MOC: Management of Change

Also referred to as Change Management

Near Hit: An abnormal event that occurred but did not cause its expected impact. It is generally accepted that these events will very likely occur in the future and cause undesired effects.

Near Miss: See: Near Hit

Normal: An Alarm State that indicates that the particular point is not in Alarm, however the alarm has not been inhibited or otherwise prevented from activation normally.

Nuisance alarm: An alarm that transitions to the alarm state but does not require action from the Operator

Occam's Razor: A mental (or thought) test that states that if there are two competing theories or explanations for something important, we should strongly lean towards believing the simpler one

OJT: On-the-job training

Operating Mode: The particular situation under which a plant is operating; generally, it is presumed that the various operating modes are sufficiently different to affect how it is managed and perhaps how the alarm system must work.

Operator Area: The specific portion of the plant that a single Operator has been assigned to operate

This would also be "Controller Area" for Controllers.

Operator Console: The equipment used by an Operator to manage aspects of production that are facilitated by remotely operated controls and/or where operational parameters are observed where they are not directly observed on the equipment

Operator Response Time: The time between the annunciation of the alarm and the time the Operator takes the (presumed) correct action in response to the alarm

Operator (in control room): The individual or individuals who operate equipment for its intended purpose by appropriate monitoring and intervention

Operator (of enterprise): The individual or entity responsible for an entire plant or facility such as a pipeline

Operator, Inside: An Operator whose primary working location is at an Operator console, usually in a sheltered area which may be within, but separate from the physical plant

Operator, Outside: An Operator whose primary working location is within the physical plant

Also referred to as a field operator

Permission-to-Operate: An immediate tactical decision whether or not the operator can continue to operate (a unit or enterprise). If permission-to-operate is lost, the only appropriate action is to render the operation safe.

See also: Safe Park

Philosophy: See: Alarm Philosophy

Plant: A coordinated or physically contiguous collection of areas. This is normally considered to be a facility that is under the management and control of an owner.

Also termed enterprise

For extensive facilities, this may require special care to delineate any differing portions that must follow different rules than laid out in a single alarm philosophy.

Plant Area: The basic part of a plant that is managed by one Operator position (but which may be actually operated by more than one physical Operator, so long as the equipment and controls interface contains all of the pieces for which the Operator is responsible and none other)

Plant Area Model: A construction by which the plant is divided up into constituent parts in a structured way that enables the ability to identify and control alarms as a group with respect to Related Alarms, Consequential Alarms, and others

Plant Enterprise: The largest entity in a corporate structure; generally consisting of multiple plants located on one or more sites

Plant Equipment: Individual items in a plant module

Plant Module: Collections of plant equipment; the smallest general entity for which alarms readiness states are managed

Plant Response Time: The time it nominally will take a plant to respond to an Operator action which is designated to bring the plant to an acceptable state of operation

Plant Site: A single, physical geographical location for manufacturing

Plant Unit: See: Unit

Priority: See: Alarm Priority

Process Safety Time PST: The amount of time between an abnormal process event and the inception of serious consequences. Effective management of the abnormal event, if done in time, can usually prevent the inception of the serious consequences.

Projection, Backwards: Treating the worst, most impactful, or most extensive implication for a weak signal that was forward extrapolated, to imagine the likely places to look for evidence

RAGAGEP: Recognized and generally accepted good engineering practice

Rationalization: The process of accepting a potential Alarm using the Alarm Philosophy, determining and documenting the rationale, and specifying all design requirements for the Alarm (including Activation Point and Priority)

Return /or/ Return to Normal: The alarm system indication that an alarm condition has cleared

RTN: Return To Normal

See Also: Normal

Safe Park: A preplanned procedure for moving a process or unit or enterprise to a responsibly safe operating status short of a full shut down

SCADA: Supervisory Control and Data Acquisition (system)

Screen: A single piece of hardware that is capable of depicting a visual graphical rendition normally associated with Operator consoles. Any number of displays may be viewed (serially or by combining views) on a given screen, up to the limit of the hardware and software controlling them.

See also: Displays

Signals, Strong: See: Strong Signals

Signals, Weak: See: Weak Signals

SIL: Safety Integrity Level – provides a universally understood safety performance rating based upon legal codes and such

SIS: Safety Instrumented System

Situation Management: The act of observing, understanding, and planning to ensure proper management, effecting action to manage, and ensuring the success of the managing activities

Small Indication: Something that is observed (from the HMI, or any other means) that seems in the moment to be of question or not quite expected

When it is decided to explore what it might mean, it is elevated to a weak signal.

See also: Weak Signal

SPOF: Single Point of Failure. A failure of any single part of something important that, when it fails, will cause a significant fault or serious abnormal situation.

Station: A single human machine interface within the Operator console. An Operator console may have a number of stations.

See: Display HMI

Strong Signals: A clear indication of some aspect of operations gone amiss. It does not include why the situation exists or what to do to remedy it.

Style Guide: A set of design recommendations for constructing HMI display screen content

SUDA: See, Understand, Decide, and Act

This is the activity of an Operator in responding to an alarm activation. It begins with the Operator seeing and recognizing an alarm activation and continues through all of the activities needed for the Operator to make what is anticipated as the proper corrective action to the process.

See: SUDA Time

SUDA Time: The time it takes for the entire SUDA activity to take place

See: SUDA

Surprise Effect: Evidence that directly confirms the reasonable truth of a worst, most impactful, or most extensive implication of a weak signal.

Sustainability: The goal of seeking activities (decisions, actions, policies) that significantly lead to better long-term outcomes

Tag: A unique label identifier that is used for BPCS data. Typical ones are temperatures, flows, and the like. Displays, alarms, logs, and other aspects are usually organized by their Tag.

Tenth Man Doctrine: A method of questioning a group decision by asking a "10th individual" to actively find all reasonable reasons why the group decision might be wrong

Time to Manage Fault: The amount of time required for the Operator to become aware of an abnormal event, understand it, take action to manage it, and the process to respond in a way that prevents serious consequences

So long as the Time to Manage Fault is less than the Process Safety Time, serious consequences can be thought of as being amenable to Operator action.

See also: Process Safety Time

Triangulation: The process of viewing the same process aspect (lab value, measurement, flow rate-of-change) by two different individuals to expose alternate information.

Two-reasons Trap: The tendency to believe something true when two reasons can be found that support it; when the second reason is significantly less supportive than the first

UI: See: User Interface; HMI

Unacknowledged: An alarm in the alarm state, that has not been acknowledged by the Operator

Unit: A general term usually used to identify a major portion of an Operator area that performs one or more important manufacturing functions in a somewhat self-contained way

User Alert: See: Alert

User Notification: See: Alert

Weak Signal: Any small indication that is being worked on to identify what it might mean. Weak signal processing is a formal work procedure.

See: Small Indication

What-Then Question: Asking the what-then question focuses a potential decision to examine all reasonable outcomes of having made the decisions, not just the one that comes to mind or is obvious

Yoking: A functionality of HMI display screen automatic navigation design for multiple displays. When a particular task requires a specific set of display screens to be visible at the same time on multiple displays, the display screens are "yoked" together.

Index

C

D

J

K

L

M

N

O

P

U

V

W

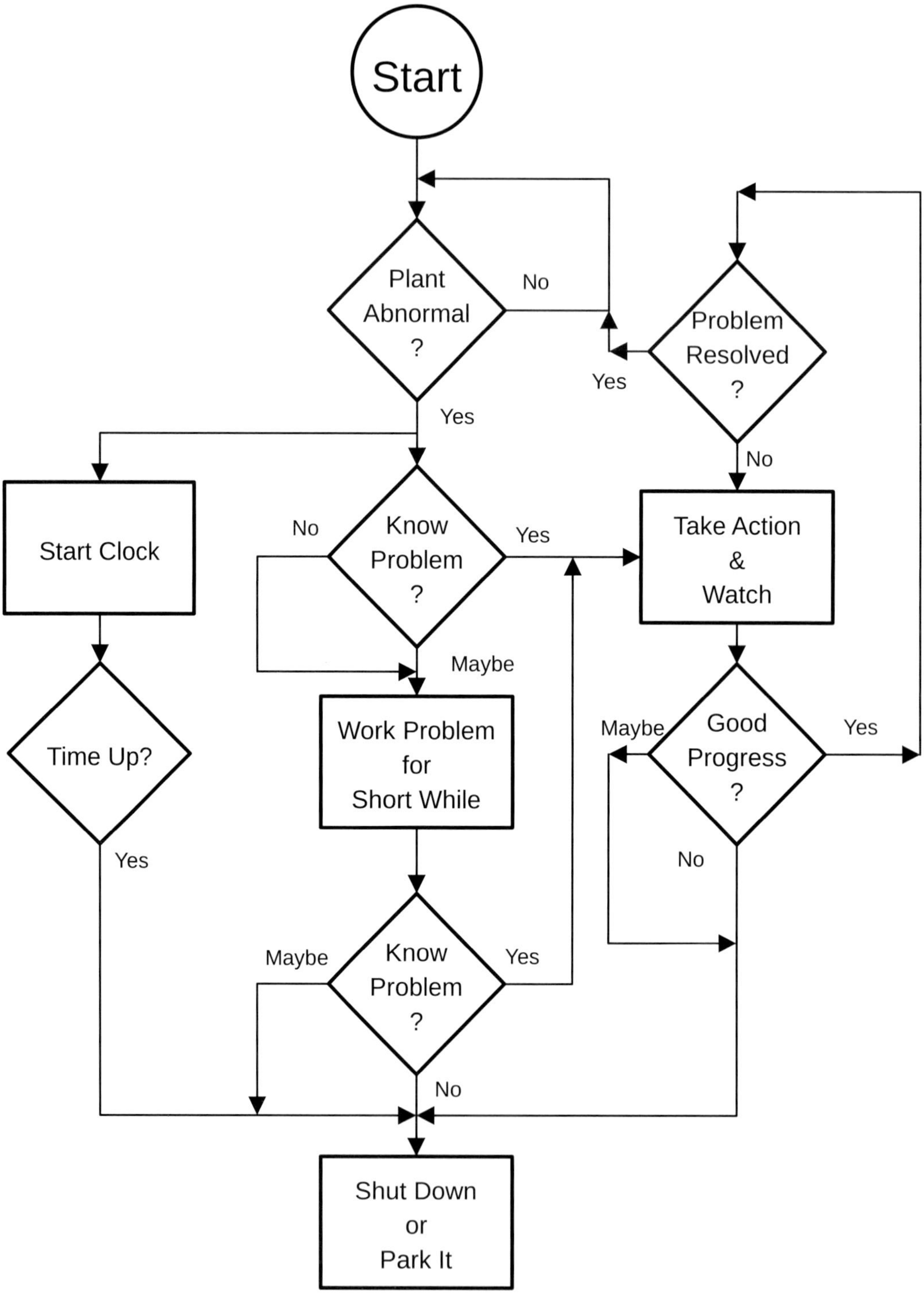
Start
Plant Abnormal ?
No
Yes
Problem Resolved ?
Yes
No
Start Clock
Know Problem ?
No
Yes
Maybe
Take Action & Watch
Time Up?
Yes
Work Problem for Short While
Good Progress ?
Maybe
Yes
No
Know Problem ?
Maybe
Yes
No
Shut Down or Park It

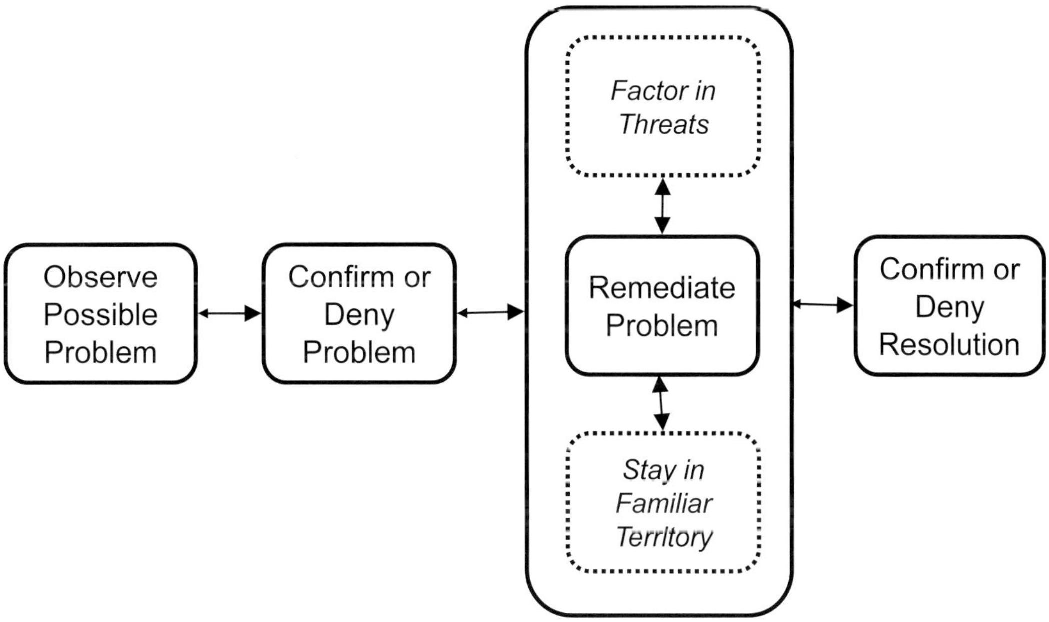
Observe Possible Problem
Confirm or Deny Problem
Factor in Threats
Remediate Problem
Stay in Familiar Territory
Confirm or Deny Resolution

Operator's Creed

As the operator, I accept responsibility for the safe, reliable, effective, and responsible operation of the facility under my control. I arrive fit for duty. I take responsible charge. I properly utilize the resources and tools at my disposal. I place safe and responsible operation above all else. When duty is done, I effectively close operations or pass on responsibility to my relief. I leave duty in a responsible manner.

Cue Card 1a

Control Room Operators' Handbook

Handbook Content Map

Control Room Operators' Handbook **Cue Cards**

Cue Card 1b

Here is the list of all 20 cards.

Card Number	Description
1	*Handbook overviews* FRONT: Handbook Map BACK: List of cue cards
2	Long-arm-of-the-operator checklist
3	*Situation management* FRONT: Basic 4 situation management components BACK: Control the Situation Management Work Process
4	Safe communications FRONT: Safe Communications Set-up BACK: Steps for Safe Communications
5	SUDA (See, Understand, Decide, & Act) FRONT: SUDA Diagram BACK: SUDA in Everyday Situations SUDA in Alarm Situations
6	Managing uncertainty
7	HMI Navigation Checklist
8	*Alarm response* FRONT: Alarm response sheet BACK: 7 Steps for using alarm response sheet
9	Shift handover tasks and sequencing
10	*Maintenance handovers* FRONT: Maintenance handover BACK: Maintenance handback
11	*Weak signals* Front: Weak signals diagram Back: Weak signals 7-step procedure with brief explanation
12	*Control room conditions* FRONT: Restricted control room BACK: Sterile control room
13	Control room codes FRONT: Code Outside (emergency outside plant) BACK: Code Inside (emergency inside plant)
14	Defensive operating Front: Defensive Operating Behaviors Back: Defensive Operating Added Guidance
15	Collaboration
16	*Escalation* FRONT: Escalation flow diagram BACK: Escalation situation diagram
17	*Permission to operate* FRONT: Permission to operate flow diagram BACK: Permission to operate situation diagram
18	Weak Signal Flags
19	Triangulation Questions
20	Crew Resource Management

Cue Card **2**

Control Room Operators' Handbook

"Long Arm" Checklist

(Coordination between Control Room and Remote Site Personnel)

1. **Operator** identifies all equipment to be observed or adjusted by remote site personnel
2. **Operator** reconfigures and adjusts current operations as needed to ensure safe and appropriate observations and adjustments by the remote site personnel
3. **Remote site personnel** verify using independent means that they have properly identified the equipment to be observed and adjusted, and it is ready for their activities
4. **Remote site personnel** take the proper steps to observe and adjust the equipment.
5. **Operator** verifies step-by-step that the equipment can actually be taken back for normal operation and is fully ready for service; or is sidelined for other work. This is an active process where the operator and remote site personnel check, test, coordinate, and otherwise ensure that the appropriate adjustments are done and equipment is operationally ready or suitably set aside.

[M4.4]

This side intentionally left blank

Four Essential Activities for Effecting Situation Management

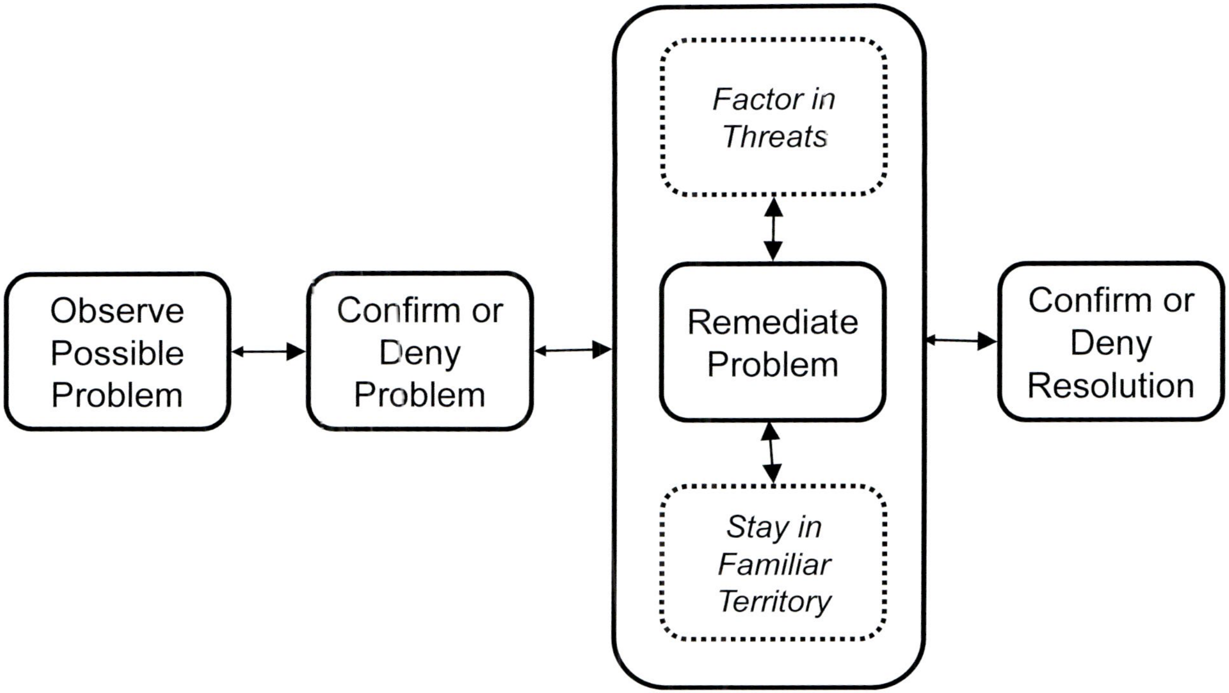

[M4.6]

Cue Card **3b** *Control Room Operators' Handbook*

Situation Management Process

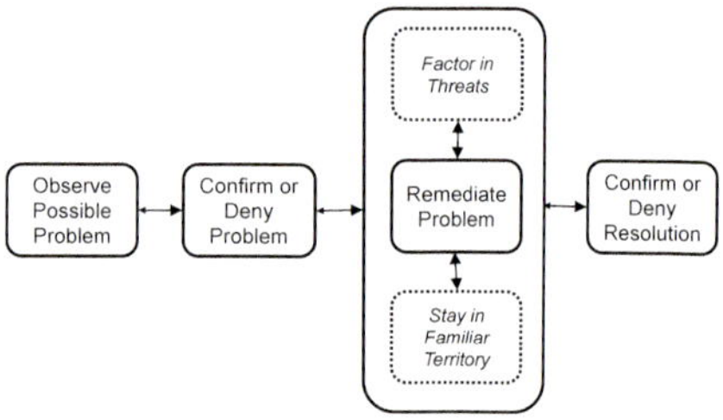

1. OBSERVE all potential problems.
2. CONFIRM the actual existence of all observed potential problems.
3. REMEDIATE all confirmed problems
   - Factor in all potential threats.
   - Never operate where you have no experience or explicit guidance.
4. CONFIRM resolution; otherwise start over

[M29.3]

Control the Situation Management Work Process

1. Use the effective operating protocols for situation awareness provided for you.
2. Ensure that there will be enough time for situation assessment and remediation activities; otherwise *permission to operate* **[Q17]** will be in effect.
3. Locate and use sufficient operational and directional information to permit successful management of the abnormal situation.
4. Utilize all safeguards and support mechanisms **[Q15]** & **[Q16]** & **[Q17]** in case your intervention is not obviously working as needed; or not going to be successful in time.

[M16.4]

Cue Card **4a**

Control Room Operators' Handbook

Mindful Conversations

(Set-up for Safe Communications)

1. **Turn Off All Devices. Focus** on only individuals necessary to conduct the conversation.
2. **Use Your Eyes** to make contact. Pick up on all needed clues about how the other person is reacting about communicating with you.
3. **Listen** to the tone of voice to pick up on what might be below the surface, hidden, or hard to say. Hear the words said, the way they are spoken, and the care and feeling they convey.
4. **Say "no" to thoughts** that have nothing to do with the conversation. Get them out of the way.
5. **Attempt to truly understand** the communication. Hear without personal judgment, critique, or your own discomfort.
6. **Watch out for external conditions** unrelated to the conversation (e.g., clothing, gender, etc). These can lead you to presuppose a message.
7. **Respond instead of react.** Give yourself the time to process what you hear. You are in the listening pathway now. Listen only. Don't plan your response before its time.

[M6.3]

Cue Card **4b**

Control Room Operators' Handbook

Steps for Safe Communications

1. **STATE WHAT IS KNOWN** and how it is known.

   Say exactly what is going to be said and how it is known.

   Repeat back to the other party succinctly what you think you heard said, and ask if there is anything more.

   Confirm that each party heard and understood what the other said. Notice and clarify any actionables, challenges or doubt. Keep a (mental) list and refer to soon later on.

2. **STATE EXPLICITLY WHAT IS THE CONCERN**

   This sets the entire reason for the conversation (but not the outcome).

3. **STATE EXPLICITLY WHAT IS EXPECTED** of each party. This sets the working relationship between each.

   Is confirmation expected or other opinion needed?

   Is actual *operational assistance* being requested (escalation, hands-on aid, collaboration) or something else? State the *something else* if so.

[M6.3]

Cue Card **5a**

Control Room Operators' Handbook

SUDA Diagram – Operator Action Timing

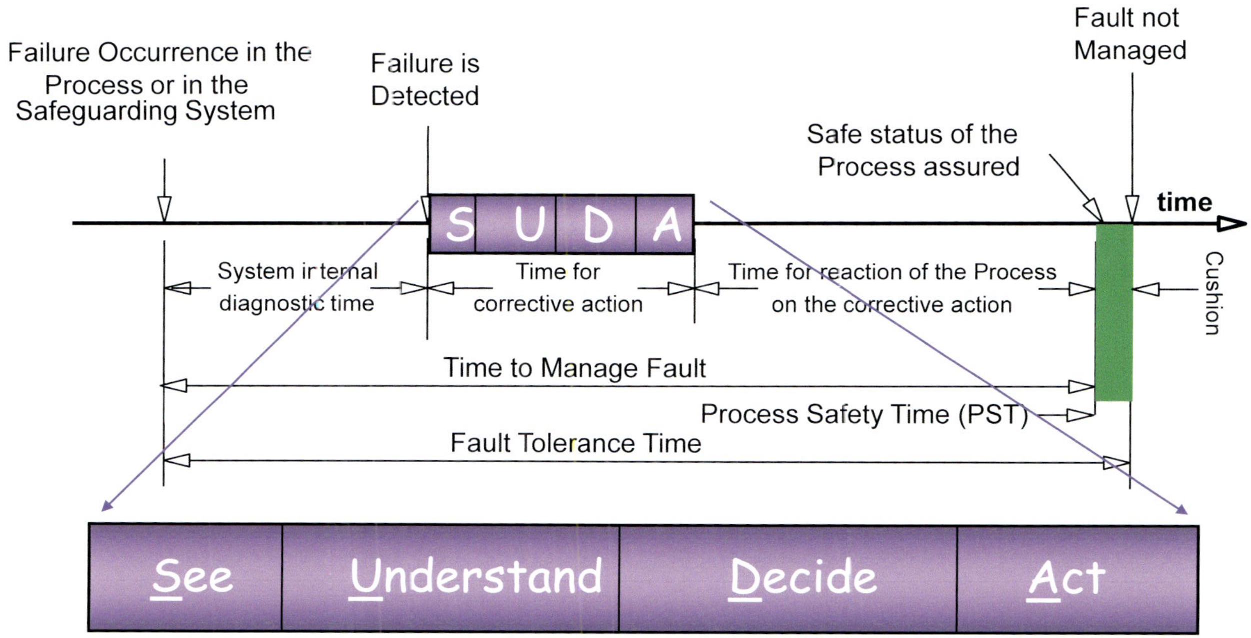

[M13.2] & [M19.2]

Cue Card 5b SUDA in Everyday Activities:

See: Be aware so you see that you have a problem

Understand: Understand the meaning of what you see and what the process will do if a remedy is not found or not found in time

Decide: Determine what to do to remedy the situation and decide how to do it

Act: Actually make all changes and other actions needed to implement your selected remedy

[M13.2]

Cue Card 5c SUDA in Alarm Situations

For activated configured alarms (implemented in the Basic Process Control System):

1. **SEE**: see it (on HMI graphics and alarm summary screen)
2. **UNDERSTAND**: be able to sufficiently understand what the alarm means
3. **DECIDE**: decide what to do to respond to it
4. **ACT**: implement your selected response action. And make sure it is working

[M19.2]

Cue Card **6**

Control Room Operators' Handbook

MANAGING UNCERTAINTY
TO UNDERSTAND, YOU NEED TO KNOW:

1. What are all SYMPTOMS of the concern?
2. What is KNOWN and what is the quality of that knowledge. Do not use what is unclear nor assumed unless and until properly verified.
3. What do you NOT KNOW but is essential to know and why do you need to know it?
4. What are the clearly perceived operational or other RISKS being faced?
5. What are the reasonable CHOICES and why?
6. What must be avoided?
7. What plan is the SAFEST and most likely to be successfully executed?
8. What is NECESSARY to execute that plan?
9. Is what is necessary to execute that plan AVAILABLE?
10. What is the BACKUP plan in case that plan cannot be selected or executed? Without a back up plan, your initial plan is unsafe.

[M15.2]

This side intentionally left blank

Cue Card **7**

Control Room Operators' Handbook

HMI Display Screen Navigation Checklist

1. **Confirm current location** – is this specific navigation starting from the place you intended?
2. Before clicking, decide **what is needed next**. (This is the purpose of the navigation: exactly what it is that you are looking for).
3. **Decide where it is located** – decide where to find exactly what you are looking for. (You know the specific "location" where to look; or you have a good idea of what is near what you are looking for, so you can get close enough to find it.)
4. **Decide how to get there** using the existing tools and infrastructure; determine the exact, specific way to navigate (use a map; do a guided search; follow a **planned idea**)
5. **Mark where you are** and then get to where you want to go.
6. **Confirm where you went** is where you need to be. Did the navigation step(s) actually take you to the intended destination?
7. **Ascertain** that what you needed is where you went to find it.
8. **Get** the information or do the actions you need.
9. **Return** to step 2 and repeat as long as needed.

[M18.5]

This side intentionally left blank

Cue Card **9**

Control Room Operators' Handbook

Operator + Supervisor + Field Operator SHIFT HANDOVER Combined Timelines

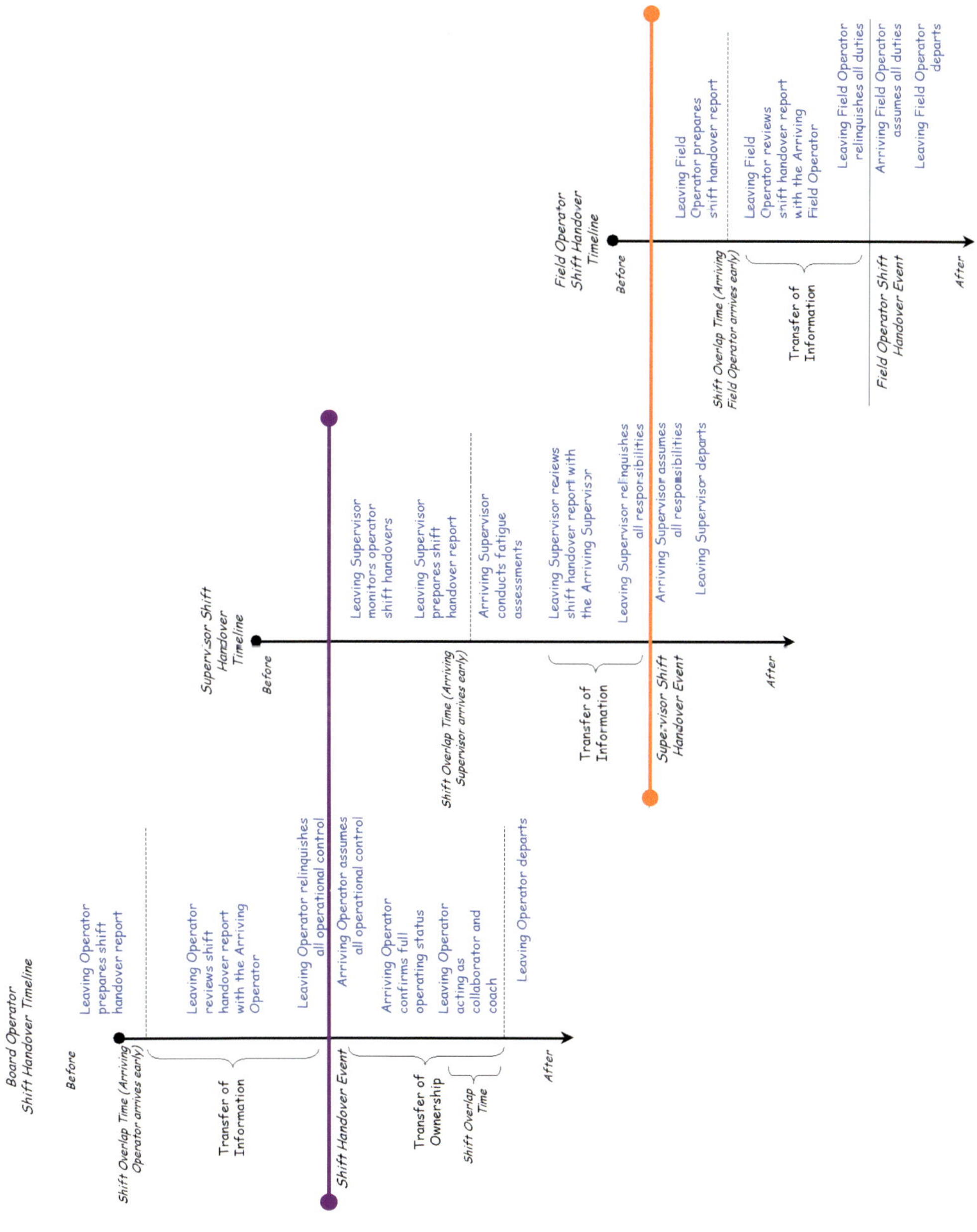

[M22.5]

This side intentionally left blank

Cue Card **10a**

Control Room Operators' Handbook

Maintenance Handover

The board operator is the responsible individual since he must properly remove equipment from service before placing it in the hands of maintenance. This is a check-list of functionality, not a work procedure.

1. Board operator identifies all equipment to be taken by maintenance
2. Board operator reconfigures and adjusts current operations to remove the need for the equipment to be taken from production. Include all ancillary parts (like piping, instruments, etc.) in this action.
3. Maintenance verifies, by using independent means, that the equipment to be taken has been properly removed from production
4. Maintenance takes (temporary) ownership of equipment (and does their work/evaluation/inspection and anything else agreed to).
5. Maintenance takes the proper extra steps (not related to operations) to render the equipment safe and ready for maintenance activities; e.g., fitting of blinds, electrical isolation, erection of physical protection barriers, etc.

[M22.6]

Cue Card **10b**

Control Room Operators' Handbook

Maintenance Handback

1. Maintenance identifies exactly what equipment is ready to be returned to service
2. Maintenance affirms that the equipment is properly repaired and/or evaluated and ready to be placed back in service; including removal of blinds, electrical restoration, clearing of physical protection barriers, etc.
3. Board operator verifies, step by step, that the equipment is actually ready for service
   a. This is an active process where the board operator and maintenance check and test and otherwise ensure that the equipment is ready to be back in service
   b. Note that this involves no verification of repairs; just that from the perspective of what the board operator can see and test, the repairs seem correct and the equipment should operate properly
4. Board operator takes back ownership of equipment
5. Maintenance leaves the area

[M22.6]

Cue Card **11a**

Control Room Operators' Handbook

Weak Signals

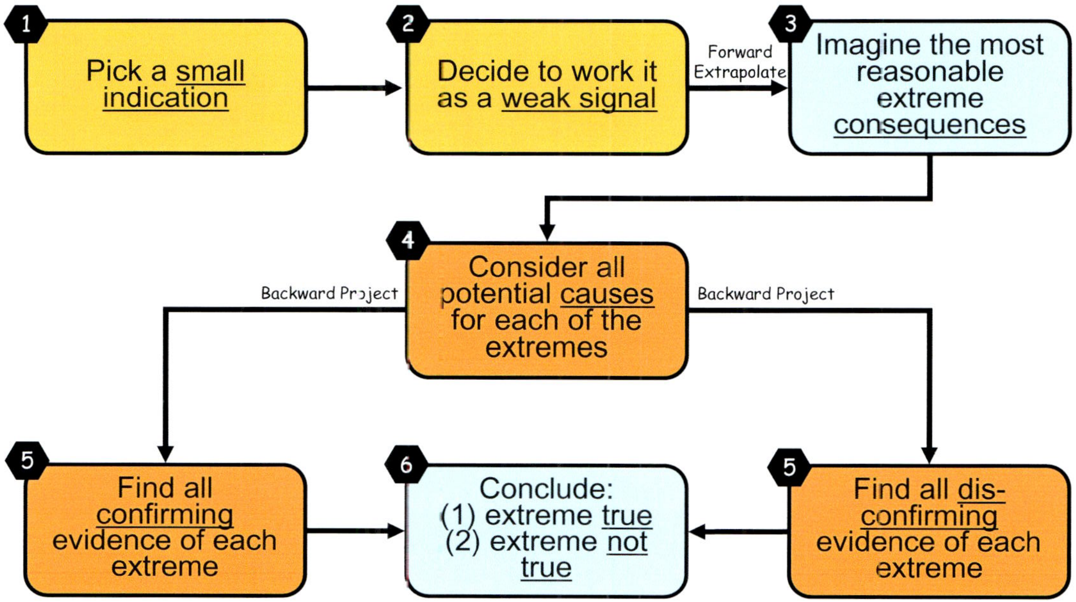

[M24.1]

Cue Card 11b Control Room Operators' Handbook
Procedure for Working Weak Signals

Selection

1. Pick a small indication that you notice and might be useful.

You've seen or thought something and are curious or concerned enough about what it might mean to consider looking more.

2. Decide the small indication might be an indicator of something you want to know more about.

Now it is to be called a weak signal.Now it is a clue of something important affecting your operating area. From now on it is a weak signal.

Forward Extrapolation

3. Use your mind's process model to imagine the reasonable but worst that might be happening if this is actually a true early warning.

You are looking for the significant thing (quite bad; really unwanted) that might be happening if not fixed or worked on. If nothing is thought of, you are forgetting something about the enterprise operation. Rethink or get help.

Backwards Projection

4. For each worst thing that might happen (possible things that go wrong), make a list of all potential causes.

In steps 1 and 2, you noticed something unusual. In step 3 you imagined a potential important (bad) outcome. Now, in Step 4, list all things that might lead to that outcome.

5. For each potential cause on the list, look for everything that confirms (or denies) each supposed thing that could actually be happening.

Use the list from step 4 to tell you exactly what to look for and where to find evidence to confirm or deny.

Confirm Problem

6. For each confirmed weak signal in step 5, you have identified a confirmed problem you will need to resolve.

Confirmed means noticed, forward extrapolated to worst bad thing, backwards projected to look for evidence, and evidence found. So the original weak signal is probably real and in need of attention.

[M24.1]

Cue Card **13a**

Control Room Operators' Handbook

CODE OUTSIDE
(Outside Plant Fence)

In this situation the focus for concern is at the enterprise boundary

A significant threat exists or is anticipated from OUTSIDE the enterprise, likely to affect what you are doing or planning to do.

Initiators:

- Severe weather
- Civil unrest (likely to affect operations)
- Significant curtailment of any utility
- Requirement exists to isolate plant from the outside community (medical, political, or any other reason potentially impacting operations)
- Significant loss of ability to store/ship product, or dramatic production increase
- Disease or risk of disease exposure

Termination:

- Authorized person declaring **CODE OUTSIDE** is terminated.

[M31.3]

Cue Card **13b**

Control Room Operators' Handbook

CODE INSIDE
(Inside Plant Fence)

In this situation the focus for concern is within the enterprise

Significant unexpected trouble is present or developing INSIDE the enterprise boundaries with potential for imminent and significant danger. Response requires coordination with proper personnel.

Initiators:

- Fire (no matter how small or if contained!)
- Explosion (including any loud sounds)
- Release (contained/uncontained, toxic or not)
- Significant equipment failure or malfunctioning including cyber attack.
- Minimum acceptable personnel not present
- Presence of unknown personnel (identity as AUTHORIZED AND SAFELY PROTECTED cannot be IMMEDIATELY verified)

Termination:

- Authorized person declaring that **CODE INSIDE** is terminated.

[M31.3]

Cue Card **14a** *Control Room Operators' Handbook*

DEFENSIVE OPERATING Behaviors

1. **Aim High**—Develop and maintain an endpoint view for what is expected and/or required during and at shift-end. *What do you need to have done as opposed to just nice to do?* Keep looking ahead: are your actions contributing to or hindering current shift objectives.
2. **Get the Big Picture**—keep alert to other aspects surrounding what is happening now. Check all operational limits. See if you are close to any, or way too far from others.
3. **Keep Your Eyes Moving**—observe current operations to the point of knowing where things are both spot on or need closer attention. Account for all details, no matter how small.
4. **Leave Yourself an Out**—always, always have a backup plan. Respond to operation that might be too near operational limits by taking steps to reduce the chances of limit crossing. Take advantage of operating space when there is still enough room that you can use.
5. **Make Sure They See You**—ensure that the rest of the enterprise knows what you are doing and reacting appropriately.

[M32.2]

Cue Card **14b**

Control Room Operators' Handbook

DEFENSIVE OPERATING—
Added Guidance:

A. **Use Extra Caution When Making Operational Changes**
   - When Maintaining current operation: like adjusting for load changes, managing disturbances and equipment malfunction, altering production rates, and the like
   - When Changing operation: like starting up, shutting down, diverting product, transiting from one product to another, and more

B. **Leave Some Space**—operating at a functional limit line may seem to be efficient use of resources, but will not be if extra is needed to account for the unexpected.

C. **Pay Attention to Current Surroundings**— be in the moment as you do your work in the control room.

D. **Trust but Verify**—confirm that what you expected is actually there.

E. **Always Yield**—teamwork and safe operation is all about working as a team and doing it safely. Yield first and find out why later.

[M32.3]

Cue Card **15**

Control Room Operators' Handbook

5 Steps for COLLABORATION

1. ***Situation***: Here's what I think we face
2. ***Task***: Here's what I think we should do
3. *Intent*: Here's why I think that is what we should do
4. ***Concern***: Here's what we should keep our eye on because if that changes, we're in a whole new situation
5. ***Calibrate***: Now, talk to me. Tell me if you don't understand, cannot do it, or see something I do not.

   Ask the following focus questions to round out step 5:

   a. What have I missed?

   b. What do you know that I don't?

   c. Where do we go from here?

[M34.3]

This side intentionally left blank

Cue Card **16a**

Control Room Operators' Handbook

ESCALATION

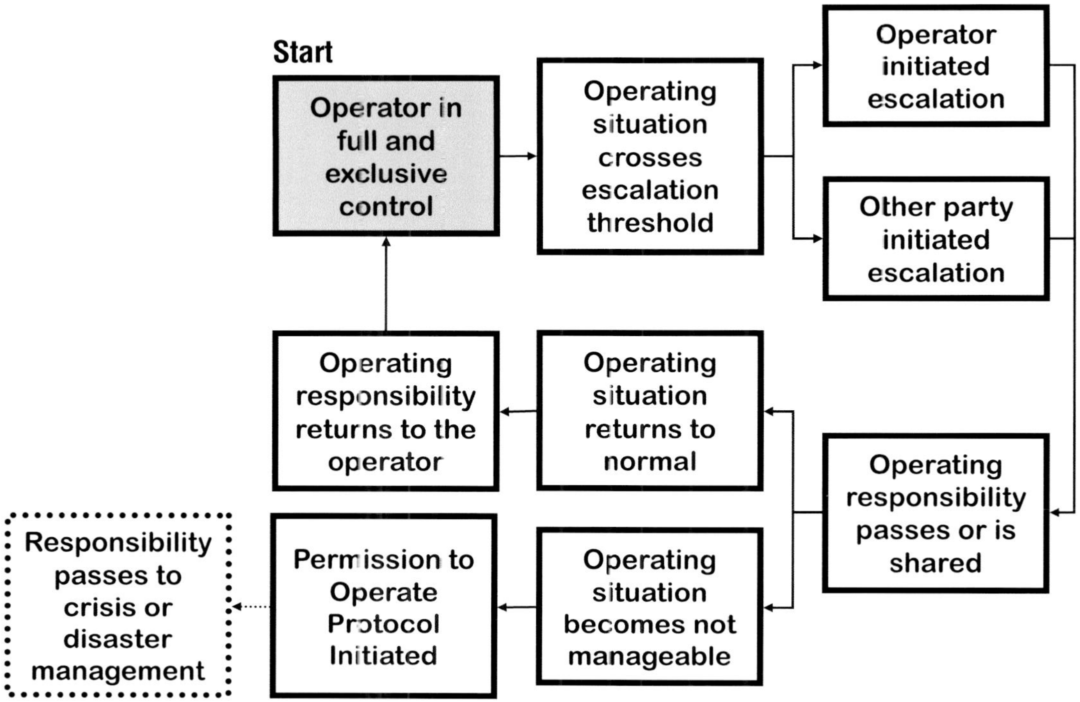

[M34.10]

Cue Card **16b**

Control Room Operators' Handbook

ESCALATION

Operational Modes:	Plant States:	Critical Systems:	Operational Goals:	Plant Activities:
Emergency	Disaster Accident	Area Emergency Response System Site Emergency Response System	Minimize Impact	Fire Fighting First Aid Rescue
	Out of Control	Physical and Mechanical Containment System Safety Shutdown, Protective Systems,	Bring to Safe State	Evacuation
Abnormal	Abnormal	Hardwired Emergency Alarms DCS Alarm System	Return to Normal	Manual Control & Troubleshooting
Normal	Normal	Decision Support System Process Equipment, DCS, Automatic Controls Plant Management Systems	Keep Normal	Preventative Monitoring & Testing

You have always lost permission-to-operate when you enter the dotted region. If you know that you have enough time to escalate, do so; otherwise shut down or safe park.

[M34.11]

Cue Card **17a**

Control Room Operators' Handbook

PERMISSION-TO-OPERATE

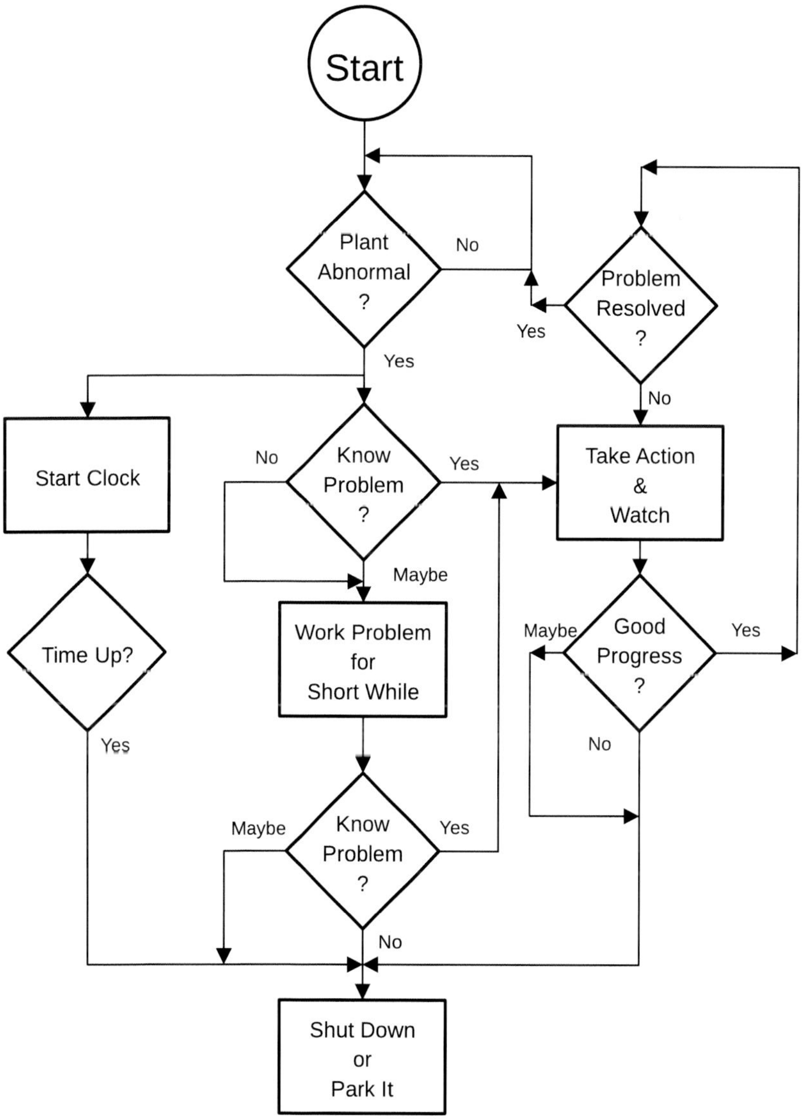

[M35.7]

Cue Card **17b**

Control Room Operators' Handbook

PERMISSION-TO-OPERATE

Operational Modes:	Plant States:	Critical Systems:	Operational Goals:	Plant Activities:
Emergency	Disaster Accident	Area Emergency Response System Site Emergency Response System	Minimize Impact	Fire Fighting First Aid Rescue
Abnormal	Out of Control	Physical and Mechanical Containment System Safety Shutdown, Protective Systems,	Bring to Safe State	Evacuation
	Abnormal	Hardwired Emergency Alarms DCS Alarm System	Return to Normal	Manual Control & Troubleshooting
Normal	Normal	Decision Support System Process Equipment, DCS, Automatic Controls Plant Management Systems	Keep Normal	Preventative Monitoring & Testing

If operating within the red box and in the operational situation of being ABOVE the dotted red line, you have immediately lost permission to operate. Shut down or safe park..

[M35.1]

Cue Card **18**

Operators' Companion Guide

Working Weak Signal FLAGS

1. INFORM supervisor whenever the number of confirmed weak signals become three or more.
2. Do FLAGS appear related
   (a) Yes– examine carefully for a new problem different from current supposed one.
   (b) No—work each problem; escalate as needed.
3. RECOMFIRM that current problem is correct.
   (a) Yes—carefully continue so long as it works.
   (b) No—for sure the current problem is wrong.
4. Current solution still working, but slowly?
   (a) Yes—Carefully look for any new problems
   (b) No—Problem is getting worse. Escalate.
5. Are ALL indicators for successful remediation still present?
   (a) Yes—Current solution needs improving
   (b) No—Wrong problem is being worked. Escalate immediately.

[M25.3]

This side intentionally left blank

Cue Card **19**

Operators' Companion Guide

TRIANGULATION QUESTIONS

1. Do you (asking the other person) see **anything unusual** or out of the ordinary?
2. Are you having any **problems** doing your work?.
3. What are the ways that anything you are doing might affect me?
   (a) Are you seeing everything you expect to see to ensure proper functioning in your area?
   (b) Does everything in your area confirm that what you are doing is proper and working?
   (c) Do you see anything amiss in your area suggestion something is not as expected?
   (d) Are you unable to see anything you need to confirm your concerns?
   (e) Do you see anything you cannot fully explain (and understand)?
4. Assuming you are doing my job:
   (a) How could what I am doing affect you?
   (b) Do you see anything unusual in my area?
   (c) Do you see everything you expect to see in my area to ensure proper function?
   (d) Do you see anything in my area that is unexpected?

(continued on reverse)

Cue Card **19**

Operators' Companion Guide

TRIANGULATION QUESTIONS

(Continued)

5. Are you aware of anything going on in the larger plant that might unduly affect your or my current operations?
6. Are you aware of anything planned in the larger plant that is not happening or not going according to plan?
7. What do you know that I do not?
8. What might I know that you do not?

Processing Answers to the Triangulation Questions:

Work down the list of 8 items only so long as you know FOR SURE that both of you have enough time to completely work all problems fully. Use this new or confirmed knowledge to understand and manage.

Otherwise ALWAYS *escalate* (if time permits) otherwise invoke *permission to operate* [**M35**].

[M34.5]

Cue Card **20**

Operators' Companion Guide

CRM – Crew Resource Management

This is the control room equivalent of "see something, say something."

The person in authority is the LEADER; the other junior individual is the FOLLOWER.

1. The FOLLOWER states the facts.
   (a) Here exactly is what I see or think.
   (b) Example: The tower temperature is way off normal.
2. If the LEADER does not affirmatively indicate hearing and understanding the facts:
   FOLLOWER issues the Challenge.
   (a) FOLLOWER restates the facts with more context and personalizes the message.
   (b) Example: [Use the name of the LEADER], are you going to continue the normal flow rate into the splitter tower? Check the low temperature at tray 22.
3. If the LEADER fails to exhibit full engagement with the situation and act appropriately:
   (a) FOLLOWER takes least impactful but adequate action on his own.
   (b) FOLLOWER identifies any other facts to LEADER but takes no further action.

[M34.4]

This side intentionally left blank